ON ARCHITECTURE

MARCUS VITRUVIUS POLLIO (*c.* 90–*c.* 20 BC) was a Roman military architect and engineer, and an expert in ballistic machines in particular. Posthumously, Vitruvius is most famous as the author of *De architectura* (On Architecture). Written during the twenties of the first century BC, it is the only complete treatise on classical architecture to have survived to modern times. Composed in ten books, it was influenced by his practical professional experiences and travels, as well as his admiration for Greek culture of the previous three centuries, which he regarded as a Golden Age for art and architecture. Throughout *De architectura* Vitruvius demonstrates his familiarity with the architecture and cities of the Greek peninsula, a Roman protectorate. He also had direct experience of the Roman province of Asia and Pontus, and of Gaul, having been recruited by Julius Caesar for his military campaign there between 58 and 51 BC. For his long and faithful service to the Roman state he was granted a pension by Augustus, the adopted heir of Julius Caesar, to whom his treatise on architecture is dedicated.

RICHARD SCHOFIELD read Classics at Oxford in the early 1970s, then architectural history at the Courtauld Institute of Art, London. After working at the University of Nottingham for many years, he moved to the Istituto universitario di architettura di Venezia in 1997, where he is Professor of the History of Architecture. He has published many articles on fifteenth- and sixteenth-century architecture and sculpture and the Counter-Reformation. He is the author of *Giovanni Antonio Amadeo: I documenti*, Como, 1989 (with J. Shell and G. Sironi), *Pellegrino Tibaldi architetto e il S. Fedele di Milano*, Como, 1994 (with Stefano Della Torre), Palladio's *The Four Books on Architecture*, The MIT Press, 1997 (with Robert Tavernor) and *Architettura e controriforma: I dibattiti per la facciata del Duomo di Milano, 1582–1682*, Milan, 2004 (with Francesco Repishti). He has edited two volumes of essays, *Nuovi Antichi: Committenti, cantieri, architetti 1400–1600*, Milan, 2004 and *La bottega dei Lombardo:*

Architettura e scultura a Venezia tra Quattrocento e Cinquecento, Marsilio, 2006 (with A. Guerra and M. Morresi).

ROBERT TAVERNOR studied architecture in London, Rome and Cambridge and practises as a consultant architect. He was Professor of Architecture at the universities of Edinburgh and Bath, and is currently Professor of Architecture and Urban Design at the London School of Economics and Political Science (LSE). He is the author of *Palladio and Palladianism*, Thames & Hudson, 1991, *On Alberti and the Art of Building*, Yale University Press, 1998, and *Smoot's Ear: The Measure of Humanity*, Yale University Press, 2007. He is co-editor (with G. Dodds) of *Body and Building: Essays on the Changing Relation of Body to Architecture*, The MIT Press, 2002; and co-translator of two English translations of architectural treatises: Alberti's *On the Art of Building in Ten Books*, The MIT Press, 1988 (with J. Rykwert and N. Leach); and Palladio's *The Four Books on Architecture*, The MIT Press, 1997 (with Richard Schofield).

VITRUVIUS

On Architecture

Translated by RICHARD SCHOFIELD
with an Introduction by ROBERT TAVERNOR

PENGUIN BOOKS

PENGUIN CLASSICS

Published by the Penguin Group
Penguin Books Ltd, 80 Strand, London WC2R 0RL, England
Penguin Group (USA) Inc., 375 Hudson Street, New York, New York 10014, USA
Penguin Group (Canada), 90 Eglinton Avenue East, Suite 700, Toronto, Ontario, Canada M4P 2Y3
(a division of Pearson Penguin Canada Inc.)
Penguin Ireland, 25 St Stephen's Green, Dublin 2, Ireland
(a division of Penguin Books Ltd)
Penguin Group (Australia), 250 Camberwell Road, Camberwell, Victoria 3124, Australia
(a division of Pearson Australia Group Pty Ltd)
Penguin Books India Pvt Ltd, 11 Community Centre, Panchsheel Park, New Delhi – 110 017, India
Penguin Group (NZ), 67 Apollo Drive, Rosedale, North Shore 0632, New Zealand
(a division of Pearson New Zealand Ltd)
Penguin Books (South Africa) (Pty) Ltd, 24 Sturdee Avenue, Rosebank, Johannesburg 2196, South Africa

Penguin Books Ltd, Registered Offices: 80 Strand, London WC2R 0RL, England

www.penguin.com

This edition first published in Penguin Classics 2009

023

Set in 10.25/12.25 pt PostScript Adobe Sabon
Typeset by Rowland Phototypesetting Ltd, Bury St Edmunds, Suffolk
Printed in England by Clays Ltd, Elcograf S.p.A.

ISBN: 978-0-141-44168-9

www.greenpenguin.co.uk

Contents

Abbreviations

Barbaro, 1567	*I dieci libri dell'architettura di M. Vitruvio tradotti e commentati da mons. Daniele Barbaro*, published by Francesco de'Franceschi and Giovanni Chrieger, Venice, 1567; cited from modern edition, with introductions by M. Tafuri and M. Morresi, Milan, 1987
Budé	*Vitruve*, Paris, Les Belles Lettres: vol. 1, ed. Ph. Fleury, 1990; vol. 2, ed. L. Callebat, P. Gros and C. Jacquemard, 1999; vol. 3, ed. P. Gros, 1990; vol. 4, ed. P. Gros, 1992; vol. 6, ed. L. Callebat, 2004; vol. 7, ed. B. Liou, M. Zuinghedau and M.-T. Cam, 1995; vol. 8, ed. L. Callebat, 1973; vol. 9, ed. J. Soubiran, 1969; vol. 10, ed. L. Callebat with Ph. Fleury, 1986
Ferri, 1960, 2002	*Vitruvius Pollio: Architettura*, ed. S Ferri, Rome, 1960, reissued with an introduction by S. Maggi, Milan, 2002
Fleury, 1993	Ph. Fleury, *La Mécanique de Vitruve*, Caen, 1993
Fra Giocondo, 1511	*M. Vitruvius per Iocundum solito castigatior factus cum figuris et tabula ut iam legi et intelligi possit*, Venice, 1511
Granger, 1931–4	F. Granger, *Vitruvius on Architecture*, 2 vols., Cambridge, Mass., 1931–4
Gros, 1997	*Vitruvio De architectura*, ed. A. Corso

	and E. Romano, introduction by P. Gros, 2 vols., Turin, 1997
Lewis, 2001	M. J. T. Lewis, *Surveying Instruments of Greece and Rome*, Cambridge, 2001
Morgan, 1960	W. H. Morgan, *Vitruvius: The Ten Books on Architecture*, Harvard, 1914; reprinted New York, 1960 (Dover)
Palladio, 1997	Andrea Palladio, *I quattro libri dell'architettura*, translated as *The Four Books on Architecture*, trans. R. Tavernor and R. Schofield, Cambridge, Mass., and London, 1997
Rose, 1867	*Vitruvii de architectura libri decem*, ed. Valentin Rose and Herman Müller-Strübing, Leipzig (Teubner), 1867
Rowland, 1999	*Vitruvius: Ten Books on Architecture*, illustrated by Thomas Noble Howe, Cambridge, 1999

Introduction

VITRUVIUS AND HIS *DE ARCHITECTURA*

Marcus Vitruvius Pollio has influenced two millennia of architectural theory and practice. His name has been on the lips of every famous architect in the Western world: many have been proud to promote themselves as heirs to the classical tradition he defined. Vitruvius completed his Latin treatise *De architectura* (On Architecture), arranged in ten books, towards the end of the first century before Christ. It became the principal touchstone for the classical Renaissance in Western architecture that emanated from Italy in the fifteenth century, and has provided aspiring architects with an essential theoretical basis, and the inspiration for many subsequent architectural design approaches, from classical to modern.

The durability of Vitruvius' text derives in part from the fact that it is open to interpretation: none of his original illustrations survives, and, without direct knowledge of many of the buildings and technical devices he refers to, his language is sometimes obscure and hard to fathom; but the very difficulty of parts of the text proved extraordinarily stimulating to its many interpreters. At the same time Vitruvius provided innumerable architects with a ready-made vocabulary for all the elements of the four principal column types – Tuscan, Doric, Ionic and Corinthian – as well as the terminology for the many component parts of temples, houses, farms and civic buildings. His is the only substantial survivor of architectural theory and practice from the ancient world, and the 'Ten Books on Architecture' have provided a key source for a practical understanding of how a Roman architect conceived, designed and implemented buildings and planned cities according to what he regarded as natural, rational, principles. Moreover, Vitruvius has shaped the modern professional architect by inspiring the notion in the Italian Renaissance that the individual architect, well versed in ancient traditions of design and building, should be regarded as a creative force for the betterment of society. His recommendations on the education of the architect established architecture as a respectable profession well embedded in the liberal arts, and writers on architecture from the Renaissance onwards have repeated them, so that eventually, two thousand years later, architecture has become a university-based discipline.

The precise date of the publication of *De architectura* is unknown, and is much debated by classical scholars, but it was probably composed during the second decade of the first century BC. The treatise is dedicated to Augustus, the title granted to Octavian in 27 BC. Vitruvius variously refers to Octavian–Augustus as 'Caesar, Supreme Ruler' (e.g. 1.1, 4.1 and 5.1), in recognition of the fact that he was the adopted heir of Julius Caesar (and Caesar was the name Octavian himself used from 44 BC), and 'Imperator' in recognition of his military successes.[1] In 27 Vitruvius would have been in his sixties, and, as his lifespan is estimated as between *c.* 90 and *c.* 20 BC, around seven years from death. The quantity of material digested in the treatise suggests he most likely began to compose it some years earlier.

His rare autobiographical remarks present him as a vulnerable, not especially distinguished, pensioner-architect: 'nature has not given me an impressive stature, age has disfigured my face and ill health has reduced my powers' (2.Introduction, 4). He reminds Augustus, and informs posterity, of the considerable experience he gained during a lifetime of service:

> I was initially known to your father [Julius Caesar] for my work in this field and was a devoted admirer of his bravery. When, therefore, the heavenly council had consecrated your father in

> the residences of immortality and had passed his power into your hands, my enduring devotion to his memory was transferred to you and found favour with you. And so, with Marcus Aurelius, Publius Minidius and Gnaeus Cornelius, I was put in charge of the supply and repair of *ballistae*, *scorpiones* and other types of artillery, and, with them, I received my reward: and it was you who granted it to me first and who continued it on the recommendation of your sister. (1.Introduction, 2)

The 'reward' to which Vitruvius refers, a pension for the remainder of his life, made on the recommendation of Augustus' full sister, Octavia minor, had presumably been earned from the long and valued service he had provided his father. Evidently Julius Caesar had recruited him as a military architect and engineer for the conquest of Gaul, and Vitruvius followed his campaigns – hence his knowledge of Caesar's 'bravery' – between 58 and 51 BC.[2] During this time he is likely to have set out military camps and repaired and refined artillery in the field of battle: he describes in detail the construction and modularity of catapults and *ballistae* (10.10–12). He experienced first-hand many different constructional techniques, including fortifications strengthened by horizontal timber framework (1.5.3), and the timber fortress at Larignum made of larch wood that proved surprisingly difficult to burn (2.9.14–17). He continues:

> Therefore, since I was indebted to you for this favour, which was such that I need have no financial anxieties for the rest of my life, I began to write this work for you since I noticed that you have built much and continue to do so now, and that for the foreseeable future you will ensure that both public and private buildings will so match the majesty of your achievements that they will be handed down in the memory of future generations. I have put in writing precise recommendations so that by examining them, you yourself may become familiar with the characteristics of buildings already constructed and of those which will be built: for in these books I have laid out all the principles of the discipline. (1.Introduction, 3)

Judging from the treatise, Vitruvius was a military engineer, an expert in ballistic machines in particular (he was perhaps an officer in charge of the *doctores ballistarum*), which he describes in minute detail in Book X, and in various types of siege engines, though even here his reliance on written sources is considerable.[3] He describes only one building of his own design, the Basilica at Fano (5.1.6–10), built under Augustus. Fanum was a modest new colony on the Adriatic, the *Colonia Iulia Fanestris*, founded by Augustus for army veterans; nothing of Vitruvius' Basilica remains. Had Vitruvius designed any other buildings, especially those belonging to the categories he describes so carefully in the treatise (temples, houses and farms, as well as forums), he would certainly have said so. Nor do we know much of his connections with other architects, though there were perhaps useful links within the extended family for employment of this kind and the exchange of knowledge: Lucius Vitruvius Cerdo, perhaps a freedman of his family clan, designed and built the Arch of the Gavii in Verona; and another Vitruvius is recorded in North Africa as having financed an arch.

Almost all the detailed information provided by Vitruvius on temples, especially the Ionic ones, comes from Greek sources.[4] A number of Greek authors, writing as early as the fourth century and as late as the second century BC, reflected on what was to be regarded by Vitruvius as a Golden Age of art and architecture, and provided him with much essential terminology as well as the largely backward-looking theme to his treatise. His own experience of buildings would also have provided him with an invaluable architectural education, and he would presumably have benefited from his travels with Caesar across the Alps. He may also have travelled east to the Roman province of Asia and Pontus, as he mentions cities there such as Sardis and Halicarnassus (2.8.10), and Tralles (5.9.1), and buildings of sun-dried brick (2.8.9–10), and he gives the reader the impression that he knew these places and their key buildings. He certainly expresses his familiarity with Athens (2.1.5, 2.8.9; 8.3.6) and buildings in mainland Greece, which is to be expected, as Greek and Roman architects were encouraged to work in both regions.

The Greek peninsula had become a Roman protectorate in 146 BC, followed by the Aegean islands and Pergamum in 133 at the bequest of its king, and Augustus organized the peninsula as the province of Achaea in 27. There was a considerable exchange of scholarship during the second century BC and into the first, and Rome was undoubtedly shaped by Greek culture of some five centuries earlier: the foundations of Roman education were derived from Greek philosophy, rhetoric and art, and their cultures were entwined. The installation of Augustus' late autobiographical declaration, *Res Gestae Divi Augusti* (The Deeds of the Divine Augustus), in various cities, with parallel Greek and Latin inscriptions, was a powerful official assertion at that time of the shared cultural inheritance.

Consequently, Romans tended to be philhellenic, and Vitruvius was no exception. In tune with officialdom, he refers in Book I of *De architectura* to the exemplary character of Greek architectural precedents based on ancient Greek advances in geometry, philosophy, music, medicine, law and astronomy; subjects essential to the education of the architect (1.1.3). His curriculum appears to be based on the Greek-derived educational model Terentius Varro had compiled in the mid-30s BC in his *Disciplinae* (in nine books) of the liberal arts, which established the *trivium* (grammar, logic and oratory) and *quadrivium* (geometry, arithmetic, astronomy and music), with medicine and architecture as additional subjects. Vitruvius succeeds in weaving all these 'arts' into his treatise, because, as he is keen to emphasize, he believes the architect

> should have a literary education, be skilful in drawing, knowledgeable about geometry, familiar with a great number of historical works and should have followed lectures in philosophy attentively; he should have a knowledge of music, should not be ignorant of medicine, should know the judgements of jurists and have a command of astronomy and of the celestial system. (1.1.3)

He felt it necessary to justify why each subject was of relevance to the architect and in the context again of the origins of Greek culture. Thus, history is important

> because architects often devise a great deal of ornament for their buildings, the meaning of which they must be able to explain to those who ask why they have made them. For example, if an architect sets up in a building marble statues of women draped in matronly robes, called Caryatids, instead of columns, and puts mutules and cornices above them, he would supply the following explanation to those who asked him why: Caria, a city in the Peloponnese, sided with their enemy, the Persians, against Greece; subsequently, the Greeks, liberated from war by a glorious victory, formed a league and declared war on the Carians. (1.1.5)

He justifies philosophy because it 'includes the study of nature, which the Greeks call *physiologia*, which must be pursued in some depth because it covers numerous different types of problems in nature' (1.1.7). Indeed, Vitruvius goes on to advocate the study and imitation of nature as one of the most important pursuits for an architect. For nature leads to beauty, which is fundamental to the practice of architecture once durability and utility have been achieved in a building (1.3.2). These three conditions – a famous triad at the root of architectural design – of durability (*firmitas*), utility (*utilitas*) and beauty (*venustas*), were to be applied through rigorous laws learnt from nature: every aspect of an architectural endeavour was to be harmonized according to such natural principles, which were truly rational according to Vitruvius. Again, he was in step with some famous intellectuals of his day. Cicero had said much the same thing in 55 BC when he argued for a natural 'harmony of style' in his treatise on oratory (*De oratore*).[5] The relationship between nature and buildings is explicit in the siting and layout of cities described by Vitruvius in Book I, and in the 'Origins of Building' with which he opens Book II, before describing the natural elements and natural characteristics of materials that are combined and located in walls to make buildings durable.

An explicit formal relationship follows in Book III, where Vitruvius describes the design of temples through the analogy of the proportions and modularity of the perfect human body. Again, he is basing this approach on an ancient Greek body-building analogy. As he puts it:

> without modularity and proportion no temple can be designed rationally, that is, unless its elements have precisely calculated relationships like those of a well-proportioned man.
>
> For nature so designed the human body that, with regards to the head, the face from the chin to the top of the forehead and the lowest roots of the hair, is a tenth; the palm of the hand from the wrist to the tip of the middle finger is the same; the head from the chin to the crown of the head is an eighth; from the top of the chest and the bottom of the neck to the lowest roots of the hair is a sixth, and from the chest to the crown of the head, a quarter. From the bottom of the chin to the lowest point of the nostrils is a third of the height of the face itself, and from the lowest point of the nostrils to a point between the eyebrows is the same; from that point to the lowest roots of the hair, including the forehead, is also a third. The length of the foot is one-sixth the height of the body: the forearm and hand, and the chest are both a quarter. The other limbs, too, have their own commensurable proportions, by adhering to which celebrated ancient painters and sculptors achieved widespread and lasting fame. (3.1.1–2)

Cicero similarly believed that an oration should be composed like a well-formed body, a notion that would have been familiar to him from Plato, who had earlier ascribed to Socrates the idea of speech as something corporeal 'with a body of its own as it were, and neither headless nor feet-less, with a middle and with members adapted to each other and to the whole'.[6]

Vitruvius' written account – or canon – of human proportions is probably also an echo or parallel of an ancient composition by the Greek sculptor Polyclitus of Sicyon (or Argos), who had canonized the proportions of his sculpture the Doryphoros (the Spear-Bearer) in the fifth century BC. His representation of the Doryphoros was to become one of the most copied of ancient sculptures in stone and bronze because it was deemed to be a beautiful – a perfect – representation of the natural male body: all the examples that survive today are Roman copies of the Greek original, and being copies of various quality they have made the recovery of the canon highly problematic. Vitruvius probably received his more detailed description from secondary sources, as he fails to mention Polyclitus by name: only fragments of Polyclitus' written account are known to us, though there are explicit references to it in the later writings of Plutarch, Philo of Byzantium, Galen and Lucian.[7] No matter; as with Polyclitus' famous statue of the ideal male form, Vitruvius' particular rendering of idealized proportions exerted a considerable influence on the notion of what constitutes physical beauty in architecture. As with a well-shaped man, writes Vitruvius, the design of a temple depends on the fundamental principles of modularity and proportion, and 'Proportion is the commensurability of a predetermined component of a building to each and every other part of a given structure, and modularity (*symmetria*) is based on this commensurability' (3.1.1). We have chosen to translate *symmetria* as 'modularity', rather than 'symmetry' as is usual, as the English 'symmetry' relates more to balance – left matching right – when Vitruvius clearly means an all-pervasive interrelation of modular parts to the whole composition.[8]

Measurements derived from the human body are also of fundamental importance when setting out a building. Vitruvius states that 'the finger, palm, foot and cubit' were the measures that architects should adopt, because they were associated with the ancient Pythagorean perfect numbers 6, 10 and 16. The Greeks had valued them for philosophical and mathematical reasons, which both Pythagoras and Plato had explained, and which Vitruvius relishes relating in some detail (3.1.5–7). They are also embodied in the parts and proportions of the human canon that Vitruvius was accepting as the natural, rational, norm. Thus, a foot is one-sixth of a man's height, the hands are composed of ten fingers (including two thumbs), and the foot is sixteen fingers long (3.1.7–8).

As well as embodying perfect numbers, the idealized human form was also circumscribed by the primary geometries of the circle and square. The geometrical diagram of the perfect male figure, known as *homo quadratus*, was famously reinterpreted fifteen hundred years later by Leonardo da Vinci when he depicted a naked man with his limbs outstretched to create two figures in one: with legs together and arms outstretched to create a square, and with legs apart to define the boundaries of a circle.[9] Leonardo based his drawing on Vitruvius' description of the geometric properties of the 'well-shaped man', which runs:

> the central point of the human body is naturally the navel. So that if a man were laid out on his back with his hands and feet spread out and compasses are set at his navel as the centre and a circle described from that point, the circumference would intersect with all his fingers and toes. In the same way, just as the figure of a circle can be traced out on the human body, so too the figure of a square can be elicited from it. For if we measure from the soles of the feet to the crown of the head and that dimension is compared with the distance between the outstretched hands, we find that the breadth is the same as the height, as in the case of plane surfaces which are perfectly square. (3.1.3)

When describing his own design for the Basilica at Fano, Vitruvius uses perfect geometry and multiples of these perfect numbers to set out its plan and section: its plan is in the form of a double square, exactly sixty feet wide and a hundred and twenty feet long (the proportion of 1 : 2); it has a surrounding aisle twenty feet wide and high (a square), with columns five feet thick and fifty feet tall (the proportion of 1 : 10); and so on (5.1.6–10). In this way, Vitruvius demonstrated that the entire building was a natural, rational form, designed with modularity and proportion, and – like the idealized human body – the proportions of the parts of his building relate to the perfect geometry of square and circle and combinations of the perfect numbers 6 and 10.

Columns too were regarded as sculptural representations – significations – of the most essential of human characteristics, and therefore of human proportional perfection. The principal column types, Doric, Ionic and Corinthian – the 'orders' (*genera*) as Vitruvius calls them in Book IV – are Greek in origin, and their proportions, intercolumniations and details, including those of the associated entablatures, door and window openings and mouldings, determine the character – what Vitruvius calls the appropriateness (*decor*) – of the various building types, of which temples are the most important. As he relates in Book IV, when the ancient Greeks defined the three main types of column, they made Doric the stockiest, imitating the figure of a man, the Ionic and Corinthian being increasingly slender, in imitation of the graceful proportions of a mature woman and young woman, respectively (4.1.6–7). Architects should express the character appropriate to the deity, through the use of the correct orders, as Vitruvius had described in Book I:

> Doric temples should be dedicated to Minerva, Mars and Hercules, since it is appropriate to provide buildings without elaborate ornament for these deities because of their warlike character. Temples of the Corinthian order built for Venus, Flora, Proserpina, the God of Springs and of the Nymphs will clearly have the right characteristics, because, given the gentleness of these deities, more graceful and florid buildings decorated with leaves and volutes will clearly enhance the appearance appropriate to them. If Ionic temples were to be built for Juno, Diana, Father Liber and all other such deities, their intermediate position will be acknowledged because their principal characteristics will be balanced between the severity of the Doric style and the delicacy of the Corinthian. (1.2.5)

These body-inspired tectonic ornaments have their own associated mythology, and Vitruvius relates a particularly poignant and poetic story about the origins of the Corinthian capital created by the ancient Greek sculptor Callimachus (4.1.9–10).

Vitruvius also regarded architectural ornamentation as an expression of the construction techniques associated with early – he uses the word 'primitive' – timber buildings, Greeks masons having translated timber and carpentry techniques into stone, turning the expression of timber to artifice. By way of example, he refers to triglyphs, the term used for the stone tablets on a Doric frieze composed of three vertically aligned rectangular projections (alternating with recessed metopes), which he likens to the ends of timber beams:

> And so ancient carpenters, [. . .] built masonry between the joists, and decorated the cornices and gables above them with carpentry of great elegance; then they cut off the projections of the joists flush with the vertical planes of the walls; but when this looked clumsy to them, they fixed wooden boards, shaped in the same way that triglyphs are now made, on the faces of the cut-off joists, and painted them with blue wax so that the cut-off ends, now covered, would not be unpleasant to look at: so it was that in Doric buildings the separation of the joists faced by the deployment of triglyphs began to provide space for the metopes between them. (4.2.2)

The plans of the different temple types, the disposition and naming of their parts, occupy Vitruvius for much of Book IV. The formal relationship in plan achieved from setting out columns in particular combinations, and the spacing between them expressed as multiples of column widths, provides a clear expression of the modularity of a relatively simple building type. In Book V, Vitruvius turns to public space and public buildings, the forum and basilica, and to more complex spatial arrangements of the treasury, prison and Senate House. This book includes a long account of the form of the theatre, Greek and Roman, as a direct response to the natural relationship of the circular emanations of sound from the stage to the audience, and natural amplification achieved by placing sounding vessels under auditorium seats. Baths and gymnasia are then described, as are harbours and shipyards. What Vitruvius omits are amphitheatres (such as those at Sutri, Capua and Pompeii), the only building type invented by the Romans; and monumental baths, which were manifestations of the later imperial city. Book VI completes Vitruvius' account of the principal building types. As he says,

> I decided that I should write a comprehensive treatise on architecture and its principles with scrupulous care in the belief that in the future it would not be an unwelcome gift for all peoples. Therefore, since I wrote at length about the development of public buildings in Book V, in this one I will explain the principles

and commensurability implicit in the modular systems of private houses. (6.Introduction, 7)

It is curious from a modern perspective that Vitruvius should wait until Book VI to describe the private house, the most fundamental component of any town and city. However, then as now, only the wealthiest would employ an architect to design their homes and Vitruvius deals with housing as a generic type. He refers to traditional large and small Roman atrium-houses, as well as Greek house designs, though here he distinguishes between regional climatic differences and how these affect the disposition and types of spaces. The design of the farmhouse is also inserted into this book.

Without the ingratiating and context-setting opening sections, the first six books of *De architectura* form a unity that focuses on architecture and urban design, its origins, component parts, relation to the natural world and ancient Greek culture, as well as the principal building types. These books provided Vitruvius' contemporaries with all the components necessary with which to plan a town or colony in some detail. As such, the first draft of this part of the treatise may have been completed prior to Augustus' rule, perhaps with the patronage of Julius Caesar in mind before the fateful Ides of March in 44 BC.

Book VII is in a category of its own, with a long introduction that provides a point of transition between the first six books and what follows. Here Vitruvius names his principal sources, whose individual works on the modularity of different temples, for example, enabled him to write his more compendious treatise: 'I have gathered together what I regarded as useful in their treatises for the present theme in one systematic work' (7.Introduction, 14). He then describes the various building finishes for floors, stucco and revetment for vaults and walls, including those in damp situations, and their decoration with materials of different coloration. In the introduction and in chapter 5 of this book he writes about perspective effects in paintings, their power to impress and the need for them to reflect reality. He disapproves of recent preferences for 'unnatural' effect; the production of 'monstrosities rather than faithful representations' (7.5.3), and with his usual certainty that nature and the ancient Greeks were right complains further that:

> new tastes have forced on us a situation in which bad judges condemn artistic excellence as incompetence. But minds obscured by faulty taste are incapable of appreciating that which really can exist in accordance with convention and criteria of appropriateness. For pictures which are unlike reality should not be approved of; nor, even if they are technically accomplished, should they immediately be judged favourably if their subjects do not conform to ascertainable criteria developed without offending established conventions. (7.5.4)

Vitruvius' indignation over those modernizers with 'faulty taste' who dared to challenge the past is one of several instances where his reactionary impulses bubble to the surface of the treatise: they no doubt provided him with the motivation to continue his arduous task to completion. The final three books, VIII, IX and X, present a miscellany of issues, and reflect Vitruvius' divisions of architecture as building, the fabrication of timepieces, and machinery.

In fact, Book VIII is concerned principally with the detection, source and quality, conveyance (via aqueducts) and storage (wells and cisterns) of water, for reasons he sets out in his introduction:

> Natural scientists, philosophers and priests maintain that the properties of water are the basis of all things, so I therefore decided that, since the principles of buildings have been explained in the previous seven books, I should write in this one about how to find water, on its qualities in relation to the characteristics of the places in which it occurs, on the methods of transporting it and, finally, on how to test its quality. For it is absolutely essential for life, pleasure and everyday use. (8.Introduction, 4)

Book IX describes how timepieces work and how a knowledge of astronomy is necessary to record time, and there is much else that is not strictly relevant – on the zodiac and planets, the phases of the moon, the constellations and astrology. Not until chapter 7 does Vitruvius finally arrive at how to construct the elaborate diagram of the *analemma* in Rome, from which the hours of the day are established from the shadow of a gnomon. Sundials are described in chapter 8, though his explanations of them are somewhat elliptical, and reconstructions of them difficult for modern scholars. More care is taken over water clocks, and he describes two main types in moderately comprehensible detail.

Book X, the longest in the treatise, reflects the breadth of the engineering experience Vitruvius had gained through maintaining and refining military engines in the field. But his attention to detail does not make this book an easy read without accompanying explanatory illustrations. His original drawings – probably ten or twelve of them – were appended to this and five other books. He starts by introducing mechanics and its everyday application; the role of the fulcrum and pulleys for hoisting weights, which leads to waterwheels and bucket-chains operated manually. The ability of humankind to manipulate water – using Archimedes' water-screw and Ctesibius' force pump – clearly fascinates him, as does the ability to turn necessity to delight through the water-organ. He then leads the reader back to measuring devices, describing land and sea odometers, before focusing on the minutiae of military engines from chapter 10 onwards. These include the artillery and siege engines, the catapults and *ballistae* with which he was hugely experienced, as well as descriptions of the methods of defence that he would have witnessed first-hand. The treatise concludes simply: 'In this book I have dealt in as much detail as I could with the mechanical systems which I thought most efficacious in time of war and peace. In the nine preceding books I have composed an account organized around single subjects and their subdivisions so that the entire work would explain all aspects of architecture in ten books.'

BUILDING ON VITRUVIUS

There are many references to Vitruvius down the centuries, though interest is focused on specific aspects of his treatise rather than his comprehensive account of architecture. In the encyclopedic *Natural History*, published in AD 77, a century or so after Vitruvius' own publication, Pliny the Elder lists *De architectura* in the bibliographies to botany and mineralogy, and Sextius Julius Frontinus made use of Vitruvius when he compiled his *De aquaeductu* (On the Aqueduct). A century and a half later M. Cetius Faventinus extracted excerpts on private dwellings and Q. Gargilius Martialis took information relevant to his book on gardens, *De hortis*, which does not survive in full. A century later still, Rutilius Palladius made use of Vitruvius for his work on agriculture, *De re rustica*. Sidonius Apollinaris referred to the treatise in the fifth century.

More significantly, Isidore, Archbishop of Seville, refers to *De architectura* in his own compendious work, the *Etymologiae* (Etymologies, *c.* 623), which brought together very many significant works of the ancient world; as did Hrabanus Maurus in his mid-ninth-century encyclopedic *De universo*. However, their interest was in accumulating knowledge and theory, and while this proved invaluable to future scholars, the substance of architecture does not come from theory alone, as Vitruvius states in relation to the education of the architect at the outset of his treatise (1.1.1–2).

Although known sporadically throughout the Middle Ages, it was the rediscovery of *De architectura* by one of the greatest Florentine humanist scholars of the early fifteenth century, Poggio Bracciolini (1380–1459), which enabled classical architectural theory and practice to be recombined and promoted so powerfully. Poggio recovered numerous ancient texts from monastic libraries and it was in the greatest treasure trove of them all, the library of St Gallen in Switzerland, that he discovered the manuscript of *De architectura* that was to inform the Florentine Early Renaissance of art and architecture *all'antica*.[10] Filippo Brunelleschi (1377–1446), architect of the dome of Florence cathedral, knew Poggio, as did Leon Battista Alberti (1404–72), whose close engagement with this circle stimulated him to write the first compendious treatise on architecture since Vitruvius. Composed in Latin in the mid-fifteenth century, Alberti's *De re aedificatoria* (On the Art of Building) combines Vitruvius' admiration for Greek antiquity with his own all-embracing scholarship and admiration for classical antiquity, both Greek and Roman.

Alberti probably began his treatise as a commentary on Vitruvius; but the obscurity of many of Vitruvius' statements about ancient buildings led him to write something entirely new, and in the process to create the first modern work of its kind.[11] His treatise was born of frustration:

> I grieved that so many works of such brilliant writers had been destroyed by the hostility of time and of man, and that almost the sole survivor from this vast shipwreck is Vitruvius, an author of unquestioned experience, though one of those whose writings have been so corrupted by time that there are many omissions and many shortcomings. What he handed down was in any case not refined, and his speech such that the Latins might think that he wanted to appear a Greek, while the Greeks would think that he babbled Latin. However, his very text is evidence that he wrote neither Latin nor Greek, so that as far as we are concerned he might just as well not have written at all, rather than write something that we cannot understand.[12]

Although Vitruvius' specialized use of Greek and Latin terminology in *De architectura* made it difficult to comprehend, Alberti had the background and determination to more than cope. He was renowned as an expert antiquarian who had studied classical literature and surveyed ancient buildings himself. Through a potent combination of scholarship and architectural practice he renewed the Vitruvian message, and produced a literary standard by which all subsequent architect-theorists would be measured. The first printed editions of Vitruvius' *De architectura* and Alberti's *De re aedificatoria* were published in 1485 and 1486, respectively, and were unillustrated.

In 1511 Fra Giovanni Giocondo of Verona produced a scholarly edition of Vitruvius' *De architectura* accompanied by 136 woodcuts and an index.[13] It proved invaluable for successive commentators on Vitruvius, not least because of the audacious brilliance of Fra Giocondo as an emendator of the ancient text. Ten years later Cesare Cesariano published a translation of Giocondo's Latin edition, supplemented by his own commentary and illustrations, which drew on the architecture of northern Italy and, most famously, Milan cathedral.[14] In 1544 a classical philologist, Guillaume Philandrier, published his Latin *Annotationes* on Vitruvius in Rome.[15] The artist and architect Sebastiano Serlio, under whom Philandrier had studied in Venice, wrote the first illustrated treatise on ancient architecture to be accessible to architects and patrons alike. His *Architettura* bears little resemblance to the treatises by either Vitruvius or Alberti, primarily because Serlio designed it as a combination of words and images (woodcuts), placed in close proximity to a descriptive text. It was subdivided into seven books, published separately and non-sequentially, each book defining a topic; the first five books – On Geometry, On Perspective, On Antiquities, On the Five Styles of Building, On Temples – were finally published together in 1551.[16] Daniele Barbaro (1514–70), a Venetian diplomat and patriarch-elect of Aquileia, learnt from Serlio's example, and when he translated Vitruvius into the vernacular, accompanied by a commentary two-thirds longer than Vitruvius' own text, he revolutionized the format of the architectural treatise.

The illustrations in the Barbaro *Vitruvius* are different in composition and character from those in Fra Giocondo's *Vitruvius*, even though Barbaro copies a number of characteristics of his predecessor's reconstructions: many were extracted from their immediate reference points in the text and given a full folio page, or an extended page. Andrea Palladio, who was the architect of Daniele and Marc'Antonio Barbaro's wonderful villa at Maser, on the Venetian *terra firma*, designed the majority of the illustrations for the folio edition of the Barbaro *Vitruvius*, published in 1556.[17] A smaller, more compact version of the folio edition was produced in separate Italian and Latin versions in 1567.[18] Palladio and Barbaro were both more positive about Vitruvius than Alberti had been: Palladio refers to Vitruvius as his principal 'master and guide' in matters architectural,[19] and Barbaro even claimed to find eloquence in Vitruvius' writing.[20]

By the time the first edition of the Barbaro *Vitruvius* was published in the mid-1550s, Palladio had started work on three of the books that were to be developed famously into *I quattro libri dell'architettura* (The Four Books on Architecture, 1570). Palladio's treatise may in fact have been intended to total ten books: he refers specifically to books 'on antiquities', 'temples', 'baths' and 'amphitheatres' – the latter having been excluded by Vitruvius.[21] In this form it would have complemented Barbaro's *Vitruvius* completely, by providing illustrated examples of every Roman building type. Barbaro even implies that their two publications were conceived as one: Palladio, he writes, 'is publishing his Book, and I am republishing Vitruvius anew'.[22]

Barbaro employs three kinds of illustrations in his *Vitruvius*: the architectural plans, sections and elevations designed by Palladio that reconstruct whole buildings; partial reconstructions or fragments of buildings that are presented as partly ruined; and drawings appropriated from other sources. Thus, Barbaro presents the principal ancient building types in their totality, in plan and exterior elevation, as Vitruvius recommends, and as if he were an architect presenting a building to his patron. But he mostly ignores Vitruvius' use of *scaenographia* – perspective drawing – preferring instead what he calls *sciografia*, or *profile*, an orthogonal section. Also, when describing a part of a building in elevation – such as the side of a temple portico in Book III, a few bays of the exterior of a theatre in Book IV, and a partial view of the front of the House of the Ancients in Book VI – he presents them as ruined fragments, their visible edge or edges having a rough finish, from which vegetation has grown.[23] Barbaro's *Vitruvius*, unlike Palladio's *Four Books*, also contains illustrations derived from other sources – constructional elements, fragments of ornament, tools, machines and devices – that are usually shown scenographically, in perspective.[24]

Barbaro describes his approach to the illustrations in an autograph addition to his manuscript for the 1556 edition explaining that they are devoid of shading and perspective so that they might be useful when designing in this manner: like the measurable templates (*sacome*) used in the construction of buildings and prepared by the master mason for the masons to follow.[25] Palladio was an obvious source for this practical way of thinking, as he refers in his professional account books to the 'sagome', or cutouts of details he provided for the builders on site. More philosophically, Barbaro was also informed by Vitruvius' description in Book III of the perfect measures and proportions of the ideal male body in relation to the proportions and symmetry of temples. In his commentary, Barbaro refers to temples as 'bodies' and their constituent elements as 'members', and describing these temples within an Aristotelian frame he invokes the 'order of human cognition'. Barbaro explains that when viewing an object from a distance 'human cognition' enables us: 'first [to] form a confused notion of it, then, coming closer, we see through its movement some of its parts, and thus we see that it is an animal. But passing closer, we know it as a man', until, from its details, the figure is recognizable as a friend.[26] Barbaro wrote his commentary on Vitruvius guided by a similar principle. He enabled his reader to construct a complete impression of an ancient building, little by little, through words and images, that lead from the general to ornamental details, and as if – to maintain the body analogy prompted by Vitruvius – each elevation were an upright man.[27] Furthermore, the bodily symmetry of a building *all'antica* obviates the need to draw both halves of an elevation or section, the missing half usually being the mirror reflection of that which is illustrated. Indeed, one may assume that in some cases a mirror may have been used to reveal a façade or interior in its entirety by placing one of its edges at right angles to the axis of symmetry.

The influence of Vitruvius abroad was stimulated by the success of Palladio's *Four Books*, and because Vitruvian and Palladian classicism became to a large extent an interchangeable concept; both promoted architecture *all'antica*, in the ancient manner. Sebastiano Serlio became architect to the French king François I and Inigo Jones imported architectural classicism to England in the early seventeenth century, encouraged by courtiers loyal to King James I. European monarchies and the establishment that supported them used the very antiquity of the classical tradition and – through Vitruvius – the clear expression of natural principles as a public statement of their authority and God-given – natural – right to rule.[28] In England, the hugely successful expansion of London during the Palladian Neoclassicism of the eighteenth century coincided with the unbroken Hanoverian monarchical span of 116 years between 1714 and 1830. Georgian architecture is characterized by buildings with regular geometry, symmetry and grand classical squares, lined with residential terraces dressed as palaces. Several versions of Vitruvius and Palladio were published during this period: the architect Colen Campbell produced a *Vitruvius Britannicus* (London, 1715–25), which catalogued recent buildings in the classical – Vitruvian – style in Britain, as well as republishing the *First Book of Palladio* (1728) and Palladio's *Five Orders of Architecture* (1728).

The success of architecture *all'antica* in Britain was extended to the colonies. It was promoted enthusiastically in America by the politician and architect Thomas Jefferson (1743–1826), who fully appreciated the power of architecture and its potential for society. Jefferson acquired many of the insights and skills that Vitruvius demanded of an architect: he sought enlightenment in the pristine sanctity of antiquity and Republicanism (he was elected President in 1801), and he was concerned with representing the purity of nature and natural form in his architecture through pure geometry, an elementalism that was by then popular too in eighteenth-century French architecture. His masterplan and designs for the University of Virginia campus – the foundation stone of which was laid in 1817 – demonstrate his consummate abilities as an architect and are a tangible testament to Vitruvian principles.[29] Meanwhile, these same principles were being challenged in France.

Under Jean-Nicolas-Louis Durand (1760–1834), professor of architecture since 1792 at the new technology-driven École Polytechnique, architectural education became decidedly anti-Vitruvian. The curriculum of the École Polytechnique was organized to create scientists and technicians with specialized skills. Students were provided with a solid general mathematical and scientific foundation before specializing in a particular discipline and entering public service. Durand's teaching epitomized the rationalizing thinking of late Enlightenment, post-Revolutionary France. The architects this institution produced were well versed in mathematical reasoning but were given little encouragement by their professors to study art and the humanities, as the arts cannot be reduced to the formulaic certainties of mathematics. In this regime, the technology of building was considered superior to the art of architecture, the main worth of which was ornament and decoration – or so it was reasoned – and architecture came to be regarded as a mere sub-discipline of civil engineering. This was the direct consequence of Napoleon's appreciation of the certitudes of mathematics and his firm belief that architecture was a luxury of marginal worth to an expanding empire: he put his trust instead in mathematicians and engineers.

Durand was at one with the curriculum of the École Polytechnique, as his lectures, published as *Précis des leçons d'architecture données à l'École Polytechnique* (1802–5), make clear. Fundamentally, he questioned the rationale of the Vitruvian tradition, particularly the analogy of body to building, and he uses Enlightenment reason to unpick almost two millennia of belief. In the introduction to volume 1 of the *Précis* Durand asks, rhetorically, whether Vitruvius' ideal proportions are true imitations of the human body. He commences with Vitruvius' description of the Doric order:

> the Greeks, it is said, defined [the Doric order] by the proportion of six diameters, because a man's foot is one-sixth of his height. First of all, a man's foot is not one-sixth but one-eighth of his height. What is more, in all Greek buildings, the proportions of

> Doric columns are endlessly varied; and, within this infinite variety, the exact proportion of six to one is not found in a single case. [. . .] The same variety is observed in the proportions of the other orders, supposedly imitated from the body of a woman and from that of a girl. It is therefore untrue that the human body served as a model for the orders.[30]

Having posed the question, Durand then delivers the hammer blow to any notion that the human analogy is appropriate to architecture:

> But even supposing that the same order always has the same proportions in the same circumstances; that the Greeks consistently followed the system attributed to them; and that the length of a man's foot is one-sixth of his height: does it then follow that the proportions of the orders are an imitation of those of the human body? What comparison is there between a man's body which varies in width at different heights, and [the column shaft] a kind of cylinder with a constant diameter throughout?[31]

By volume 2 of the *Précis*, Durand had clarified his rational response to these ancient myths. Analogy and imitation are appropriate to other arts but not architecture:

> For, if any pleasure is to result from imitation, the object imitated must be an object of nature, beyond which we know nothing, and beyond which, in consequence, nothing can interest us. The imitation of the object must, furthermore, be perfect [. . .] the human body, bearing no formal analogy to any architectural body, cannot be imitated in its proportions.[32]

The separation of the human body analogy from architectural discourse was decisive for the development of modern architecture. Attempts would be made to reconnect them, but, for the most part, architectural expression became the servant of a new kind of rationality and science.

The modern age brought with it many new building types, and the expansion of empires led to the ornaments of different expressions of architecture being applied to them. Architecture became global, even universal in outlook, and the Vitruvian architectural tradition abandoned as a totality. Certainly some aspects of Vitruvius' classicism, and its variants through Palladianism and Neoclassicism, have survived: particularly the column types and their countless descendants, which were developed by architects across two millennia, and are still used to lend authority to the outward appearance of buildings. But there are few, if any, architects today who can match the holistic integrity and beauty of more distant interpretations of Vitruvius' classicism.

If Vitruvius' treatise has been largely abandoned in practice, his message retains its potency. It is no exaggeration to suggest that the potential of architecture would have been less understood and less valued without Vitruvius. The relevance of *De architectura* is therefore undiminished by time, and this treatise should remain a fundamental starting point for architects if architecture and society are to prosper in harmony with the natural world, rather than attempt vainly to impose order on it.

NOTES

1. The fact that Vitruvius addressed his patron as 'Caesar' and 'Imperator', rather than Augustus, does not prove he was writing before 27, because writers rarely used this title in the 20s BC.
2. Vitruvius may have seen action with the Legio VI Ferrata under Julius Caesar in Gaul between these dates, at the battles at Avaricum, Gergovia, Alesia (52 BC), Uxellodunum (51), as well as at Dyrrachium (48) in Albania, Pharsalus (48) in Greece, Zela (47) in Turkey, and Thapsus (46) in Africa.
3. He lists thirteen writers on military machinery of various kinds.
4. He lists some twenty-five Greek writers on temples, including his favourite, Hermogenes, but just three Roman writers on architecture: Fuficius, Varro and Publius Septimius. The dearth of Roman authors provides him with a primary motive to write his own Latin treatise on architecture.
5. Cicero, *De oratore* 44.149, 49.164 and 65.220.

6. Plato, *Phaedrus* 264c. See also Aristotle, *Poetics* 1459ᵃ 17 ff.
7. See R. Tavernor, *Smoot's Ear: The Measure of Humanity*, New Haven and London, 2007, pp. 23–7.
8. See R. Tavernor, *On Alberti and the Art of Building*, New Haven and London, 1998, pp. 43–4.
9. Tavernor, *Smoot's Ear*, p. 10.
10. F. Pellati, 'Vitruvio nel medio evo e nel rinascimento', *Bollettino reale dell'istituto archeologia e storia*, 5/4–5 (1932), pp. 111–18; L. A. Ciapponi, 'Il *De architectura* di Vitruvio nel primo umanesimo', *Italia medievale e umanistica*, 3 (1960), pp. 59–99; P. N. Pagliara, 'Vitruvio da testo a canone', in S. Settis (ed.), *Memoria dell'antico nell'arte italiana*, Turin, 1986, vol. 3, *Dalla tradizione all'archeologia*, pp. 5–85.
11. Tavernor, *On Alberti*, pp. 15–23.
12. *Alberti, On the Art of Building in Ten Books*, ed. and trans. J. Rykwert, N. Leach and R. Tavernor, Cambridge, Mass., and London, 1988: vol. 1, p. 154.
13. See L. A. Ciapponi, 'Fra Giocondo da Verona and his Edition of Vitruvius', *Journal of the Warburg and Courtauld Institutes*, 47 (1984), pp. 72–90.
14. *Di Lucio Vitruvio Pollione De Architectura libri dece traducti del Latino in Vulgare affigurati*, ed. Cesare Cesariano, Como, 1521; facsimile: A. Bruschi (ed.), Milan, 1981.
15. *Gulielmi Philandri Castilionii Galli Civis Ro. In decem libros M. Vitruvii Pollionis de Architectura Annotationes*, Rome, 1544.
16. *Sebastiano Serlio on Architecture*, vol. 1 (Books I–V of *Tutte l'opere d'architettura et prospettiva*), trans. V. Hart and P. Hicks, New Haven and London, 1996.
17. D. Barbaro, *I dieci libri dell'architettura di M. Vitruvio. Tradotti et commentati da Monsignor Barbaro*, Venice, 1556 (hereafter Barbaro, 1556), I.6.40(a).
18. Barbaro's commentary was simplified and abridged, and Palladio's illustrations reduced in size and recut, though in the process some images were reversed, their proportions and details changed, and the fine lines of the original engravings were lost in the coarser woodcuts. See M. Morresi, 'Le due edizioni dei commentari di Daniele Barbaro', in *I dieci libri dell'architettura tradotti e commentati da mons. Daniele Barbaro*, Venice, 1567, modern edition with introductions by M. Tafuri and M. Morresi, Milan, 1987 (henceforth Barbaro, 1567), pp. xli–lviii; and M. M. D'Evelyn, 'Word and Image in Architectural Treatises of the Italian Renaissance', unpublished Ph.D. thesis, Princeton Univer-

sity, 1994 (hereafter D'Evelyn, 1994), 4 vols., UMI Microfilm no. 94–29175, pp. 248 ff., where the illustrations provided by Palladio are identified. See also the numerous recent publications since 1998 on this topic by Louis Cellauro listed in our Appendix below.

19. Andrea Palladio, *The Four Books on Architecture*, trans. R. Tavernor and R. Schofield, Cambridge, Mass., and London, 1997 (hereafter Palladio, 1997), p. 5.
20. Barbaro, 1556, V, p. 127; and cf. Barbaro, 1567, p. 203.
21. Palladio, 1997, p. 161; for other citations see p. 348 n. 25; see also L. Puppi, *Palladio Drawings*, New York, 1989, pp. 11–27 and especially p. 13.
22. Barbaro, 1556, VI.10, p. 179; and Barbaro, 1567, p. 303.
23. These are illustrated on pp. 139, 250 and 281, respectively, of Barbaro, 1567.
24. For example, a Tuscan roof taken from Philandrier is shown in perspective in Book IV, chapter 7; see *M. Vitruvii Pollionis de Architectura libri decem ad Caesarem Augustum, omnibus omnium editionibus longe emendatiores, collatis veteribus exemplis. Accesserunt, Gulielmi Philandri Castilionii, civis Romani, annotationes castigatiores & plus tertia parte locupletiores*, Geneva, 1552, p. 163. Two ancient fireplaces are taken from Francesco di Giorgio Martini. In both these cases the redrawings were by Palladio; see D'Evelyn, 1994, pp. 257 and 265. Some of the illustrations Barbaro uses, however, are by unknown artists; see his sequence of six engravings of pumps and mills: G. Scaglia, *Francesco di Giorgio: Checklist and History of Manuscripts and Drawings in Autographs and Copies from ca. 1470 to 1687 and Renewed Copies (1764–1839)*, Bethlehem, Pa., London and Toronto, 1992, p. 117. Barbaro also includes an illustration of a spiral staircase taken from Albrecht Dürer's *Underweysung der Messung* (Nuremberg, 1525) for Book IX, chapter 2.
25. Barbaro, MS Italian 5106: III.1, f. 95^{v} (Venice, Biblioteca Nazionale Marciana, cod. 4., Cl. IV, 152 (5106); and see D'Evelyn, 1994, p. 324 n. 24 and pp. 360–63.
26. Barbaro, 1556, III.2, p. 74; cf. Barbaro, 1567, p. 124; and D'Evelyn, 1994, p. 321 n. 13.
27. D'Evelyn, 1994, pp. 291–3, 300–301.
28. See R. Tavernor, *Palladio and Palladianism*, London and New York, 1991.
29. Ibid., pp. 188–209.
30. J.-N.-L. Durand, *Précis of the Lectures on Architecture*, trans.

D. Pritt, Los Angeles, 2000, p. 81; and see Tavernor, *Smoot's Ear*, pp. 112–17.

31. Durand, *Précis*, p. 81.
32. Ibid., p. 133.

Translator's Note

The aim of this translation is to make Vitruvius' text, the most important and influential treatise on architecture in the Western world, available to those who are starting to study the classical and Renaissance traditions in architecture and who have no prior knowledge of the author or of the language in which he wrote. The uniquely rich terminology Vitruvius used to describe many kinds of buildings, and especially the complexities of the orders, conditioned the vocabulary that has been used to discuss architecture from the Renaissance onwards more than that of any other author. Accordingly I have tried to make this translation serve both as an introduction to the Latin text itself for those who may want to turn to the original, and as a series of aides-memoires as to what the original says and how it says it. This is why a selection of Latin technical terms is included in the text in brackets after the translation in English when they first occur; many of these words are also keyed into the captions of the plates and figures. Vitruvius also included a number of Greek words in his text, frequently with their Latin synonyms; at other times he includes Greek terms when an exact Latin equivalent was not available. The Greek words have been retained in the text and are followed by a transliteration in all cases, and sometimes a translation when the Greek word is not interchangeable with a Latin term already presented by Vitruvius. A glossary of words in Greek, Latin and English is provided at the end of the book with transliterations and translations where necessary. In a book designed for general readers in the tradition of the Penguin Classics series, heavy annotation or commentary is not required, and indeed, is unnecessary, given the availability of the more than ample recent commentaries in English, French and Italian, mentioned below, all provided with a plethora of technical drawings. In the notes I have indicated just a few divergences from the text I have usually followed (see below) and just occasionally presented alternative interpretations when they occurred to me.

From the point of view of the translator, however, it is impossible not to agree that 'Very few students and admirers of the great literary legacy of the Augustan age could put their hands on their hearts and thank their lucky stars that they had been granted the opportunity to read the Ten Books of Vitruvius on architecture'.[1] While it is supremely informative, the text is at the same time unlovely to read, and includes a number of interpolations, innumerable obscurities of expression and many technical words that appear nowhere else in Latin literature, or appear here for the first time or used in novel senses.

I have tried to turn everything in his text into English, except for certain words that have no modern, plain language equivalents, such as *apophysis*, *chalcidicum*, *displuviatus* or *tablinum*, or those words that have, in any case, ended up in English without much change, such as *atrium, exedra, pycnostyle* or *vestibulum*. I have also tried to make the English as clear as possible, and shake off, or at least soften and smooth over, a number of aspects of Vitruvius' mannerisms without losing fidelity to the sense of the original; the translation therefore gives a fairly diluted idea of the knotty texture of the original prose. Translating Vitruvius is no easy task, since the text is frequently very compressed, particularly when he describes natural phenomena, such as the constellations and their orbits or the genesis of natural materials, where, often, mental re-punctuation is needed to make the right set of phrases attach themselves to the right nouns; glutinous too are the descriptions of machines, especially of the *ballista* and *scorpio*, where he seems to be copying out technical specifications. Vitruvius also has some unnerving habits that can embarrass the translator by forcing him to present inanities; his frequent use of 'but' when the sense required is 'and'; his persistent use of pleonastic descriptions of things already perfectly described by a single noun – a 'circular base which is equal in all directions', 'a cube is a body with flat, square surfaces of which all sides are of equal breadth'; and statements of the blindingly obvious, such as 'I will explain this so that it will be clear and not obscure to the reader . . .', 'so-and-so is not different but exactly the same', 'foundations . . . should be dug out of solid ground, if it can be found, and as deep down into such solid ground as . . .'.

An attempt to translate Vitruvius into English for the fourth time since 1914 suggests that there must be good reasons for doing so. The translation by William Hickey Morgan (1914), which is still in print, is certainly the best in English and deserves its longevity. He based himself on Valentin Rose's second edition of 1899 (the less good of the two by that distinguished classicist), and I doubt if his dignified and intelligent prose could be surpassed, even though here and there it is faintly dated; he also occasionally uses terms that do not seem to be current in English, such as 'beetles' for wooden rams, or *congé*, the French term for *apophysis*.[2]

The translation by Frank Granger of 1931–4, frequently reprinted and also still available, is based on only one manuscript, MS Harley 2767 in the British Library, to which Granger gave too much importance, and this warps his text at a number of points by precluding other readings equally as likely from other manuscripts.[3] But his translation remains very useful; he frequently broke down to very good effect the long periods used by Vitruvius into short sentences, and was particularly adept at going right to the point and cutting through the circumlocutions. He was also very much at home with the technical terminology, which comes in a number of different varieties in Vitruvius' text depending on the source he was copying out. At the same time, it has to be said that his own English was curiously archaic in patches, even considering its date, and sometimes grammar and even sense break down.[4]

A translation into English by Ingrid Rowland is included in the immensely useful volume *Vitruvius: Ten Books on Architecture*, Cambridge, 1999, which has a valuable introduction and commentary as well as a wealth of line drawings by Thomas

Noble Howe. This book constitutes the only commentary in English and is the most copiously illustrated edition of Vitruvius in any language. The translation is often very acute and adroit, sometimes droll,[5] but on many occasions reveals an attitude towards the translation of a Latin classic which the present translator does not share – that is, there is often too strict an adherence to the word- and phrase-order of the original at the expense of comprehension, as well as the inappropriate retention of Latinate words;[6] but this a question of taste.

In the last decade or so there have been other important developments relating to the text and its exegesis; in 1997 there appeared a magnificent text, translation and commentary by two Italian scholars, Elisa Romano and Antonio Corso (who dealt with alternate books), and a long introduction by Pierre Gros, in a 1,563-page book published by Einaudi at Turin – a *summa* of knowledge about Vitruvius in all departments. Here all aspects of the treatise are examined in vast detail and the full weight of classical archaeology is applied to the whole of the work; appendices by Maria Losito deal also with the *analemma* and the history of attempts to construct the Ionic volute. This is the text that I have used for the present translation, and on those occasions when I have ventured to disagree with their readings or translations, I console myself that I have found myself agreeing often with the best single-volume text of Vitruvius, Valentin Rose's remarkable first edition of 1867.[7] I have also consulted continuously the indispensable Budé series of one-volume editions of each book, started as long ago as 1969 and still proceeding (only Book V is lacking now), and here and there I have followed readings and translations from them against Rose, Corso and Romano. The Budé texts are supplied with rich commentaries and introductions, a text with translation, and particularly, an apparatus criticus at the bottom of the page, which, unfortunately the Einaudi edition does not include (where, instead, divergences from Rose are listed in vol. 2, pp. 1437–53). Between them, the later Budé texts particularly and Corso and Romano present a mass of new suggestions and considerations of all types relating to the whole of the text, and this is one of the main reasons why it seems reasonable to attempt to take account of these researches in another English version.[8]

Editions consulted most frequently:

Vitruvii de architectura libri decem, ed. Valentin Rose and Herman Müller-Strübing, Leipzig (Teubner), 1867.
Budé texts, published at Paris by Les Belles Lettres: *Vitruve*, vol. 1, ed. Ph. Fleury, 1990; vol. 2, ed. L. Callebat, P. Gros and C. Jacquemard, 1999; vol. 3, ed. P. Gros, 1990; vol. 4, ed. P. Gros, 1992; vol. 6, ed. L. Callebat, 2004; vol. 7, ed. B. Liou, M. Zuinghedau and M.-T. Cam, 1995; vol. 8, ed. L. Callebat, 1973; vol. 9, ed. J. Soubiran, 1969; vol. 10, ed. L. Callebat with Ph. Fleury, 1986.
Vitruvio De architectura, ed. A. Corso and E. Romano, introduction by P. Gros, Turin (Einaudi), 1997.

Indispensable too are the following:

Vitruve. De architectura. *Concordance*, ed. L. Callebat, P. Bouet, Ph. Fleury and M. Zuinghedau, Hildesheim, Zurich and New York, 1984.
Ph. Fleury, *La Mécanique de Vitruve*, Caen, 1993.
Dictionaire des termes techniques du De architectura *de Vitruve*, ed. L. Callebat and Ph. Fleury, Hildesheim, Zurich and New York, 1995.

I dedicate the translation to Christine, who has put up with this and so much else in the last thirty-five years.

NOTES

1. L. D. Reynolds and S. F. Weiskittel, in Reynolds (ed.), *Text and Transmission: A Survey of the Latin Classics*, Oxford, 1983, p. 440.
2. *Vitruvius: The Ten Books on Architecture*, Cambridge, Mass., 1914; reprinted New York, 1960. A revised edition of Morgan's

translation of Books I and III–VI appeared recently, with sumptuous illustrations by T. Gordon Smith, with S. Kellogg and M. A. Rosenshine, *Vitruvius on Architecture*, The Monacelli Press, New York, 2003.

3. *Vitruvius on Architecture*, Cambridge, Mass., 1931–4, 2 vols., facing text and translation.
4. e.g. vol. 2, p. 275 (10.1.2): 'But the design which gains an impulse by the power of moving air reaches neat results by the scientific refinement of its expedients.'
5. e.g. p. 113: 'The Serpent Handler . . . with his left foot stomping Scorpio full in the face' (9.4.3).
6. e.g. p. 99: 'For just as a bronze vessel, filled with water not to the brim but to two-thirds capacity, and with a lid set in place, when touched by the intense fervor of fire, will force the water to become superheated, and the water in turn because of its natural rarefaction, as it absorbs the powerful inflation of the boiling, will not only fill the vessel, but raise the lid on its gusts as it grows and overflows, and then with the lid removed and its inflations emitted into the open air, recede back to its place again . . .' (8.3.3).
7. The first Teubner edition (1867) was an extraordinary achievement, particularly given the fact that Rose was working without the vastly increased knowledge of archaeology and technology available today. His sane and cautious text with full apparatus is certainly the best of all the one-volume editions of the nineteenth century; his second edition of 1899 (again, Teubner) includes a number of needless emendations, while the third Teubner by F. Krohn of 1912, while always interesting because he draws attention to difficulties (though sometimes difficulties only he perceives) and often sees things from a new angle, includes much needless interference with the text.
8. I have also consulted the text and limpid translation of the acute and combative Silvio Ferri, which unfortunately includes only a selection from Books I–VII: *Vitruvius Pollio: Architettura*, ed. S. Ferri, Rome, 1960, reissued with an introduction by S. Maggi, Milan, 2002.

Further Reading

The importance of Vitruvius today can be indicated by a single observation; one of the biggest websites, used by architectural historians in particular, is the combined catalogue of the libraries of the German Art Historical Institutes at Florence, Rome and Munich (kubikat.org): it lists more than 230 contributions about him or his influence after 1997, when the fundamental edition by A. Corso and E. Romano with an introduction by P. Gros appeared.

The Einaudi and Budé editions, as well as the commentary and translation by Rowland and Howe, include all the classical and technical bibliography up to the late 1990s (see Translator's Note). We therefore give here a brief list of works, mostly published in the 1990s and the majority in English, that include more panoramic treatments of a number of the themes for which Vitruvius is important.

Ackerman, J. S., *The Villa: Form and Ideology of Country Houses*, London, 1990.

Adam, J. -P., *L'arte di costruire presso i romani*, Milan, 1996.

Campbell, D. B., with B. Delf, *Greek and Roman Artillery 399 BC–AD 363*, Oxford, 2003.

—— *Greek and Roman Siege Machinery 399 BC–AD 363*, Oxford, 2003.

Clarke, G., *Roman House – Renaissance Palaces: Inventing Antiquity in Fifteenth-Century Italy*, Cambridge, 2003.

Germann, G., *Vitruve et le vitruvianisme: Introduction à l'histoire de la théorie architecturale*, trans. M. Zaugg and J. Gubler, Lausanne, 1991.

Gros, P., *L'architettura romana dagli inizi del III secolo a.c. alla fine dell'alto impero: I monumenti publici*, Milan, 2001.
—— *Vitruve et la tradition des traités d'architecture: Fabrica et ratiocinatio; recueil d'études*, Rome, 2006.
Hart, V., and P. Hicks (eds.), *Paper Palaces: The Rise of the Renaissance Architectural Treatise*, New Haven and London, 1998.
Kruft, H. W., *A History of Architectural Theory from Vitruvius to the Present*, New York, 1994.
Lewis, M. J. T., *Surveying Instruments of Greece and Rome*, Cambridge, 2001.
McEwen, I. J., *Vitruvius: Writing the Body of Architecture*, Cambridge, Mass., 2003.
Onians, J., *Bearers of Meaning: The Classical Orders in Antiquity, the Middle Ages and the Renaissance*, Princeton, 1988.
Rowland, I. D. (ed.), *Ten Books on Architecture: The Corsini Incunabulum. Vitruvius with the Annotations and Autograph Drawings of Giovanni Battista da Sangallo*, Rome, 2003.
Rykwert, J., *The Dancing Column: On Order in Architecture*, Cambridge, Mass., 1996.
—— *On Adam's House in Paradise: The Idea of the Primitive Hut in Architectural History*, 2nd edn., Cambridge, Mass., 1997.
Schuler, Stefan, *Vitruv im Mittelalter: Die Rezeption von* De architectura *von der Antike bis in die Frühe Neuzeit*, Cologne, 1999.
Stamper, J. W., *The Architecture of Roman Temples: The Republic to the Middle Empire*, Cambridge, 2005.
Welch, K. E., *The Roman Amphitheatre: From its Origins to the Colosseum*, Cambridge, 2007.
Wilson Jones, M., *Principles of Roman Architecture*, New Haven and London, 2000.
Yegül, F., *Baths and Bathing in Classical Antiquity*, Cambridge, Mass., 1995.

Conferences

Munus non ingratum: proceedings of the international symposium on Vitruvius' *De architectura* and Hellenistic and Republican architecture, ed. H. Geertman and J. J. De Jong, Leiden, 1989.

Le Projet de Vitruve: Objet, destinataires et réception du 'De Architectura', École Française de Rome, 1994.

Vitruvio nella cultura architettonica antica, medievale e moderna: Atti del convegno internazionale, scritti in onore di Claudio Tiberi, ed. G. Ciotta with M. Folin and M. Spesso, Genoa, 2003.

Vitruviuscongres, ed. R. Rolf, Heerlen, 1997.

List of Plates

List of Figures

The second group of illustrations consists of versions of modern technical drawings made by Torsten Schroeder (London School of Economics). This category of illustration is particularly important where Barbaro fails us entirely, above all in reconstructing the machines described in Books IX and X, where all modern readers are indebted to the editions of Book IX by J. Soubiran and Book X by L. Callebat and Ph. Fleury, and the essential contribution of the latter, *La Mécanique de Vitruve*, Caen, 1993 (see p. xliii). The Figures are to be found at the end of the book, pp. 323–60.

ON ARCHITECTURE

BOOK I

Introduction

1. Caesar, Supreme Ruler: while your divine intelligence and supernatural power were acquiring the mastery of the whole world and Roman citizens were glorying in your triumph and your victory, once all your enemies had been obliterated by your indomitable bravery, and all the peoples you had conquered awaited your command, and the Roman People and Senate, freed from fear, began to be governed by your far-ranging plans and decisions,[1] I did not dare, when you were so occupied with such important matters, to publish my writings on architecture and the ideas I had developed after long reflection, for fear that by interrupting you at an inopportune moment I might incur your displeasure.

2. When, in fact, I realized that you have taken in hand not only the everyday lives of all our citizens and the organization of the state, but also the development of public buildings so that not only has the state been enriched, thanks to you, with new provinces, but also the majesty of its power is already being demonstrated by the extraordinary prestige of its public buildings, I thought that I should not let slip the opportunity to publish my writings on this subject, dedicated to you, as soon as possible, particularly since I was initially known to your father for my work in this field and was a devoted admirer of his courage. When, therefore, the heavenly council had consecrated your father in the residences of immortality and had passed his power into your hands, my enduring devotion to his memory was transferred to you and found favour with you.[2]

And so, with Marcus Aurelius, Publius Minidius and Gnaius Cornelius, I was put in charge of the supply and repair of *ballistae*, *scorpiones* and other types of artillery, and, with them, I received my reward: and it was you who granted it to me first and who continued it on the recommendation of your sister.[3]

3. Therefore, since I was indebted to you for this favour, which was such that I need have no financial anxieties for the rest of my life, I began to write this work for you since I noticed that you have built much and continue to do so now, and that for the foreseeable future you will ensure that both public and private buildings will so match the majesty of your achievements that they will be handed down in the memory of future generations. I have put in writing precise recommendations so that by examining them, you yourself may become familiar with the characteristics of buildings already constructed and of those which will be built: for in these books I have laid out all the principles of the discipline.

CHAPTER I

The Education of the Architect

1. The architect's professional knowledge is enriched by contributions from many disciplines and different fields of knowledge and all the works produced by these other arts are subject to the architect's scrutiny. This expertise derives from practice and theory. Practice consists of the ceaseless and repeated use of a skill by which any work to be produced is completed by working manually with the appropriate materials according to a predetermined design. Theory, by contrast, is the ability to elucidate and explain works created by such manual dexterity in terms of their technical accomplishment and their proportions.

2. So architects who have struggled to achieve practical proficiency without an education have not been able to achieve recognition commensurate with their efforts: by contrast, those who have relied only on theory and book-learning were evidently chasing shadows rather than reality. But those who have mastered both, like men supplied with all the necessary weapons, have achieved recognition and fulfilled their ambitions more quickly.

3. For all fields, and especially architecture, comprise two aspects: that which is signified and that which signifies it. That which is signified is the object under discussion, while that which signifies is an explanation of it conducted according to scientific principles. Therefore it is evident that a man who wants to proclaim himself an architect must be proficient with regards to both aspects. So he must be naturally gifted and ready to learn, for neither natural aptitude without learning nor learning without natural aptitude will make a perfect practitioner. He should have a literary education, be skilful in drawing, knowledgeable about geometry and familiar with a great number of historical works, and should have followed lectures in philosophy attentively; he should have a knowledge of music, should not be ignorant of medicine, should know the judgements of jurists and have a command of astronomy and of the celestial system.

4. These are the explanations for all this. The architect must have a literary education so that he can leave a more dependable record[4] when writing up his commentaries. Then he must have the expertise in drawing which will enable him to represent more easily the appearance of the work he wishes to design with painted models.[5] Geometry as well is extremely helpful in architecture, and teaches, first, the use of straight lines and then of the compasses, thanks to which, above all, the plans of buildings on their designated sites can be prepared very readily with the correct alignments of right angles, horizontals and verticals. Again, a knowledge of optics enables light to be drawn correctly from well-defined areas of the sky. A knowledge of arithmetic enables us to calculate the cost of buildings precisely and apply the techniques of mensuration correctly, and difficult problems relating to modular systems [*symmetriae*] are resolved by the application of the laws of geometry.[6]

5. He should also have a wide knowledge of history because

architects often devise a great deal of ornament for their buildings, the meaning of which they must be able to explain to those who ask why they have made them. For example, if an architect sets up in a building marble statues of women draped in matronly robes, called Caryatids, instead of columns, and puts mutules [*mutuli*] and cornices [*coronae*] above them, he would supply the following explanation to those who asked him why: Caria, a city in the Peloponnese, sided with their enemy, the Persians, against Greece; subsequently, the Greeks, liberated from war by a glorious victory, formed a league and declared war on the Carians.[7] They captured the town, killed the men and auctioned the citizens; the women were taken into slavery and were not allowed to remove their robes or other such ornaments indicating their married status: in this way they would not only be dragged along in a triumphal procession but, as eternal examples of slavery crushed by appalling humiliation, would seem to pay the penalty for the whole city. This is why architects of the period devised images of them placed in load-bearing positions in public buildings so that the punishment for the crime of the Carians would be known to posterity and remain in history [Plates 1, 2].

6. Similarly the Spartans, after defeating the countless hordes of the Persian army at the battle of Plataea with a handful of soldiers led by Pausanias, the son of Agesipolis, celebrated a glorious triumph with the spoils and booty, and built a Persian portico with the proceeds as proof of the bravery and glory of their citizens, and as a trophy of victory for future generations.[8] There they placed images of the prisoners dressed in barbarian clothing supporting the roof, their arrogance punished with well-deserved humiliation, so that enemies would be terrified at the results of the Spartans' courage, and their own citizens, contemplating this paradigm of bravery and buoyed up by the glory, would be ready for the defence of their liberty. So from then on, many architects put up statues of Persians supporting architraves [*epistylia*] and their mouldings[9] and have consequently introduced striking variations on the theme in their buildings. And there are other stories from history of the same kind which architects should be aware of.

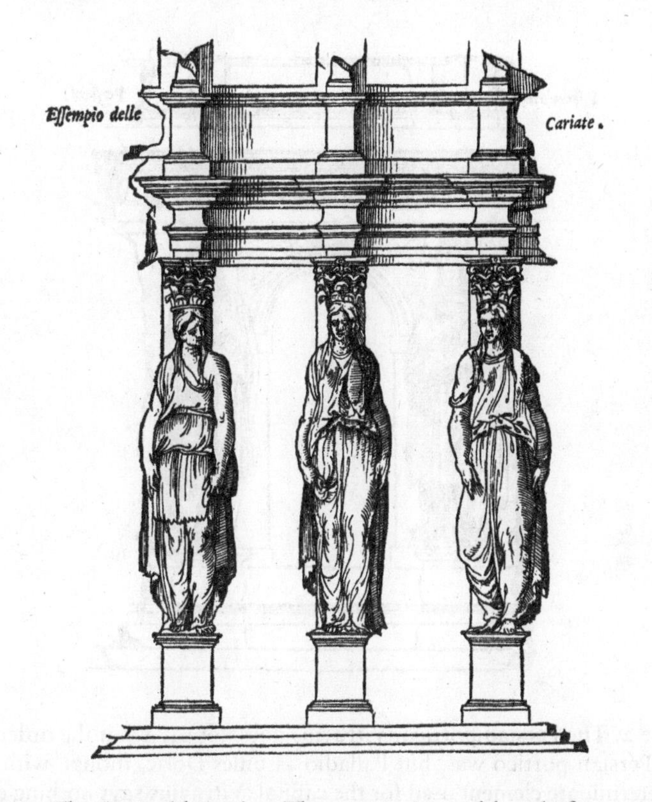

Plate 1. The Caryatid portico. The proportions of female figures used as substitutes for columns at, for example, the Treasuries of the Cnidians and Siphnians at Delphi, the Erechtheum, the Perikles shrine at Limyra, presumably the Agrippan version of the Pantheon and in the upper storey of the forum of Augustus had the disadvantage that they were too short for their breadth, causing ancient architects to add the *polos*, a circular element under an echinus, and other devices, such as pedestals, to raise their height. Here Palladio provides backing pilasters for the figures, which shows that they are not free standing as in ancient exemplars. To fill out Vitruvius' bare description, Palladio has derived much from an illustration by Marc'Antonio Raimondi of the 1520s which showed the male statues in the Doric ground floor and the female figures above surmounted by Ionic (not Corinthian) capitals, as was usual in the Renaissance. Vitruvius in fact places his Caryatids in a Doric context.

Plate 2. The Persian portico. Vitruvius does not say what the order of the Persian portico was, but Palladio assumes Doric, though with an indeterminate element used for the capital. Vitruvius says nothing else about the architecture, so Palladio uses elements he knows from a variety of other sources: the pulvinate *basamento* of the type used for the Arco di Portogallo, introduced into sixteenth-century architectural vocabulary by Raphael at the Villa Madama and taken up by Galeazzo Alessi at the Villa Giustiniani (Albaro, Genoa) and the Palazzo Marini (Milan); *bucrania* in metopes probably following the frieze of the circular temple at Tivoli, which he also used in his own courtyard of the Carità in Venice; the great figures from the celebrated Trajanic statues of captured Dacians, known as the Farnese Prisoners, then in the Palazzo Colonna at SS Apostoli, Rome (now in the Museo Archeologico Nazionale, Naples). The most impressive Renaissance version of the portico with male prisoners was the façade of the House of Leone Leoni (Milan) where mighty male half-figures (terms), representing barbarians defeated by Marcus Aurelius, are substituted for parts of the pilasters.

7. Philosophy in fact makes the architect high-minded and ensures that he will not be arrogant, but rather flexible, fair and trustworthy without greed, which is the most important quality, since no work can be carried out satisfactorily without loyalty or integrity; and [philosophy also ensures] that he will not be avaricious or preoccupied with receiving rewards, but will safeguard his own standing rigorously by maintaining his own good name; this is what philosophy teaches us. Furthermore, philosophy includes the study of nature, which the Greeks call φυσιολογία [*physiologia*], which must be pursued in some depth because it covers many different types of problems in nature, such as how to conduct water. For pressure-pockets naturally form in one place or another in conduits that run downhill and are led round circuitously and ascend after a horizontal stretch, and anybody who has not learnt about the laws of nature from a study of philosophy will be unable to avoid the resulting damage. So whoever reads the works of Ctesibius or Archimedes and of other authors who wrote manuals of the same type will not understand these laws unless he has been instructed in these matters by philosophers.

8. The architect should also understand music so that he is conversant with the system of harmonic relationships and mathematical theory, and, consequently, is capable of adjusting the tuning of *ballistae*, catapults and *scorpiones* correctly. For to right and left of the frames of these machines are the spring-holes through which ropes made of twined cord are put under tension with windlasses and hand-spikes; they are not secured and tied off unless they produce sounds which are both identical and correct to the ear of the engineer. For when the arms inserted in these springs are put under tension, they should deliver the same thrust equally on each side, because if they are not identically tuned, they will interfere with the straight trajectory of the projectiles.

9. So too in theatres, bronze vessels, which the Greeks call ἠχεῖα [*echeia*], are placed in small chambers under the steps following the mathematical principles of the musical intervals. These bronze vessels are arranged around the circumference of the theatre at regular intervals so as to provide musical concords

or harmonies with intervals of a fourth, a fifth and from the octave up to the double octave. The aim is that when the voice of the actor resonates with the vases distributed around the theatre as it strikes them, it is amplified and reaches the ears of the audience more clearly and pleasantly. Similarly nobody can construct hydraulic organs and other such instruments without a knowledge of musical principles.

10. The architect should also have some knowledge of medicine, because of the problems posed by the latitude, which the Greeks call κλίματα [*klimata*], by the properties of the air, by locations which are healthy or infected, and by water use: for without such knowledge no healthy house can be built. He should also know the mandatory legal regulations for the construction of buildings with party walls, for the distribution of eaves and sewers, and of windows and water-pipes around buildings. And architects should know about other similar things, so that, before starting building, they can ensure that no legal disputes are left for householders after the work has been completed, and that in drawing up contracts due caution is observed with regard to the interests of both the employer and the contractor: if the contract has been well drafted, either party could release himself from the other without specious wrangling. And through a study of astronomy, the architect recognizes east, west, south and north as well as the principles governing the sky, the equinoxes, the solstices and the courses of the stars. If someone is ignorant of these subjects it will be impossible for him to understand the principles behind sundials [*horologia*].

11. Consequently, since such a wide discipline should be enriched, and overflow with many different kinds of expertise, I do not think that people can justifiably profess themselves architects at the drop of a hat, unless, having climbed the steps of these disciplines from their youth and having been brought up with a wide literary and technical knowledge, they have reached the highest sanctuary of architecture.

12. But perhaps laymen will find it unbelievable that a man's intellect enables him to understand and retain such a large number of disciplines. But when they realize that all disciplines are connected with, and feed into, each other, they will readily believe that this can happen. For a general education is like a single body composed of these different limbs. This is why those who are instructed in various subjects from a tender age recognize the ground common to all areas of study and the complementary relationships between all the disciplines, and for that reason can more readily understand all of them. This is why one of the ancient architects, Pytheos, who designed the Temple of Minerva at Priene so brilliantly,[10] says in his *Commentaries* that an architect should be able to achieve more in all the arts and sciences than those who, by hard work and constant practice, have taken their individual specialities to the highest level of sophistication. But in reality this does not happen.

13. The architect need not and cannot be a grammarian of the stature of Aristarchus, though he must not be illiterate; nor a musicologist such as Aristoxenus, though he must not be ignorant of music; nor a painter like Apelles, though he should not be incompetent as a draftsman; nor a sculptor on the level of Myron or Polyclitus, though he should not be ignorant of the techniques of sculpture; nor, again, a doctor such as Hippocrates, though he should not be entirely ignorant of medicine; nor indeed should he be outstanding in any one of the other sciences, though not incompetent in any of them.[11] For, given the vast variety of these subjects, nobody can attain mastery in each because it is hardly possible for anyone to absorb and assimilate their theoretical principles.

14. In any case, it is not only architects who are unable to achieve the greatest proficiency in all fields; not even all of those who pursue a specialism in one field of the arts can attain the highest level of excellence. Therefore, if individual specialists – and not all of them, but only a few – have, with great difficulty, achieved enduring distinction across the whole of time in individual disciplines, how can the architect, who should be competent in many fields, not only avoid being deficient in these subjects, a really remarkable feat in itself, but even surpass all the specialists who have devoted unremitting industry to their individual fields?

15. Accordingly, it seems that with respect to this point Pytheos was mistaken in not taking into account the fact that each discipline comprises two aspects: the work produced and the theory implicit in that work. The first of these, that is, the production of the work, is the province of those who have trained themselves in single fields. The other, theory, is common to all cultivated men; for example, the rhythm of the pulse in our veins and that of the locomotion of our feet is the province of doctors and musicians respectively; but if one needs to cure a wound or snatch some sick person from danger, a musician would not make the house-call, since this will properly be the doctor's work. So too in the case of a musical instrument, it will not be a doctor but a musician who will play it so that our ears receive the appropriate pleasure from its music.

16. Similarly, astronomers and musicians share common ground for discussions of the harmonies of the stars and of musical concords, of squares and triangles, and of fourths and fifths; and astronomers share with geometricians the subject of vision, called λόγος ὀπτικός [*logos optikos*] in Greek; and in all other sciences, many, or even all points are common ground, at least as far as debating is concerned. In fact, undertaking works which are perfected by manual dexterity is the province of those who are specially trained in the exercise of just one art. It therefore seems that anyone who has an average understanding of those aspects of the separate disciplines and their theoretical foundations essential for architecture has done more than enough to ensure that he would not be left floundering if he had to judge and evaluate one of these subjects or techniques.

As for those on whom nature has bestowed so much ingenuity and acuteness and such powers of recall that they have a profound understanding of astronomy, geometry, music and the other subjects, they go beyond the competence of architects and become great intellectuals; so they can easily argue the pros and cons of such disciplines since they are armed with so many dialectical weapons from within those same disciplines. People like this, however, are rarely found; but at various times there were men like Aristarchus of Samos, Philolaus and Archytas of Tarentum, Apollonius of Perge, Eratosthenes of Cyrene,

Archimedes and Scopinas from Syracuse, who left for future generations many inventions in the fields of mechanics and sundials which they had developed and explained using mathematics and the laws of nature.[12]

17. Since, then, such intellectual gifts deriving from innate talent are not granted indiscriminately to all mankind, but to only a few, and since the architect's profession requires proficiency in all branches of learning, and our intellectual faculties certainly do not allow us a consummate, but rather a middling knowledge of these disciplines because of the vastness of the field, I ask you, Caesar, and others who will read these books, to overlook anything in my exposition that falls short of the highest literary standards. For it is not as a consummate philosopher, nor as a fluent rhetorician, nor as a grammarian practised in the highest skills of his art that I have laboured to write these books, but as an architect with only a passing acquaintance of these cultural themes. But with respect to the possibilities presented by the art and its theoretical foundations, I promise, and hope, that by means of these books I will provide an unequivocally authoritative account not only for those who build but also for all men of culture.

CHAPTER II

The Principles of Architecture

1. Architecture consists of planning [*ordinatio*], called τάξις [*taxis*] in Greek, projection [*dispositio*], which the Greeks call διάθεσις [*diathesis*], harmony [*eurythmia*], modularity [*symmetria*], appropriateness [*decor*] and distribution [*distributio*], called οἰκονομία [*oikonomia*] in Greek.

2. Planning consists of adapting each individual element of a building to the right dimensions and establishing its overall proportions by reference to modularity. It comprises quantity, ποσότης [*posotes*] in Greek: quantity consists of the selection of modules from elements of the building itself and the harmonious

execution of the whole work starting from the individual components of its parts.

Projection consists of the appropriate placement of the components of a building and the elegant completion of the work based on a combination of the parts appropriate to the characteristics of the work.[13] The types of projection, ἰδέαι [*ideai*] in Greek, are these: the ground-plan [*ichnographia*], the orthogonal[14] elevation [*orthographia*] and the perspectival drawing [*scaenographia*]. The ground-plan involves the correct use to scale of the compasses and the ruler and the plans of buildings at ground level are derived from it. But the orthogonal is an elevation drawing of a façade and a representation of it rendered to scale following the proportions of the future building. And a perspective drawing consists of a sketch of a façade and its receding sides with all the lines converging at the centre of a circle. These three result from analysis [*cogitatio*] and invention [*inventio*]. Analysis is the most concentrated attention, full of enthusiasm, hard work and alertness, aimed at the pleasurable outcome of a project, while invention is the resolution of intricate problems and the discovery of solutions thanks to intellectual versatility. These are the terms implied by projection [Plates 3, 4].

3. Harmony consists of a beautiful appearance and harmonious effect deriving from the composition of the separate parts. This is achieved when the heights of the elements of a building are suitable to their breadth, and their breadth to their length, and, in a word, when all the elements match its modular system.

4. Modularity is the appropriate agreement of the components of the building itself and the correspondence of the separate parts to the form of the whole scheme based on one of those parts selected as the standard unit. Just as in a human body the nature of its harmony is modular[15] and derives from the forearm, the foot, the palm, the finger and other small parts, the same happens when it comes to the construction of buildings. First: in the case of sacred buildings, their modular systems are derived either from the diameters of columns or a triglyph [or from an *embater*],[16] but also in the case of the

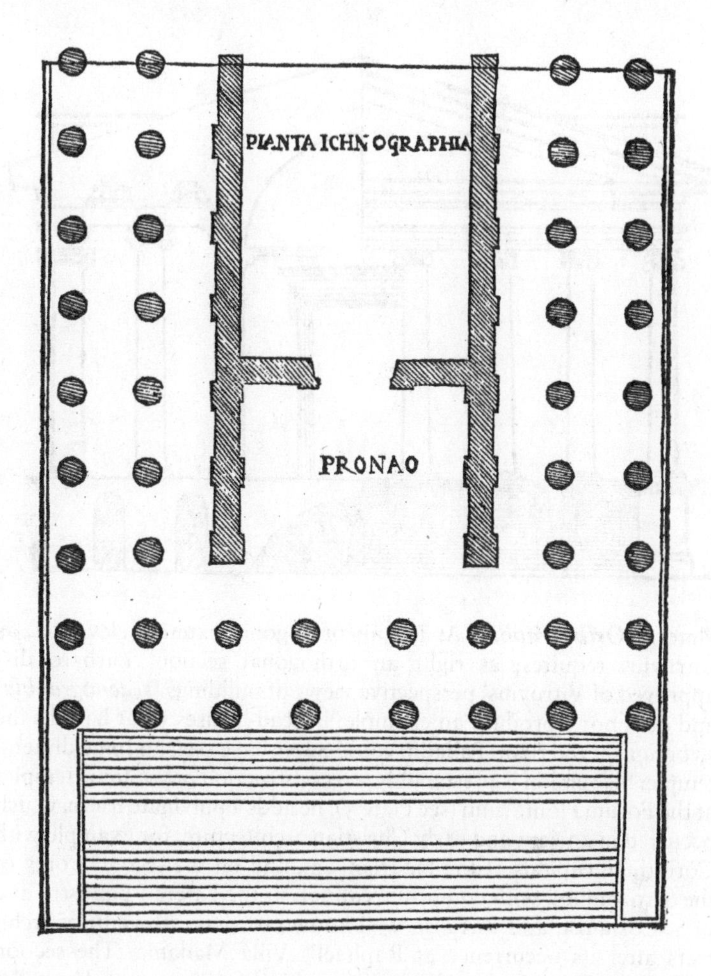

Plate 3. *Ichnographia.* To illustrate the ground-plan, Palladio shows the front half of a dipteral temple with octastyle façade and wider central aperture surmounting a podium on three sides, an illustration he reuses to illustrate the dipteral temple. The plan probably owes something to the central, Ionic, peripteral hexastyle temple (dedicated to Juno Sospita?) or the hexastyle Ionic temple without *posticum* at the north (dedicated to Janus?) in the Forum Holitorium at S. Nicola in Carcere, and bears a remarkable resemblance to that published by Palladio in his *Four Books* (1570) (see Palladio, 1997, p. 283) for the octastyle Temple of Vespasian (IV.71).

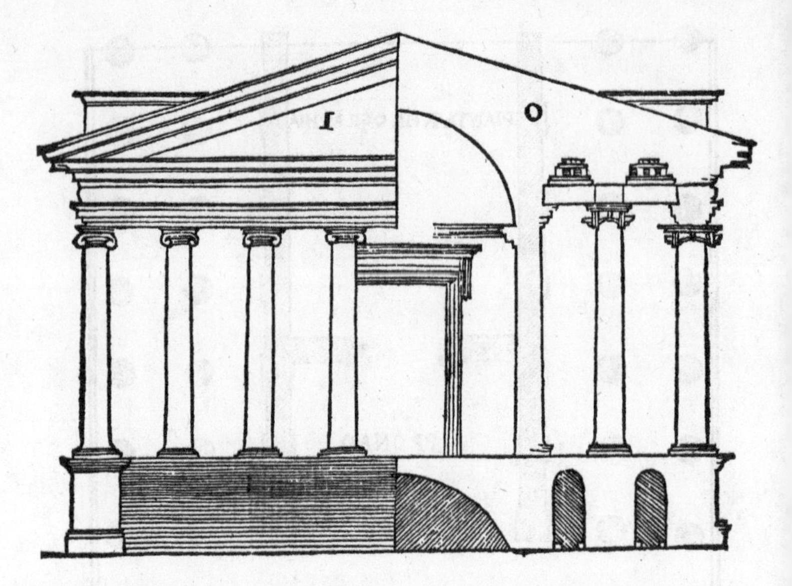

Plate 4. *Orthographia.* At left an orthogonal exterior elevation, as Vitruvius requires; at right an orthogonal section. Barbaro disapproved of Vitruvius' perspective views of buildings (*scaenographia*) and does not reproduce an example; instead he uses what he calls the *sciografia* at the right. Palladio could never have seen an Ionic dipteral temple: his version is presumably a reconstruction of the Ionic temples at the Forum Holitorium (see Plate 3): he adds a pulvinate frieze, which occurred in ancient and early Christian architecture, for example with Corinthian capitals at the Hadrianeum, and on the second storey of the Settizonium, with composite capitals at the Lateran Baptistry and in S. Costanza, and was also well known to sixteenth-century architects after its occurrence at Raphael's Villa Madama. The section includes an inner order of columns without architraves and so taller than that on the exterior. Perhaps the low vault in the centre of the *stereobate* derives from those of the temples of Minerva in the Forum of Nerva and of Antoninus and Faustina, also drawn by Palladio (IV.25, 33; Palladio, 1997, pp. 235, 245). The door, appropriately, is Ionic with consoles, as recommended by Vitruvius.

ballista, from the hole in it which the Greeks call the περίτρημα [*peritrema*], and in the case of boats, from the distance between the thole-pins, which is called the διάπηγμα [*diapegma*]; and so on, for all other works, the calculation of the modular systems is derived from the elements that compose them.

5. Appropriateness consists in the perfect appearance of a work composed using the correct elements in accordance with precedence. This is achieved by following a rule, θεματισμός [*thematismos*] in Greek, or a custom or nature. One follows a rule when roofless buildings open to the sky [*hypaethra*] are built to Jupiter, Creator of Lightning, Caelus and the Sun and Moon: for the appearances and manifestations of these deities are visible to us in the sky when it is clear and bright. Doric temples should be dedicated to Minerva, Mars and Hercules, since it is appropriate to provide buildings without elaborate ornament for these deities because of their warlike character. Temples of the Corinthian order[17] built for Venus, Flora, Proserpina, the God of Springs and of the Nymphs will clearly have the right characteristics, because, given the gentleness of these deities, more graceful and florid buildings decorated with leaves and volutes will clearly enhance the appearance appropriate to them. If Ionic temples were to be built for Juno, Diana, Father Liber and all other such deities, their intermediate position will be acknowledged because their principal characteristics will be balanced between the severity of the Doric style and the delicacy of the Corinthian.

6. Appropriateness in accordance with custom is demonstrated when, for example, suitable and elegant vestibules [*vestibula*] matching magnificent interiors will be built for buildings; for if the interiors have elegant finishes but the entrances are ordinary and shabby, they will not conform to what is appropriate. Again, if dentils [*denticuli*] were to be carved in the cornices of Doric entablatures or if triglyphs were to be designed along with pulvinate capitals[18] and Ionic architraves, the appearance would be disconcerting, since elements particular to one system have been transferred from one type of building to another, when the conventions already established for the system[19] are different from this.

7. Appropriateness will conform to nature if, at the outset, very healthy sites are selected for all sacred enclosures and adequate sources of water are chosen for the locations in which sanctuaries [*fana*] are to be built; this applies particularly to Aesculapius, Salus and other deities by whose healing powers great numbers of sick people are evidently cured. For when their ailing bodies are transferred from some unhealthy location to a healthy one and are treated with waters deriving from curative springs, they will recover more rapidly. The result of this would be that because of the nature of the site, the deity will enjoy both greater fame and enhanced prestige. Again, appropriateness to nature will be observed if the light for bedrooms [*cubicula*] and libraries [*bibliothecae*] is derived from the east, for bath-rooms [*balnea*] and winter apartments from the west, for picture galleries [*pinacothecae*] and rooms which require steady light from the north, because that zone of the sky is not dazzled or obscured by the trajectory of the sun, but the light remains constant and regular all day long.

8. Distribution consists of the appropriate management of resources and of the site, and the prudent control of the finances during construction thanks to careful calculation. This will be achieved if, at the outset, the architect does not try to acquire materials that either cannot be found or can only be acquired at high cost. For supplies of quarry-sand, quarry-stones [*caementa*], fir, deal and marble are not available everywhere, since one occurs in one area and another elsewhere, and importing them is difficult and costly. Instead, where quarry-sand is not available, we must use river sand or washed marine sand, and similarly, the absence of supplies of fir or of planks of deal will be obviated by the use of cypress, poplar, elm and pine, and other difficulties will be resolved in the same way.

9. The second level of distribution consists of planning buildings which differ depending on whether they are intended for use by the heads of families[20] or people with great wealth or political power. For it is obvious that urban houses should be constructed differently from those whose owners derive their income from country estates, and again, those for moneylenders should differ from those of the rich and cultivated; and then

houses for the powerful, whose deliberations govern the state, will be organized to suit their needs: in brief, the layouts of buildings must be appropriate to each class of person.

CHAPTER III
The Divisions of Architecture

1. Architecture has three divisions: the construction of buildings, of sundials and of machines. Construction in turn is divided into two parts, one of which consists of the deployment of city-walls and civic buildings on public sites, the other, of the development of private buildings. Public buildings are divided into three categories: first, those destined for defence, second, those for religious use, and third, those for public utility. Defence involves devising walls, towers and gates capable of resisting enemy onslaughts at any time: under religion comes the planning of sanctuaries and other sacred buildings dedicated to the immortal gods; and public utility is concerned with the arrangement of communal areas for use by the people, such as ports, squares, porticoes, baths, theatres, covered walks [*inambulationes*] and everything else that is designed on the same principles for public sites.

2. And all these buildings must be executed in such a way as to take account of durability, utility and beauty. Durability will be catered for when the foundations have been sunk down to solid ground and the building materials carefully selected from the available resources without cutting corners; the requirements of utility will be satisfied when the organization of the spaces is correct, with no obstacles to their use, and they are suitably and conveniently orientated as each type requires. Beauty will be achieved when the appearance of a building is pleasing and elegant and the commensurability of its components is correctly related to the system of modules.

CHAPTER IV

The Choice of Healthy Sites for Cities

1. These will be the principles to follow with regard to city-walls. First, the choice of a very healthy site: this will be in a high place, without mists or frost, and exposed to weather conditions that are neither sweltering nor freezing, but temperate; moreover, proximity to marshy terrain is to be avoided. For when the morning breezes blow towards the town at sunrise, and these are joined by the mists that have sprung up, and the noxious breath of marsh animals mixes with the mist and wafts into the bodies of the inhabitants – all this makes the site unhealthy. Again, if the city-walls are to be on the sea and face south or west, they will not be good for the health, because in the summer the zones exposed to the southern sky heat up with the rising sun and burn hot at midday, while a site facing west is warmed when the sun has risen, is hot at midday and swelters in the evening.

2. So the result is that these oscillations between hot and cold will damage the bodies of people living in these places: one also notices the same thing in the case of inanimate objects. For example, nobody takes the light for covered wine-stores from the south or west, but from the north, because that orientation is never subject to changes in temperature, but is always stable and unchanging. The same applies to granaries [*granaria*] exposed to the trajectory of the sun which rapidly damages the qualities of the produce, and the foodstuffs and fruit which are not stored in a location facing away from the course of the sun will not keep for long.

3. For when heat cooks the consistency out of things and extracts their innate qualities by sucking them out with its hot vapours, it always dissolves them and makes them soft and weak with its intensity. We can see this in the case of iron: although it is naturally hard, it becomes so soft when heated through by the fiery vapour in furnaces that it can easily be forged into any shape; and the same iron, if it is cooled down by being dipped in cold water while soft and incandescent, will become hard again and resume its original properties.

4. We can confirm the truth of this by considering the fact that during the summer, all organisms grow weaker because of the heat, not only in unhealthy but also in healthy locales, and that during the winter even the most noxious zones become healthier because they are reinvigorated by the cold. In the same way, too, even organisms transported from cold to warm zones have no capacity to resist, but become completely drained, while those that pass from hot regions to cold northern zones not only do not suffer ill health because of the change of locale, but in fact grow stronger.

5. Accordingly, we must avoid locating city-walls in regions where unhealthy vapours can infiltrate the bodies of the inhabitants because of the heat. For while all our bodies are composed of elements, which the Greeks call στοιχεῖα [*stoicheia*], namely, heat, humidity, earth and air, so too the characteristics of all living creatures in the world, depending on the species, are governed by the mixture of elements according to natural equilibria.

6. Therefore, in those organisms in which heat predominates over the other elements, it will kill and destroy all the others with its intensity. This damage is produced by the torrid climate of certain zones when heat infiltrates open pores in an amount greater than the natural balance of the elements mixed in the body can tolerate. In the same way, if moisture has invaded the pores of our bodies, destroying their equilibrium, the other elements dissolve when broken down by the liquid, and the normal characteristics of their composition are destroyed. Again, defects of this type infiltrate organisms because of the cooling effect of the moisture in winds and breezes. In the same way, an increase or diminution of the natural proportion of air or earth in our organisms weakens the other elements; the predominance of earth is due to too much food, that of air to the density of the atmosphere.

7. But whoever wants to understand these principles more deeply by personal observation should take note of and study the natures of birds, fish and land animals, and so will identify

the differences in their composition. For the species of birds is composed of one mixture of elements, that of fish of another, and that of land animals is very different again. Birds have less earth and moisture, a moderate amount of heat and a lot of air: therefore they can soar more easily against the currents of the air since they are composed of lighter elements. In the case of the aquatic nature of fish, since they are moderately supplied with heat and composed mainly of air and earth but very little moisture, it follows that the less moisture their bodies include in relation to the other elements they comprise, the more easily they survive in water; consequently, when they are brought to land, they lose their lives along with the moisture. So too, land animals cannot survive long in water because the humid components preponderate since they are composed of moderate amounts of the elements of air and heat, and have less earth but a lot of water.

8. Therefore, if these facts are as I have explained them and we can see for ourselves that the bodies of living things are composed of these elements, and if we are convinced that they suffer and become debilitated from excesses or deficiencies of certain elements, then we can be in no doubt that we must take the greatest care to select the most temperate zones, since good health is the essential prerequisite when siting city-walls.

9. So I am convinced that we must resort to the ancient method more and more; for our ancestors sacrificed animals grazing on the sites where they wanted to build towns or permanent camps, and examined their livers; if, according to the first test, the livers were bluish and damaged, they sacrificed other animals because they were not certain whether the livers had been damaged by some disease or by bad alimentation. After they had made a number of tests and demonstrated that the consistency and solidity of the livers resulted from the local water and pasturage, they set up their fortifications on that spot; if, however, they kept on finding that the livers were infected, they would deduce that the supply of water and food occurring naturally in these areas would be just as damaging to human beings, and so they moved on and changed to another area in search of conditions which were healthy in every respect.

10. That the qualities of healthy land can be determined from its pasturage and food can be observed and investigated by an examination of the fields in Crete around the river Pothereus, which flows between the cities of Cnossus and Gortyn on that island. For cattle graze on both banks of the river, but those which feed near Cnossus have spleens, while those that feed on the other side near Gortyn do not seem to have them. Consequently, the doctors investigating this phenomenon found a plant growing in the area that the cattle chew, which caused the reduction of the spleen. So they collect the plant and treat the splenetic with this medicine, which the Cretans in fact call ἄσπληνον [*asplenon*; without spleen]. This fact shows that we can learn from the food and water whether the characteristics of sites are naturally healthy or unhealthy.

11. Again, if city-walls are to be built in marshy areas lying along the seashore and facing north or north-east, and the marshes are higher up than the seashore, then the walls will clearly have been laid out sensibly. For the marsh-water can be released onto the shore by digging ditches, and when the sea gets up because of storms, the surge of water is driven by the turbulence into the marshes, where, because of the admixture of salt water, it prevents the marsh animals from reproducing, while the creatures which swim down from higher areas and arrive in the vicinity of the shore are killed by the unaccustomed level of salinity. This phenomenon can be exemplified by the Gallic marshes around Altinum, Ravenna, Aquileia[21] and other municipalities in similar areas near marshes because they are incredibly healthy, for the reasons given.

12. But in areas where there are standing marshes with no exit-channels through rivers or canals, like the Pomptine marshes, they stagnate and go bad, giving out heavy and unhealthy vapours in the surrounding areas. In Apulia, for example, the ancient town of Salpia, founded by Diomedes on his way back from Troy or, according to other writers, by Elpias of Rhodes, was built in an area of this type:[22] for this reason the inhabitants suffered for years from various complaints and eventually went to Marcus Hostilius and requested him, with a public petition, to search for and select a suitable site for them to which they could relocate their city-walls. Then without delay, he immediately made exhaustive investigations, bought a site in a healthy area by the sea and petitioned the Senate and Roman People for authorization to transfer the town there. He then built the walls, divided up the lots and sold the rights for each one to every citizen for a *sestertius*. After this, he opened up the lake towards the sea and turned it into a port for the city. This is why the inhabitants of Salpia, who moved themselves four miles from their old town, now live on a healthy site.

CHAPTER V

City-walls, Towers and Ramparts

1. When, following these procedures, the question of the healthiness of the site of the city-walls to be laid out has been resolved, and when areas of the country rich in provisions for supplying the city have been selected, and the construction of roads or the exploitation of navigable rivers or access for maritime traffic through ports have made the transport of goods to the city practicable, then the foundations of the towers and walls must be constructed like this: one should dig down to solid ground, if it can be found, then down into it as far as seems proportionate to the size of the building; the breadth of the excavation should be greater than that of the walls to be built above ground; the hole should then be filled in with the strongest possible masonry [*structura*].

2. Again, the towers should project from the walls in such a way that when the enemy wants to approach the wall for an assault he can be attacked with missiles on his exposed sides from the towers to left and right. Above all, it is obvious that one must ensure that there will be no easy approach by which to attack the wall, but that the wall will be laid out around crevasses and so devised that the approaches to the gates are not straight, but come in from the defenders' left: when the

approaches are arranged like this, the right flank of those advancing, which will not be protected by shields, will be closest to the wall. Towns should not be laid out as squares or with salient angles but with curved perimeters, so that the enemy can be spotted from as many points as possible. For when city-walls have salient angles, they are difficult to defend because an angle offers more cover to the enemy than to the citizens.

3. With regard to the thickness of the wall, my view is that it should be built in such a way that armed men meeting each other on the rampart can pass each other without obstruction. Then in the space [between the two walls] whole beams of charred olive-wood should be set together as closely as possible, so that the inner faces of the two walls, bound together by the beams as though by metal braces, last indefinitely; for this is a wood that neither decays nor is damaged by bad weather or the passage of time, and even if it is buried in the ground or submerged in water remains unscathed and permanently fit for use. So not only city-walls, but also substructures and those types of walls which must be built as thick as city-walls will not deteriorate rapidly if they are tied together using this method.

4. The distances between towers are to be set so that they are no further than a bowshot apart, so that if any of the towers is attacked, the enemy could be driven back by *scorpiones* and other projectile-throwing machines fired from the towers located at right and left. Moreover, the walls corresponding to the interiors of the towers are to be interrupted by gaps as wide as the towers themselves in such a way that the walkways across the interiors of the towers are made of planks of wood, but are not fixed down with nails;[23] for if the enemy takes some part of the wall, the defenders could cut away the walkways, and, if they manage to do so quickly, could prevent the enemy from making his way to other areas of the towers and walls unless he is prepared to take a really bad fall.

5. Towers should be built in circular or polygonal[24] form. Siege engines destroy square towers more rapidly because the hammering of the battering-rams pounds the corners off: but in the case of circular towers they can do no damage because

they merely drive the stones to the centre like wedges. Defensive systems with walls and towers in conjunction with ramparts are the most protected because neither battering-rams, mining nor any other machinery is capable of damaging them.

6. The rampart system, though, is not to be used in all locations, but only where there is an approach from high ground outside the wall from which the walls can be attacked at the same level. So in places like this, we must first dig ditches as wide and deep as possible, and then the foundation of the wall must be sunk into the bed of the excavation and built thick enough to contain the earthwork easily.

7. Then another foundation should be built on the inside of the substructure and separated from it by a wide area, large enough for the cohorts to take up position on the broad platform of the rampart and make their defence lined up as though in combat formation. When the foundations have been built separately from each other in this way, then cross-walls should be laid out between them connecting the exterior and interior foundations, and arranged in a zigzag, as the teeth of a saw usually are. When the substructure is built like this, the enormous weight of earth distributed over small compartments cannot conceivably force the substructures of the wall to bulge outwards since it will not bear down on them with all its weight.

8. With regards to the wall itself, we should not decide in advance the material of which it will be constructed or surfaced, because we cannot procure the materials we prefer in all areas. But where blocks of stone [*saxa quadrata*], hard limestone [*silex*], quarry-stone [*caementum*], fired [*coctus later*] or unfired brick [*crudus later*] are available, they should be used. For not all regions are like Babylon, where, given the abundant supply of liquid pitch which can be used instead of lime and sand, they have a wall made of fired bricks; similarly all regions or localities with particular characteristics may have useful resources of their own which, once exploited, enable them to build an excellent wall without defects and which would last indefinitely.

CHAPTER VI

The Winds and City-planning

1. When the walls have been built around the city, the lots for housing inside them must be allocated and the main avenues [*plateae*] and narrow cross-streets [*angiporta*] orientated so that they take account of climatic conditions. The streets will be laid out correctly if care is taken to keep the winds out of the cross-streets: if the winds are cold they damage the health, if hot, they are infectious, and if humid, they are noxious. So it seems best to avoid this fault and make sure that what often happens in a number of cities does not recur. For example, Mytilene, on the island of Lesbos, is a town that was built magnificently and elegantly but orientated unwisely. When the south wind blows in the city, men fall ill: when the north-west wind blows, they cough: when the north wind blows, they return to good health but cannot stand about in the side streets and avenues because of the biting cold.

2. For the wind is a wave of air flowing in unpredictable directions. It is generated when heat encounters moisture and the violence of the impact forces out a powerful blast of air. The truth of this can be confirmed by an observation of bronze *aeolipilae*: one can elicit the divine truth from the hidden laws of the universe by means of such ingenious inventions. For *aeolipilae* are made in the form of empty bronze vessels with a very small hole by which they are filled up with water: they are put on a fire and, before they heat up, emit no steam, but as soon as they begin to get very hot they release a strong blast of air at the fire. So from this small and rapid demonstration we can understand and evaluate the vast, grand laws of nature governing the heavens and the winds.

3. Excluding the winds will not only make a place healthy for people who are well, but also, if diseases should happen to develop from other infections which in healthy places elsewhere would be treated by antidotes,[25] they will be cured more rapidly in these areas because of the moderate climate created by the exclusion of winds. For there are diseases which are difficult to cure in the areas which I have written about above, which are these: inflammation of the trachea, coughs, pleurisy, consumption, spitting of blood and all others that are cured not by purging but by building the body up. Such illnesses are hard to treat, first, because they are created by the cold; then, because the patients' resistance has already been lowered by illness, the air, agitated by the turbulence of the winds, rarefies and simultaneously sucks the vital fluid out of their diseased bodies, making them much weaker. By contrast, air that is mild and dense, which does not generate currents or constant changes of direction, strengthens the limbs thanks to its permanent stability and so restores to health people who are afflicted by such diseases.

4. Some liked to believe that there are four winds, the Solanus from the equinoctial east [due east], the Auster from the south, the Favonius from the equinoctial west and the Septentrio from the north. But those who have looked into the matter more carefully have taught us that there are eight winds – above all, Andronicus of Cirrhos, who even built a marble octagonal tower in Athens as a demonstration of this fact.[26] He decorated each side of the octagon with the images in relief of the individual winds, each facing the direction from which it blows: above the tower he placed a marble pinnacle on which he set a bronze Triton holding a rod in his right hand, which he designed so as to swing around with the wind and, always facing into the current, to hold out the rod as a pointer above the image of the wind that was blowing.

5. And so the Eurus was placed at the south-east between the Solanus [E] and the Auster [S]; Africus in the south-west between Auster and Favonius [W]; Caurus, which many call Corus [NW], between Favonius and Septentrio [N]; and Aquilo, between Septentrio and Solanus [NE]. In this way there was evidently a representation in relief intended to include the number, names and directions from which the blasts of the winds habitually blow.

When one has tested the theory satisfactorily like this, one should proceed as follows to discover the directions and points of origins of the winds.

6. A sheet of marble should be placed horizontally in the middle of the city or an area levelled with ruler and level so that the marble sheet is not needed: at the point in the centre, place a bronze gnomon which tracks shadows, called a σκιαθήρης [*skiatheres*; shadow-hunter] in Greek. At about the fifth hour before midday, take the very end of the shadow cast by the gnomon and mark it with a point: then, with the compasses opened as far as the point which indicates the length of the shadow of the gnomon, describe a circle from the centre. In the same way, watch the lengthening shadow of the gnomon in the afternoon, and when it touches the circumference of the circle, producing an afternoon shadow equal in length to the morning shadow, mark it with a point.

7. From these two points, describe intersecting arcs with the compass and draw a line passing through the point of intersection of the two arcs and the centre of the circle as far as the circumference of the circle in order to mark the north–south zones. Then take a sixteenth of the entire circumference of the circle [as a diameter] and put the compass point on the line at the south where it touches the circumference, and mark the points on the circumference at right and left on both the south and north sides. From these four points, draw intersecting lines through the centre from one side of the circumference to the other: in this way an eighth of the circumference will be allocated to the Auster and [another eighth to] the Septentrio. The other segments, three to the right and three to the left, are to be distributed equally around the whole circumference, so that equal segments for the eight winds are marked on the diagram. It is obvious then that the avenues and cross-streets should be laid out along the angles between the sectors of any two winds.

8. The unpleasant force of the wind will be excluded from houses and smaller streets if these methods are followed; for if the principal streets are aligned with the direction in which the winds blow, their fierce and frequent blasts, coming from the vast spaces of the sky, will rush here and there with even greater force when enclosed in the restricted openings of the cross-streets. For these reasons, the rows of houses should be aligned away from the directions in which the winds blow, so that when they arrive, they buffet the corners of the blocks of houses and so are repelled and dissipate themselves [Plate 5].

9. Perhaps those who know a number of names for winds will be surprised that in our account there are only eight. But if they bear in mind that Eratosthenes of Cyrene,[27] analysing the course of the sun, the equinoctial shadows of the gnomon and the latitude with mathematical calculations and geometrical methods, discovered that the circumference of the earth is 252,000 stadia, which is 31,500,000 paces, and that an eighth of that circumference, each occupied by a wind, is evidently 3,937,500 paces, they should not then be surprised that a single wind roaming around in such a vast area could blow with variable intensity as it goes back and forth.

10. So Leuconotus and Altanus usually blow to the right and left of Auster [S]; Libonotus and Subvesperus to the right and left of Africus [SW]; Argestes blows around Favonius [W], and at certain times, the Etesians; to the sides of Caurus [NW], Circias and Corus; on either side of Septentrio [N], Thracias and Gallicus; to right and left of Aquilo [NE], Supernas and Caecias; around Solanus [E], Carbas, and at a certain period, the Ornithiae; Eurocircias and Volturnus in the distant regions of Eurus [SE], which is in the middle. There are also many other names for other winds derived from localities, rivers or mountain storms [Fig. 1].

11. Besides these there are the morning breezes: for where the sun, emerging during its rotation from below the earth,

Plate 5. The Vitruvian city. The octagonal city has round towers with wooden, removable walkways across them and ramparts between them; as Vitruvius recommends, the main streets and subsidiary cross-streets are not aligned with the principal winds but along the lines of the angles between them, and the blocks of housing therefore present their corners to the main winds. The plan of the city is tilted so that north is at the left. **T** Septentrio [N] **G** Aquilo [NE] **+** Solanus [E] **S** Eurus [SE] **O** Auster [S] **G** Africus [SW] **P** Favonius [W] **M** Caurus, Corus [NW] **T** Moat **V** Tower **X** Gates **Y** Forum **o** Basilica **I** Streets

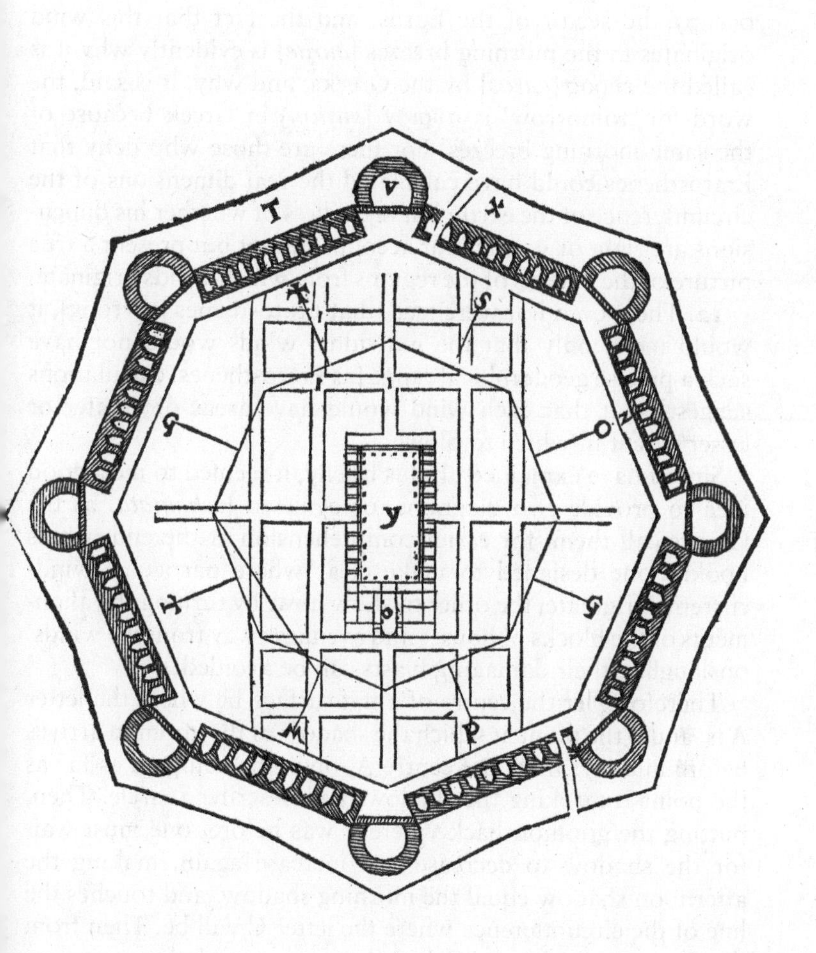

T
V
X
y

strikes the moisture in the air and pushes it on with the impetus of its ascent, it forces out currents from the breezes blowing before first light. Those that remain when the sun has risen occupy the sector of the Eurus, and the fact that this wind originates in the morning breezes [*aurae*] is evidently why it is called the εὖρος [*euros*] by the Greeks, and why, it is said, the word for 'tomorrow' is αὔριον [*aurion*] in Greek because of the same morning breezes. For there are those who deny that Eratosthenes could have calculated the real dimensions of the circumference of the earth; but regardless of whether his dimensions are right or wrong, our account cannot but present a true picture of the extents of the regions from which winds originate.

12. Then, even if it were true [that Eratosthenes is wrong], it would mean only that the individual winds would not have such a precise geographical range [as Eratosthenes' calculations suggest], but that each wind would have areas of greater or lesser extent in which to blow.

Since I have explained all this briefly, it seemed to me a good idea to provide two diagrams, or σχήματα [*schemata*] as the Greeks call them, for easier comprehension at the end of this book:[28] one designed to make clear where particular wind-currents originate; the other to show how, by turning the alignments of the blocks of houses and avenues away from the winds' onslaughts, their damaging blasts can be avoided.

Therefore, let the centre of a flat surface be where the letter A is, and B the point at which the shadow of the gnomon arrives before midday: from the centre A, open the compass as far as the point B marking the shadow, and describe a circle. Then, putting the gnomon back where it was before, one must wait for the shadow to decrease and increase again, making the afternoon shadow equal the morning shadow, and touches the line of the circumference where the letter C will be. Then from the points marked B and C describe two arcs with the compasses intersecting at the point to be marked D; then through the point of intersection D a line should be drawn through the centre [A] to the circumference on which the letters E and F will be marked. This line will indicate where the southern and northern sectors are located.

13. Then, using the compasses, take a sixteenth of the entire circumference, and put the point of the compasses on the southern line where it touches the circumference at the letter E, and mark it with the letters G and H at left and right. Similarly, in the northern sector put the point of the compasses on the point of intersection between the circumference and the northern line at letter F, and mark points to right and left where the letters I and K are; lines should then be taken through the centre [A] from G to K and from H to I. Thus the distance from G to H will demarcate the sector of the Auster and the southern area, and the distance from I to K will be that of the Septentrio [N]. The other segments are to be divided equally into three at the right and three at the left, with those to the east including the letters L and M, and those to the west including the letters N and O. Intersecting lines should be drawn from M to O and from L to N. In this way there will be eight equal sectors for the winds around the circumference. When the diagram has been constructed like this, at each of the angles of the octagon beginning from the south the letter G will be marked between the Eurus [SE] and the Auster [S]; between Auster and Africus [SW], H; between Africus and Favonius [W], N; between Favonius and Caurus [NW], O; between Caurus and Septentrio [N], K; between Septentrio and Aquilo [NE], I; between Aquilo and Solanus [E], L; between Solanus and Eurus, M. This done, the gnomon should be set on the angles of the octagon and the directions of the cross-streets laid out accordingly [Fig. 2; cf. Plate 5].

CHAPTER VII

Sites for Public Buildings and Temples

1. Once the cross-streets have been laid out and the main streets planned, we must deal with the selection of areas in the city with respect to their development for communal use and the convenience of the citizens, for sacred buildings, the forum and

other communal spaces. If the city-walls lie along the sea, a site near the port should be chosen for the location of the forum: but if the city is inland, the forum should be built in the middle. With respect to temples dedicated to the deities regarded as the chief protectors of the city, and to Jupiter, Juno and Minerva, their sites should be located in the highest possible positions from which most of the city-walls can be surveyed: but the temple dedicated to Mercury should be placed in the forum or in the marketplace, along with those of Isis and Serapis; those to Apollo and Father Liber near the theatre; that to Hercules, at the circus in cities where there are no gymnasia or amphitheatres: the temple for Mars, outside the city and near the training camp, and so too that for Venus, but at the port. According to the ritual books encapsulating the wisdom of the Etruscan diviners, the dedications are as follows: temples of Venus, Vulcan and Mars should be located outside the walls so that young men and mothers will not become habituated to erotic pleasures in the city; and buildings would evidently be spared the fear of fire once the power of Vulcan has been summoned outside the walls by sacrifices: and when a temple is dedicated to the divinity of Mars outside the walls there will be no armed struggles between the citizens, but he will defend the city from its enemies and save it from the dangers of war.

2. Again, the temple for Ceres should be placed on a site outside the city where men have absolutely no need to go except to make sacrifices, and this place must be looked after with devotion, purity and high morals. Sites for the temples of other gods should be laid out in a manner appropriate to the nature of the sacrifices offered to them.

I shall explain in Books III and IV the principles for the construction of temples themselves and their modular systems, because I thought it best to discuss first, in Book II, what the properties and uses are of the supplies which should be collected for buildings; then to deal with the commensurability of buildings, their layouts[29] and different types of proportional systems, explaining each of them in separate books.

BOOK II

Introduction

1. When Alexander was master of the world, Dinocrates, an architect full of confidence in his own ideas and ingenuity and very eager for royal patronage, left Macedonia for the army. He brought from home letters from relatives and friends addressed to the highest military officials and courtiers to gain easier access to the king, and having been received courteously by them, requested that he should be introduced to Alexander as soon as possible. Although they had promised to do so, they were fairly slow, waiting for the right moment. So Dinocrates, thinking that they were playing games with him, decided to help himself, for he was very tall, nice-looking and endowed with great physical presence and personal dignity. So, confident in these natural gifts, he undressed himself in his inn, anointed all his body with oil, crowned himself with a garland of poplar leaves, covered his left shoulder with a lion-skin, and, holding a club in his right hand, proceeded towards the tribunal where the king was dispensing justice.

2. When this remarkable vision had attracted the attention of the crowd, Alexander also noticed him, and, very taken with him, ordered the people to make way for him to approach and asked who he might be. He replied: 'I am Dinocrates, a Macedonian architect, who offers you projects and designs worthy of your celebrity. For I have made a project for Mount Athos in the form of a male statue, in whose left hand I have designed the walls of a vast city, and in his right, a bowl to collect the water of all the streams on the mountain so that it may be poured from there into the sea.'

3. Alexander, entranced by the idea behind the project, immediately asked if there were any fields in the vicinity which could maintain the city with a regular supply of corn. When he discovered that it would be impossible without transporting it across the sea, he said, 'Dinocrates, I appreciate the ingenious design of the project and am delighted with it, but I note that the judgement of someone who was to found a colony there would be heavily criticized. For just as a newborn baby cannot grow without the nurse's milk nor be led up the steps of its growing life, so a city without fields and their produce arriving inside the walls cannot expand or support a large number of inhabitants without an abundance of food, or maintain its population without a good supply. Therefore, though I believe your design has much to recommend it, at the same time I regard the site as unsuitable: but I want you to stay with me because I want to make use of your services.'

4. From then on Dinocrates never left the king and followed him to Egypt. There, when Alexander had noticed the naturally protected harbour, the outstanding trade centre, the cornfields all across the country and the immense advantages presented by the mighty river Nile, he ordered Dinocrates to found the city of Alexandria in his name. So it was that Dinocrates, recommended by his good looks and imposing presence, arrived at such a level of fame.[1] But in my case, Supreme Ruler, nature has not given me an impressive stature, age has disfigured my face and ill health has reduced my powers. So, since I have been deprived of these advantages, I will, I hope, gain your approval with the help of my expertise and my writings.

5. Now since I wrote extensively on the role of architecture and the scope of the art in Book I, but also about city-walls and the planning of sites within the walls, next in order should be an explanation of sacred temples, public and private buildings and the proportions and modular systems they should exhibit: but I thought it best not to give precedence to this discussion without having dealt first with the availability and selection of the materials with which buildings are brought to completion

using masonry and wood [*materia*], the qualities these materials should possess when put to use, or without having spoken of the natural elements of which they are composed.

But before I begin to explain matters regarding nature, I will present, first, a discussion of the origins from which the methods of building arose and the development of inventions related to them; I will follow the advances made by the ancients with respect to nature and by those who committed to writing what they had learnt regarding the beginnings of civilization and the inventions which they had studied. So I will lay out my material in the way that I have learnt from them.

CHAPTER I

The Origin of Buildings

1. As was usual in ancient times, men were born like wild animals in the forests, caves and woods, and spent their lives feeding on fodder. Somewhere or other in that period, the trees grew together densely, and, being tossed about by storms and gales, burst into flames when their branches rubbed together; those who lived in the area were terrified by the raging flames and ran away. Then, after the fire had died down, they moved closer and, when they noticed that being near the heat of the fire was very beneficial for their bodies, they threw wood on it and kept it going; then they called the others over and, showing it to them with sign-language, made clear the advantages it provided them with. In such gatherings, when sounds with many different meanings were emitted when they uttered, men began to form words as they happened to arise through daily use; then by indicating frequently the things they used, they began to talk in a haphazard way and so generated a common language.

2. So when the initial impetus for men's social gatherings, for their councils and communal life, occurred because of the discovery of fire, and more and more of them assembled in one

place, and they were endowed by nature with the advantage over other living creatures that they could walk upright without facing the ground and could observe the splendours of the earth and heavens, as well as being able to handle any object as readily as they wished with their hands and fingers – it was then that some of them from these first groups began to make shelters of foliage, others to dig caves at the foot of mountains and yet others to build refuges of mud and branches in which to shelter in imitation of the nests of swallows and their way of building. Next, by observing each other's shelters and incorporating the innovations of others in their own thinking about them, they built better kinds of huts day by day.

3. Since men were naturally imitative and quick to learn, they would show each other the results of their building, proud of their own inventions, and so, sharpening their wits in competition, became more competent technically every day. At first, after putting up stakes with forked ends and laying branches across them, they covered the walls with mud. Others built walls by drying out lumps of mud, binding them together with wood and covering them with reeds and foliage to avoid the showers and heat. When the roofs proved incapable of resisting showers during winter storms, they made gables, and, having covered the inclined roofs with mud, led the rain down them.

4. And we can see for ourselves that these practices developed from the origins which I have written about because to this day buildings are constructed of these materials in foreign countries: for example, in Gaul, Spain, Lusitania and Aquitaine, houses are roofed with oak shingles or thatched. Because of the abundance of woods at Pontus, in the country of the Colchians, whole trees are laid flat on the ground to right and left, separated by a distance equal to their length; and then two other trees are placed on their ends at right angles, so marking out the interior of the house. Then, by binding together the corners by placing alternate beams on all four sides, and so constructing walls with trees on the same vertical plane as those at the bottom, they build up tall towers, and block up the spaces between beams resulting from the thickness of the timbers with wood-shavings and mud. Again, they lay the roofs across these structures by cutting off the ends of the cross-beams [*transtra*] and progressively shortening them, and so build up pyramidal roofs high in the middle from all four sides, which they cover with leaves and mud, and so complete the pointed roofs of their towers in barbarian style.

5. But the Phrygians, who live on the plains, have very little timber because of the lack of forests: they choose natural mounds, and, cutting trenches through the middle and digging out passages, enlarge the interior space as much as the nature of the site allows. Above, they bind stout branches together to form conical roofs which they cover over with reeds and brushwood, then pile up great heaps of earth on top of their homes. These types of roof keep them very warm in winter and very cool in summer. Some peoples build huts with roofs made of sedge from the marshes. Among other nations as well, and in other locales, the structures of the houses are built using the same, or similar methods. So too at Marseilles we can see roofs without tiles made of earth worked with straw.[2] Even today an exemplar of this ancient type at the Areopagus is roofed with mud.[3] Again, the hut of Romulus on the Capitol and the sacred buildings covered with thatch in the citadel can provide us with significant evidence of practices in ancient times.[4]

6. So after a consideration of these remains we can reasonably infer that these were the ancient innovations in construction.

Besides, when they had daily become more manually adept at building and had, by cleverly exercising their ingenuity, arrived at a mastery of these arts, then the habit of hard work instilled in their minds also enabled those pursuing these practices most assiduously to proclaim themselves craftsmen. So when these arts had been developed like this at the beginning, and nature had not only enriched human beings with senses like other living beings, but had also reinforced their intellectual capacities with powers of reasoning and common sense and had subjected other animals to their power, then men, progressing gradually from the construction of buildings to other arts and disciplines, moved themselves on from their wild, rustic lives to gentle civilization.

7. Then, growing in self-confidence and looking forward with greater ambition on the basis of the variety of the arts, they began to construct houses, rather than huts, with foundations and walls built of brick or stone and roofed with timberwork and tiles; next, as a result of the observations they made during their investigations, they progressed from vague and imprecise ways of thinking to the ascertainable rules of modularity. After they had noticed that nature had given birth to copious resources of timber and an abundant supply of building-materials, they developed them carefully when they used them and enhanced the elegance of their lives, already improved by the arts, with luxuries.

Therefore I shall now discuss, to the best of my ability, the materials which are suitable for use in houses and what properties and qualities they should have.

8. If, however, someone wanted to argue with the position of this book, thinking that it should have been placed first, I would explain my rationale like this, lest he thought I had made a mistake: when I started writing a complete treatment of architecture, I decided to explain in the first volume the expertise and disciplines which the subject involves, to define its subdivisions with the technical terminology, and to state what it was that caused it to develop. So there I declared [in Book I] what skills the architect should possess; accordingly in the first book I talked about the scope of the art,[5] but in this I shall discuss the use of naturally occurring materials. For this book does not pronounce on the origin of architecture, but why buildings were originally constructed as they were,[6] and with what principles they were developed and gradually progressed to this level of perfection. **9.** Therefore the organization of this book will be in the right order and place.

Now I shall return to the theme I set myself, and shall discuss the materials which are suitable for the construction of buildings, how they seem to have been created by nature, and with what combinations of elements these compounds are composed, in such a way that these subjects should not be obscure but absolutely clear to my readers. For no kinds of matter, living organisms or objects can come into existence or be perceived without the combination of elements; and, otherwise, natural phenomena are not susceptible of valid explanations when discussed in the light of the doctrines of natural scientists unless the nature and character of the inherent causes of these phenomena are made clear with sophisticated analyses.

CHAPTER II

The Elements

1. First of all, Thales thought that water was the basic element of all matter. Heraclitus of Ephesus, who was nicknamed Σκοτεινός [*Skoteinos*; obscure] by the Greeks because of the impenetrability of his writings, thought it was fire. Democritus and his follower Epicurus thought that it was atoms, which our writers called 'uncuttable bodies', and others, 'indivisibles'. The school of the Pythagoreans added air and earth to water and fire. Democritus, then, though he did not name these objects specifically but merely defined them as 'indivisibles', seems, in fact, to have said exactly the same thing, because, since they are separate entities, they do not suffer damage or destruction and cannot be subdivided into smaller units, but retain their integrity for all time.[7]

2. Since, then, it seems that all things coalesce and are generated by the combination of such elements and have been distributed by nature amongst an infinity of types of objects, I thought it appropriate to discuss the varieties and differences in their uses and the qualities they have with respect to buildings, so that when these things have been understood, those who are thinking of building will not make mistakes, but will collect the right supplies for their constructions.

CHAPTER III

Bricks

1. First, then, with respect to unfired bricks [*lateres*], I will talk about the type of earth from which they should be moulded. They should not be moulded from sandy or pebbly clay or fine sand because they immediately become heavy when made from these types, and then, when splashed by the rain in walls, dissolve and fall to bits and the straw in them does not hold them together because of their coarseness. Instead they should be made of chalky-white or red clay, or even of rough sand: these types are durable because they have good consistency; they are not heavy to work and can be laid easily in walls.

2. Bricks should be moulded in the spring and autumn so that they dry uniformly. For those prepared in midsummer become faulty because the sun makes the brick look dry by fiercely cooking the outer layer prematurely, while the inside is not cooked. Afterwards, when the brick contracts on drying, the parts that have dried out previously break up; and so bricks full of fissures become weak. But bricks will be much more usable if they have been made more than two years before, since they cannot dry out completely in less time. When new bricks that have not dried properly are laid in a wall, and the plaster revetment [*tectorium*] has been applied, and the bricks have hardened into a solid mass, they settle and cannot maintain the same level as the plaster, and, displaced by contraction, do not adhere to it, but are pulled away from the points of contact. So the plaster separates from the wall and cannot stay upright by itself because it is so thin, but breaks in pieces, and the walls themselves are damaged by settling inconsistently. This is why the people of Utica use bricks in structures with walls only if they are dry and have been made more than five years before, provided that they have been certified as such by the decision of a magistrate.[8]

3. Now, three kinds of bricks are produced. The first, called the Lydian in Greek, is the one our people use, and is a foot

and a half long and a foot wide. Greek buildings are built with the two other kinds, of which one is called the πεντάδωρον [*pentadoron*; five palms], the other the τετράδωρον [*tetradoron*; four palms]. The Greeks call the palm δῶρον [*doron*] because they call the offering of gifts δῶρον and this is always done with the palm of the hand. So a brick which is five palms square is called a *pentadoron*, and that four palms square, a *tetradoron*; public buildings are constructed with πεντάδωρα [*pentadora*] and private buildings with τετράδωρα [*tetradora*; Fig. 3].

4. Half-bricks are also made along with these full-size bricks: when they are laid in a wall, courses of full-size bricks are laid on one side and courses of half-bricks on the other:[9] so that when they are laid accurately on the two faces, the walls are tied together by the alternating courses and the mid-points of the bricks placed over the vertical joints below provide the wall with strength and a not unattractive appearance on both faces.

Now in Further Spain, there are the cities of Maxilua and Callet, and in Asia, Pitane,[10] where the bricks, once moulded and dried, float when thrown into water. It seems that they can float like this because the earth from which they are made is porous. So because it is light, it does not take in or absorb liquid after being hardened by exposure to the atmosphere. Therefore, since these bricks have the property of being light and of low density and do not let the effects of moisture penetrate them, they are compelled by the laws of nature to float on water just like pumice, regardless of their weight. They therefore present great advantages, since they are not heavy to handle in building and are not dissolved by storms when assailed by them.[11]

CHAPTER IV

Sand

1. Instead, with regard to walls built with quarry-stones [*structurae caementiciae*],[12] one must find out first whether the sand would be suitable for the preparation of mortar and that it has no earth mixed in with it. These are the types of quarry-sand: black, grey, light and dark red. The best of these will be the one that squeaks when rubbed in the hand, while the one which has earth in it will not be rough enough: the sand will be suitable if, when wrapped up in a white cloth and then shaken out, the cloth is not stained and no earth is left on it.

2. If, however, there are no sandpits where sand can be dug up, then it must be sifted out from river sands or gravel, or, of course, from the seashore. This kind, however, has the following defects when used in masonry: it dries with difficulty, nor can a wall made of it tolerate continuous loading without pauses in the work, nor can it support vaults. But sea sand has the added difficulty that the walls reject its salinity and fall to bits when the revetments have been applied to them.

3. When used with masonry, quarry-sands dry rapidly, but only that freshly taken from the quarries; the revetments stay in place, and the wall can support vaults; for if they lie unused too long in the open after extraction, they are weathered by the sun, moon and frost and deteriorate and become earthy. So when they are used with masonry, they cannot bind the stones, which consequently settle, and the loads, which the walls cannot support, collapse. Recently dug sands, though, although they have many virtues when used with masonry, are not well adapted for use in revetments because, by virtue of its richness, the lime mixed with straw made from them cannot dry without cracking because of its strength.[13] But because of its fineness, river sand becomes perfectly solid in plaster revetments when worked over with plasterers' floats, like Signian sand.[14]

CHAPTER V

Lime

1. Having discussed supplies of sand, we must now turn our attention to lime and how to cook it out of white or hard stone. When it is derived from thick, dense stone it will be useful in

masonry, and that derived from porous stone will be suitable for revetments. When it has been slaked, the mortar should be mixed in such a way that if quarry-sand is to be used, three parts of sand to one of lime should be poured in; if using river or sea sand, two parts of sand should be mixed with one of lime; this will be the right adjustment of the proportions of the mixture. Furthermore, if anyone using river or sea sand were to add in a third of ground-up and sifted fired brick [*testa*], this will produce a mortar better mixed for use.

2. But the reason why lime strengthens masonry when it takes in water and sand seems to be that, like all other bodies, stones too are mixtures of the elements. And those which have more air are soft, those with more water are pliant because of the moisture, those with more earth are hard, and those with more fire are more brittle. And so if, before being fired, these stones are minutely ground up and mixed with sand and inserted into masonry, they do not solidify nor can they hold it together. But when they have been thrown into a furnace and have been enveloped by the fierce heat of the fire, they lose their naturally hard properties: then, with all their strength burnt out and exhausted, they are left empty and with open pores. So, when the moisture and air in the body of the stone are burnt out and extracted, it will retain residual heat: if the stone is immersed in water before it can absorb the intensity of the fire, it heats up strongly when the liquid penetrates the empty pores, and then, cooled down in this way, ejects the heat from the body of the lime.

3. Accordingly, the weight of these stones when thrown into the furnace cannot correspond to their weight when extracted; their volume remains the same once the liquid has been burnt out, but when weighing them one finds that about a third of the weight has been lost. Therefore, when their cavities and fissures are open, they absorb the sandy mixture and so coalesce with it and, on drying, create strong masonry by binding with the rubble.

CHAPTER VI

Pozzolana

1. There is also a kind of powder which naturally produces extraordinary results. It occurs in the region of Baiae in the countryside belonging to the municipalities around Mount Vesuvius. When mixed with lime and rubble, this powder not only ensures the durability of different types of construction, but even when masonry piers are built with it in the sea, they set hard under water. The explanation seems to be this: under these mountains there is hot earth and a large number of springs which would not be there but for the fact that deep down there are enormous fires burning because of the presence of sulphur, alum or pitch. In the depths, therefore, the fire and the heat of its flames diffuse themselves through the fissures in the earth and the heat makes the earth there light, and the tufa created there emerges free of moisture. So when these three substances [lime, rubble, pozzolana], all generated in a similar way by the intensity of fire, mix together, they coagulate as soon as they have absorbed liquid, and, hardened rapidly by the moisture, solidify so that neither the waves nor the force of water can dissolve them.

2. That there are powerful heat-sources in these areas can again be demonstrated by the fact that in the hills around Baiae, in Cumaean territory, there are places which have been dug out as sweating-rooms, in which the hot steam, generated deep down, emerges; it penetrates up through the earth because of the force of the fire and, spreading through it, springs up in these areas, and thus creates very effective conditions for sweating-rooms. In the same way, too, it is said that in ancient times fires developed and spread widely under Mount Vesuvius and that from there they spewed out flames across the fields. So that seems to be why porous stone, or Pompeian pumice as it is known, was reduced to this particular consistency after another variety of stone had been cooked by fire.

3. However, the type of porous stone which is extracted there is not found in all areas, but only around Etna and in the hills of Mysia, which the Greeks call κατακεκαυμένη [*katakekaumene*; burnt earth], and wherever the properties of other regions are similar. So if, in these regions, one finds hot-water springs and heated vapours wherever there are excavations, and the ancients record that these same areas had flames raging across the countryside, it seems certain that the moisture has been extracted from the tufa and earth by the intensity of the fire, just as it is from lime in furnaces.

4. In which case, when different and heterogeneous materials have been subjected to fire and reduced to the same condition, and in their hot, dry state are suddenly soaked with water, they grow hot because of the heat latent in the masses now joined together, and this causes them to combine strongly, and acquire rapidly the advantage of great solidity.

But one would still like to know why, since there are plenty of hot-water springs in Etruria, one cannot find there a powder with which masonry solidifies under water in the same way. Accordingly, I thought I had better explain – before the question crops up – what seems to be the explanation of these phenomena.

5. The same kinds of soil or rocks are not found in all locales and regions, but some are earthy, others have coarse sand or gravel; in other places again the soil is sandy and ligneous[15] and the characteristics of the soil in the earth are absolutely unlike and varied in different types of places. It is particularly worth considering the fact that where the Apennines encircle the regions of Italy and Etruria, there are sandpits practically everywhere, but there are none at all in the area across the Apennines and along the Adriatic; and the term itself is unheard of in Achaea, Asia and indeed anywhere across the sea. So the same favourable conditions cannot occur simultaneously in the same way in all the regions where lots of hot-water springs boil up, but all things are created randomly and in different ways as nature determines and not for the satisfaction of men.

6. Therefore, in those regions where the hills are not earthy but ligneous in nature, the force of the fire, coming out through their fissures, burns out the soft and yielding elements, but leaves the hard. The result is that in Campania the soil that is burnt turns into powder, and in Etruria the ligneous earth that is burnt turns into lignite. Both of them are excellent in masonry, but one has advantages for buildings on land, the other for piers in the sea. Again, the characteristic of this material is that it is softer than tufa and harder than the earth, and it produces the type of sand known as lignite in a number of regions when it is burnt deep down by the intensity of the heat.

CHAPTER VII

Stone

1. I have spoken of the different types and qualities of lime and sand. Next in order of discussion is the subject of the stone-quarries from which squared slabs and supplies of stone for buildings are extracted and prepared. We find, however, that [stones from] these quarries[16] have different and dissimilar characteristics. For some are soft, like those in the areas around the City[17] at Saxa Rubra, Pallia, Fidena and Alba; others are moderately hard, like those at Tivoli, Amiternum and Soracte, and others of this type; and some, such as hard limestone, are really tough. Again there are many other kinds, such as red and black tufa in Campania, white tufa in Umbria, Picenum and Venice, which can even be cut with a saw like wood.

2. But all these types of soft stone have the advantage that when blocks of them are extracted they can be handled easily during construction. If they are under cover, they take loads well; but if they are in open and unprotected locations, they crumble and fall to bits when compacted by frost and ice. Again, along the sea-coast, they crumble when corroded by salt, and cannot withstand the surges of the sea. But stone from Tivoli [travertine] and all other stones of that type are resistant to damage from heavy loads or bad weather, but they are entirely at the mercy of fire: as soon as fire touches it, it cracks

open and falls to bits because its composition naturally includes little moisture, and again, it does not include much earth but a great deal of air and fire. Since there is little moisture or earth in it, then the fire, driving the air out by its hot and fierce assault, penetrates deep inside and fills up the hollow fissures, and makes other parts of the stone burn in the same way on contact with its particles.

3. There are also a number of quarries, called Anician, in the territory of the Tarquinii, of which the stone is similar in colour to those at Alba: the workings are found mainly around Lake Bolsena and again, in the prèfecture of Statonia. This stone has innumerable virtues; neither the frosty season nor exposure to fire can damage it, but it stays dense and lasts for a very long time because it is naturally composed of small amounts of air and fire, a moderate amount of moisture and a great deal of earth: so it is made solid by this dense texture and is not damaged by storms or the onslaught of fire.

4. We can appreciate this best from the tombs around the municipality of Ferentum made with stone from these quarries. For they have large, excellently carved statues, and elegantly carved figurines, flowers and acanthus plants. Although they are ancient, they look as new as if they had only just been made. And the bronze-workers, too, making up moulds from stones from these quarries for melting bronze, derive great advantages from them when they cast it. If they were near the City, it would be convenient if all such works were carried out with stone from these sites.

5. But since necessity forces us to use the quarries at Saxa Rubra and Pallia and supplies nearest the City because of their proximity, anyone who wants to complete a building without defects should prepare the materials like this. The stone should be quarried no less than two years before the time for building arrives and in the summer rather than the winter, and it should remain on the ground in the open. Stone that has been damaged by weathering during the two-year period should be thrown into the foundations. The rest, which will not have been damaged, has been tested by nature and will be capable of lasting when built into structures above ground. These precautions

must be followed not only with respect to stone slabs but also with regard to walls built with stones.

CHAPTER VIII

Different Kinds of Walls

1. These are the types of masonry: *opus reticulatum*, which everyone uses now, and the ancient type called *opus incertum*. *Opus reticulatum* is the more attractive of the two, but it is apt to form cracks because it comprises disconnected bed- and vertical joints in all parts of the wall.[18] But the rough stones [of *opus incertum*], lying one above the other and interlocking, create masonry that is not pleasing to look at but is stronger than *opus reticulatum*.

2. In any case, both types should be constructed with the smallest stones so that the walls, fully saturated with mortar made of lime and sand, will be bound together for longer. For since these stones are soft and thin in consistency, they dry out the mortar by absorbing its moisture. But when the amount of lime and sand is great and very plentiful, the wall will not give way quickly since it contains more moisture, but is held together by these materials. As soon as the component of moisture has been sucked out of the mortar by the porousness of the stones, and the lime separates from the sand and disintegrates, then the stones cannot bind with them, and will ruin the wall in the course of time.

3. And we can observe this phenomenon in several tombs in the environs of the City built of marble or stone blocks and packed with masonry[19] in the middle: in the course of time the mortar has lost its strength and has been sucked dry by the porousness of the stones, so the tombs are falling down and disintegrating since the points of connection between the vertical joints have failed because of the collapse.

4. But anyone who wants to avoid this pitfall should leave a cavity in the middle behind the outer facing-stones [*ortho-*

statae], and build walls two feet thick with courses of red dressed stone, fired brick or ordinary hard stone in the cavity: then the outer faces should be bound to these with iron clamps and lead. In this way the wall, which has not been built with just a pile of material but in courses, will last indefinitely without defect, because the bed- and vertical joints sit on each other and are bound together by the ties, and would not let the structure belly out or allow the collapse of the stone of the facings, which are bound to each other.

5. This is why Greek methods of building masonry walls are not to be disparaged. For they do not use masonry faced with soft quarry-stone; but when they do not use squared slabs, they lay courses of hard limestone or some other common hard stone, and, as though building with bricks, bind their vertical joints in alternating courses, and so achieve maximum durability for an eternity. These structures are built in two ways: one of them is called *isodomum*, the other, *pseudisodomum*.

6. A wall is called *isodomum* when all the courses are equal in height, *pseudisodomum*, when rows of courses of different heights and lengths are laid out. Both types are durable, first, because the stones themselves are thick and solid in character and cannot absorb the moisture from the mortar, but rather keeps it moist for a very long time; and the bed-joints of the stones laid flat and level from the beginning prevent the mortar from settling and hold the walls together for a very long time because they are tied together throughout their entire thickness.

7. Another type of wall is that which they call ἔμπλεκτον [*emplekton*; woven], which our country-folk use as well. The faces of these types of walls are dressed, but they bind the other stones, which they put in place with mortar just as they were found, with alternating vertical joints.[20] But our workmen, wanting to work fast, set the blocks on their edges and concentrate only on the vertical outer faces of the walls, and fill up the space between them with broken-up stone and mortar thrown in separately. So three vertical surfaces are incorporated in the wall, two comprising the outer surfaces, and the other the infill between them. The Greeks do not build walls this way, but lay the stones horizontally and set alternate stones lengthwise into

the thickness of the wall; they do not fill the space in the middle with infill, but build them solid with a continuous uniform thickness between one face and the other. As well as this, they insert single stones through the whole thickness of the wall with faces at either end, called διάτονοι [*diatonoi*; cross-pieces, through-pieces], and these, binding the walls together powerfully, reinforce their stability [Fig. 4].[21]

8. So, anyone who wanted to identify and select a particular type of masonry on the basis of these notes would be fully informed of its capacity to last. For those which are made of soft quarry-stone with a thin and pleasing facing cannot but collapse after a long period. That is why, when experts are summoned to evaluate party walls, they do not assess the cost of their construction, but, having established the original price of the tenders from the contracts, deduct an eightieth for each year which has passed, and announce their decision that the remaining sum should be paid for the walls and, in effect, that they cannot last more than eighty years.

9. But with respect to brick walls nothing is deducted so long as they remain perpendicular, and they are always valued at their original price. So it turns out that in some cities we can see public buildings, private houses and even palaces built of brick: first, the wall at Athens facing Mounts Hymettus and Pentelicus; the brick *cellae* of the Temple of Jupiter and Hercules at Patras, around which the architraves and columns of the temple were built of stone; an ancient wall of exceptional workmanship at Arezzo in Italy; the palace built for the Attalid kings at Tralles which is always given to the man who holds the office of Chief Priest of the city to live in. In Sparta, paintings have even been cut out of certain walls by sawing through the bricks, and have then been enclosed in wooden frames and transported to the Comitium as an ornament to honour the aedileship of Varro and Murena.[22]

10. Then there is the palace of Croesus which the people of Sardis have dedicated as a meeting place for the College of Elders and as a place of rest for them in their old age. Again, at Halicarnassus, the palace of the immensely powerful King Mausolus, though decorated throughout with Proconnesian

marble, has walls built of brick which even today exhibit exceptional strength and of which the revetments are so highly finished that they seem to have the translucency of glass. Nor did this king use brick because of a lack of funds; he was awash with enormous revenues since he ruled over the whole of Caria.[23]

11. But one can appreciate his acuity and ingenuity with respect to the construction of buildings from this story; though he was born at Milasa, he had noticed that the site at Halicarnassus was naturally fortified, the trading-centre convenient and the port viable, so he established his residence there. Now the site has a curvature like that of a theatre: so the forum was built in the lowest part alongside the port. About halfway up the incline, at the intersection with the transverse walkway, as it were, a very wide square was laid out, in the middle of which was built the Mausoleum with a series of works so exceptional as to be nominated one of the Seven Wonders of the World. At the top of the hill in the middle is the sanctuary of Mars containing a colossal acrolithic statue made by an accomplished master:[24] some think that Leochares made this statue, others, Timotheus. At the top of the right outcrop is the Sanctuary of Venus and Mercury next to the spring of Salmacis.

12. It is wrongly held that this spring makes those who drink from it prey to erotic fixations:[25] so it will certainly not be tedious to explain why this view spread around the world as a result of garbled reporting. It is impossible, as some claim, that people become degenerate and lubricious because of this water, for the characteristic of the fountain is that it is very clear and has an excellent taste. When Melas and Arevanias led a joint colony from Argos and Troezen to this area, they drove the barbarian Carians and Lelegans out. But they fled to the mountains, joined forces and ravaged the Greeks brutally in a series of raids. Later on, one of the colonists established a well-supplied inn as a commercial enterprise at the fountain because of the excellence of the water: the way he ran the inn attracted the barbarians. They began to come down to it, one at a time, and then, mixing with groups of Greeks, were converted from their hard and fierce habits and were willingly brought round to their customs and agreeable behaviour. So this water acquired its peculiar reputation not because it caused some shameful erotic obsession but because the barbarians had been tamed by the gentleness of civilization.

13. Since I have already embarked on a description of the city-walls at Halicarnassus, it remains for me to describe all the rest. Just as there is the Sanctuary of Venus and the spring mentioned above at the right, so on the left promontory there is the royal palace which King Mausolus located on the site following his own personal project. For at the right it commands a view over the forum, the whole of the port and of the circuit of the walls; down below at the left is the port hidden below the hills in such a way that nobody could see or know what was going on inside it, and the king himself could give orders from his own palace to his sailors and soldiers about what action to take without anyone knowing about it.

14. When after the death of Mausolus his wife Artemisia succeded to the throne, the Rhodians were outraged that a woman governed the cities across the whole of Caria, so they armed their fleet and set out to take over the kingdom. When this was reported to Artemisia, she gave orders that her fleet, with the oarsmen hidden and the marines ready for action, should be concealed in the harbour, but that the rest of the citizens should station themselves on the walls. When the Rhodians deployed themselves in the larger port with their well-equipped fleet, Artemisia ordered the people on the wall to applaud them and promise to hand over the town. When the Rhodians made their way inside the walls, leaving their fleet unmanned, Artemisia, who had had a canal dug to the sea, rapidly led her fleet through it out of the smaller port to the sea and so brought it into the larger harbour. Having disembarked her soldiers, she took the unmanned Rhodian fleet out to sea. So the Rhodians, with nowhere to hide, found themselves surrounded, and were massacred in the forum itself.

15. Then Artemisia embarked her own soldiers and oarsmen on the ships of the Rhodians and set out for Rhodes. But when the Rhodians spied their own ships arriving covered with laurel, they supposed that their citizens were returning in triumph and took the enemy in. Then, after seizing Rhodes and executing the leaders, Artemisia set up a trophy to her victory in the city itself and had two bronze statues made, one of the city, the other an image of herself, showing her branding the city with a mark of shame. Afterwards, the Rhodians, inhibited by religious scruples to the effect that it was sacrilege to remove trophies once they had been dedicated, erected a building around the site and protected it with the construction of a Greek guard-post so that nobody could see it, and ordered that it be declared ἄβατον [*abaton*; inaccessible].

16. Since kings of such enormous power did not turn up their noses at structures made with brick walls, even though, given their revenues from taxation and plunder, they could often afford to build them with either quarry-stone or squared stone blocks but also with marble, I do not think, therefore, that one should reject buildings made with brick walls, so long as they are properly covered on top. But I will explain why this kind of structure should not be built by Romans in the City, not forgetting the causes and reasons for it.

17. Public laws forbid that walls thicker than a foot and a half should be built on public land; other [i.e. interior] walls should be constructed of the same thickness lest the [interior] spaces become too narrow. But brick walls a foot and a half thick cannot support more than one storey, yet they can if they are two or three bricks thick. Yet given the enormous size of the City and the great density of its population, it is essential to provide innumerable houses. Accordingly, since one-storey housing could not provide living-space for such a mass of people, this very fact forces us to resort to high-rise buildings. So tall buildings constructed with stone piers [*pilae*], walls of fired brick and rubble party walls, and tied together by a great number of wooden floors, provide partitions of great practical use.[26] This is why the Roman people have excellent housing because of the upwards proliferation of many storeys of various configurations without obstruction within the city-walls.[27]

18. Now we have explained the reason why walls of dried brick [*latericius paries*] cannot be built like this in the City because of restrictions of space; when one needs to use them outside the City, they should be built as follows to avoid defects and last a long time. On top of the walls, a fired brick parapet [*structura testacea*] about a foot and a half high with projecting cornices should be inserted under the roof-tiles. In this way the walls will not suffer from the defects that habitually afflict them; for when the roof-tiles break or are thrown off by the wind, which can let the rainwater fall inside, the fired brick breast-work would prevent damage to the dried bricks [of the walls below], and the projection of the cornices would throw the spray of water out beyond the vertical face, in this way keeping the structures of the dried brick walls intact.

19. But with respect to fired brick, nobody can tell on the spot whether it would be excellent or damaging when used in walls, because only when it is put on a roof [in the form of tiles] is its durability tested by storms and the summer heat. For if it has not been made of good clay, or has not been baked enough, it will show its faults when it is weathered by ice and frost there. So tiles that will not withstand weathering on roofs are not strong enough to bear loads in walls. This is why walls built of fired brick taken from old roof-tiles, especially, will have great strength.

20. But I wish that half-timbering had never been invented: despite its advantages with respect to speed of construction and space-saving, it is equally prone to cause devastating and widespread disasters since it is predisposed to burn like torches. So a higher expenditure on fired bricks as part of the overall outlay seems more sensible than putting oneself in danger through cutting expenses by using half-timber. Half-timber also creates cracks in plaster revetments because of the arrangement of its uprights and cross-pieces, because, when it is plastered, it swells on absorbing moisture and contracts when drying, and breaks up the solid revetment when it shrinks.

But since lack of time or money or the need for partition-walls in spaces without structural support force many people to use half-timbering, it should be made like this. The base should be built high enough that it has no contact with the concrete or the pavement, because eventually the half-timbering rots when it is fixed in them, and when it settles, leans out and destroys the surface of the revetment.

I have now discussed walls as best I could along with the preparation of the different kinds of material for them and their advantages and disadvantages; now, in line with what nature shows us, I will deal with wooden floors and the materials of which they should be made so that they will not deteriorate with the passage of time.

CHAPTER IX

Timber

1. Timber should be felled between early autumn and before the time when the Favonius [W] begins to blow. For in the spring all trees become fertile and discharge their nutritious properties into their leaves and their annual fruit. So when they are empty and swollen because of the demands of the seasons, they become weak and feeble because of their lack of density. In the same way, women's bodies, when they have conceived, are not regarded as healthy until the foetus is born, and pregnant slave-girls are not rated as healthy during sales, because the embryo, growing in the body with all the goodness of the food, takes the nourishment for itself, and the stronger it becomes as the moment of birth approaches, the weaker it makes the body in which it is created. So once the baby is born, the body, now liberated by the separation of birth, absorbs what had previously been extracted for the creation of another creature through its empty and open veins, and sucking up the moisture, grows more robust and regains its natural, original strength.

2. For the same reason, with the maturation of the fruit and the wilting of the leaves in the autumn, the roots of trees recover and are restored to their old density by absorbing moisture from the earth. But in fact the force of the winter air compresses and reinforces them during the period mentioned above.

Consequently, if timber is felled following the methods and timing already mentioned, it will be done at the right time.

3. Now a tree should be cut by slicing to the central core of the trunk and should be left to dry with the sap dripping out from the core. In this way the useless liquid in the veins, flowing out through the sapwood, will prevent the sap inside from decomposing and destroying the quality of the timber. But then, when the tree is dry and does not drip any more, it should be felled and so will be perfectly good to use.

4. One may note that the situation is the same with respect to shrubs. When these are perforated at the base and pruned at the appropriate time, they pour out from the centre all the superfluous and infected liquid which they contain through the holes, and so dry out and acquire great durability. But when liquids have no means of escaping from trees, they coagulate inside them, go bad and make them hollow and diseased. Therefore, if trees which are erect and vital do not age when they are drained, there is no doubt that when they are felled for timber, they will be extremely useful in buildings for a very long time when they have been treated in this way.

5. There are various notable differences, however, between the characteristics of trees, for example of oak, elm, poplar, cypress, fir and others which are best suited for use in buildings. For oak does not have the same properties as fir, nor cypress those of elm, and neither do the others naturally have the same qualities: but individual types are endowed with the particular characteristics of their constituent elements and are efficacious in building, each in their different ways.

6. First of all fir, having a great deal of air and fire and a minimum of water and earth, is endowed with lighter natural properties, and so is not heavy. Consequently it does not bow readily under a load since it is held together by its natural rigidity, and stays straight in flooring. Yet, because it contains a lot of heat, it generates and nourishes rot and is damaged by it: it also catches fire rapidly because of the porousness caused by the air in it, and opening up, takes in fire, and so emits fierce flames.

7. Before it is cut down, the part of the tree nearest to the

ground absorbs the humidity in the vicinity via its roots, and becomes moist and free from knots. But when the upper part has sent branches out into the air through its knots because of the vigour of the heat in it, and is then cut off about twenty feet from the ground and sawn up, it is called 'knotwood' because of the hardness of its knots. When cut down, the lowest part is divided along the grain into four sections and, once the sapwood from the tree has been thrown away, is prepared for fine carpentry [*opera intestina*]: the innermost part of the wood is called deal.

8. By contrast, when oak, which is very rich in earthy elements but contains very little water, air or fire, is sunk in structures underground, it lasts for ever. For this reason, when it is exposed to moisture it cannot absorb liquid into its tissue because it does not have the porousness produced by apertures on account of its density; but it resists the moisture by evading it and warping, which causes cracks in whichever structure it is used.

9. But the winter oak, which is composed of a balance of all the elements, is immensely useful in buildings; when put in damp locations, it absorbs liquid through its pores right to the centre, and rots once the air and fire are forced out of it by the element of humidity. Turkey oak and beech contain an equal mixture of water, fire and earth, but a great deal of air; they are porous and decay rapidly when they absorb liquid to their centre. White and black poplar, as well as willow, linden and agnus castus, which contain a great deal of fire and air, a moderate amount of humidity but very little earth, are composed of a relatively light mixture and so seem to possess remarkable rigidity when put to use. So since they are not made hard by the admixture of earth, they are white in colour because of their porousness and provide a material which is easily handled when used for sculptures.

10. The alder, which grows near river-banks and seems entirely useless as timber, has, in fact, some excellent characteristics, since it is composed of a mixture of a great deal of air and fire, but not much earth and very little water. And so when densely packed in the form of piles under the foundations of buildings in marshy areas, it absorbs water, since its own tissues contain very little themselves, and, remaining imperishable for an eternity, bears the huge loads of buildings and preserves them from damage. So this wood, which cannot survive long outside the earth, lasts indefinitely when it is immersed in wet sites. One can observe this phenomenon best at Ravenna because there all the public and private buildings have piles of this type of wood under their foundations.

11. Elm and ash contain great quantities of moisture, but minimum levels of air and fire, as well as a moderate admixture of earth. When they are prepared for use in buildings they are pliant and because of the weight of moisture in them they soon sag. At the same time, when they become dry with age, or die when they stand in the country after the liquid in them has been exhausted, they become more robust and can make strong dowels in joints and joins because of their flexibility.

12. So too the hornbeam, which contains a minimal admixture of fire and earth but a very high proportion of air and moisture, is not fragile and is extremely easy to handle. Hence the Greeks call it *zygia* because they make the yokes for draught-animals from this wood, and they call yokes ζυγά [*zyga*]. No less remarkable are the cypress and pine which tend to sag when used in buildings because of a superfluity of moisture since they comprise a great quantity of humidity and an equal admixture of other elements; but they last a very long time without defects because the liquid deep inside the trunks has a bitter taste which, because of its pungency, prevents rot from penetrating it, as well as destructive insects. So the result is that buildings made with these types of wood last for an eternity.

13. Cedar and juniper have the same characteristics and uses, and just as a resin derives from the cypress and the pine, so too an oil called *cedrium* comes from the cedar; and when various objects, such as books, are coated with it, they are not damaged by grubs or rot. The foliage of this tree is similar to that of the cypress and the wood is straight-grained. The statue of Diana in the temple at Ephesus, as well as the coffered ceilings there and in other famous sanctuaries, are made from it because of

its immense durability. These trees grow particularly in Crete, Africa and in various districts of Syria.

14. The larch, however, is known only in those municipalities along the banks of the river Po and on the shores of the Adriatic, and is not only left undamaged by rot or grubs because of the extreme bitterness of its sap, but is not ignited even by the flames of fires; nor can it catch fire by itself unless it is burnt like a stone with other types of wood in a furnace for cooking lime. Even then it does not catch fire or produce charcoal but is eventually burnt away after a long time. Because it is composed of minimal levels of the elements of fire and air, but is densely packed with a mass of earth and water, it does not have pores with apertures through which fire can penetrate; it repels the fire's onslaught and resists being damaged by it quickly; it does not float on water because of its weight, but when transported is placed on board ships or barges made of silver fir.

15. It really is worth knowing how this wood was discovered. When the Deified Caesar had deployed his army near the Alps and ordered the local municipalities to provide supplies, the inhabitants of a well-defended fort in the area, called Larignum,[28] refused to submit to his command since they trusted their natural defences. So the general ordered his forces to move up to the attack. But there was a tower in front of the gate of the castle constructed of beams of this wood laid alternately at right angles to each other, as in a funeral pyre, and built high so that besiegers could be repelled from the top with stakes and rocks. Then, when the Romans noticed that the townsfolk had no other missiles apart from the stakes and that they could not hurl them very far from the wall because of their weight, the command was given to approach and set bundles of tied-up brushwood and burning torches against the defences, which the soldiers soon piled up.

16. When the flame had kindled the brushwood around the wooden structure and had risen into the sky, it made everybody think that the whole mass had already collapsed. But when the fire had burnt itself out and subsided, the tower emerged to view intact; Caesar was incredulous and gave orders that the defenders should be surrounded with a rampart outside the range of their missiles. And so when the fear-stricken townsfolk had surrendered, they were asked where the wood left unscathed by fire came from. They then showed Caesar the trees in question, of which there were great supplies in the region; so the fort is called Larignum, just as the wood is known as larch [*larigna*]. This wood is transported to Ravenna along the Po and is supplied to Fano, Pesaro, Ancona and the other towns in that region. If there were some means of transporting it to the City, it would be extremely useful in buildings, though not necessarily in all of them; if the boards in the eaves around housing-blocks were constructed of it, buildings would be free from the danger of fires jumping between them because this kind of wood cannot be set on fire by flames or burning embers, or ignite of its own accord.

17. These trees have leaves similar to those of the pine and their fibres are long:[29] the wood from them is as manageable as deal for carpentry and contains a runny resin, the colour of Attic honey, which is administered to consumptives as a medicine.

I have discussed the individual types of trees, the properties with which they seem to be endowed by nature and the conditions in which they grow. There follows a consideration of why it is that highland fir, as it is called in the City, is of lower quality, and of why lowland fir fulfils many functions in buildings extremely well over very long periods; I will explain in this context why it is that they seem to have defects and virtues deriving from the characteristics of the regions where they are found, so that the subject will be clearer for those interested in it.

CHAPTER X

Highland and Lowland Fir

1. The first foothills of the Apennines extend from the Tyrrhenian Sea between the Alps and the most distant areas of Tuscany. But the ridges of the mountains curve and in the middle of their arc nearly touch the shores of the Adriatic, then complete their circuit by extending along the strait. So the lower part of the curve which heads towards the regions of Etruria and Campania is characteristically sunny, since it is constantly exposed to the trajectory of the sun. But the further section, sloping towards the upper sea [the Adriatic], is exposed to the north and is continuously covered with darkness and shadow. That is why the trees growing in that area, nourished by the fertile moisture, not only grow to enormous sizes, but their veins swell as well when filled with a mass of moisture and become saturated with an abundance of liquid. But when they have been felled and sawn up, they lose their life force and dry out because the stiffness of their veins persists; consequently they cannot last very long when used in buildings after they have become brittle and hollow because of their porousness.

2. But trees which grow in places exposed to the trajectory of the sun do not have porous internal channels and become dense when desiccated by dryness, because the sun not only absorbs moisture from the earth but also extracts it from trees. So trees which grow in sunny regions are made rigid by the tough consistency of their fibres since they lack the porousness produced by water: when they are sawn up for timber, they provide great advantages over long periods. Therefore the lowland firs which are transported from sunny regions are better than those brought from the gloomy highlands.

3. Considering the matter to the best of my ability, I have discussed the materials which are essential for the construction of buildings, the mixture of the various amounts of the elements they naturally comprise, and the advantages and defects of each kind so that they will not remain unfamiliar to those engaged in building. Consequently those who can follow the advice contained in this body of instructions will be better informed and equipped to choose how to use the particular types for construction. Therefore, since the preliminaries have now been explained, the buildings themselves will be discussed in the other books; and first, as the order of my argument requires, I will write about the sacred temples of the immortal gods and their modularity and proportions in the next book.

BOOK III

Introduction

1. Delphic Apollo announced through the responses of the Pythian priestess that Socrates was the wisest of all men.[1] For he is recorded as having said, sagely and with the greatest acuteness, that men's breasts should have windows in them and be open so that their thoughts would not remain concealed but open for inspection. If only, in fact, nature had followed his view and had made men's thoughts open and easily seen: if that had happened, not only would the merits and defects of men's hearts be immediately visible, but also their knowledge of the sciences, once available for visual inspection, would not be subject to fallible methods of analysis, but the learned and wise would be credited with illustrious and permanent authority. But since things were not arranged like this, but as nature wanted, the result is that it is impossible for people to determine the quality of men's knowledge of works of art when their intellects are hidden deep down in their breasts: artists may recommend their own skills if they are poor, but if they have made a name for themselves because of their studios' long standing or, indeed, are endowed with forensic and rhetorical skills, they really can acquire impressive reputations in reward for their assiduous study such that people believe that they have the knowledge they lay claim to.[2]

2. We can see this above all in the careers of ancient sculptors and painters because only those of them who acquired high status and enjoyed generous patronage remain permanently in the memory of posterity, such as Myron, Polyclitus, Phidias,

Lysippus and others, who achieved fame through their art: they acquired it by working for great cities, kings or important citizens. But those who were equally endowed with powers of application, talent and skill, and executed works brilliantly for citizens of low standing, are hardly remembered, because fortune, not their dedication or artistic skill, deserted them: for instance, Hellas the Athenian, Chion the Corinthian, Myager the Phocaean, Pharax the Ephesian, Boedas of Byzantium and indeed many others. In the same way painters of no less talent, such as Aristomenes of Thasos, Polycles, Andramithes, Niteon and others lacked neither application, enthusiasm nor skill in their art, but either reduced family circumstances, misfortune or the successes of their rivals in professional struggles prevented them from becoming famous.[3]

3. We should not be surprised that artistic excellence goes unrecognized because of public ignorance of it; but we should be extremely indignant when, as often happens, networking at social gatherings seduces men away from a disinterested evaluation [of artistic talent] in favour of baseless approbation [of worthless art]. Therefore if, as Socrates wanted, thoughts, opinions and knowledge enhanced by study were clear and transparent, neither sycophancy nor intrigue would have any effect, but commissions would automatically be assigned to those who had arrived at the summit of their professions by valid and demonstrable attention to their studies. But since these qualities are not clear and in plain view as we think they should be, and since I notice that the uneducated rather than the educated gain favour, I have decided not to compete with the pushiness of ignoramuses, but rather to demonstrate the great value of our discipline by publishing this body of instructions.

4. So, Supreme Ruler, I explained for you in Book I the art of building, its virtues and the disciplines in which the architect should be educated; I included the reasons why he should be expert in them and distributed the principles of the whole of architecture into various sections and completed it by defining them. Then I explained scientifically how, first and foremost, healthy sites for walled cities should be selected, and I showed,

using geometrical designs, what winds there are and from which direction each one blows, and demonstrated how the main streets and cross-streets inside the city-walls should be laid out correctly: and there I concluded the theme of Book I. Then I dealt comprehensively with the uses of building materials in construction and their natural characteristics in Book II. Now I will talk in Book III about the temples of the immortal gods and describe them with all appropriate detail.

CHAPTER I

Modularity in Temples and in the Human Body: Perfect Numbers

1. The design of temples depends on modularity, the principles of which the architect must adhere to rigorously: modularity originates in proportion, which is called ἀναλογία [*analogia*] in Greek. Proportion is the commensurability of a predetermined component of a building to each and every other part of a given structure, and modularity is based on this commensurability. For without modularity and proportion no temple can be designed rationally, that is, unless its elements have precisely calculated relationships like those of a well-proportioned man.

2. For nature so designed the human body that, with regards to the head, the face from the chin to the top of the forehead and the lowest roots of the hair, is a tenth; the palm of the hand from the wrist to the tip of the middle finger is the same; the head from the chin to the crown of the head is an eighth; from the top of the chest and the bottom of the neck to the lowest roots of the hair is a sixth, and from the chest to the crown of the head, a quarter. From the bottom of the chin to the lowest point of the nostrils is a third of the height of the face itself, and from the lowest point of the nostrils to a point between the eyebrows is the same; from that point to the lowest roots of the hair, including the forehead, is also a third. The length of the foot is one-sixth the height of the body: the forearm and hand, and the chest are both a quarter. The other limbs, too, have their own commensurable proportions, by adhering to which celebrated ancient painters and sculptors achieved widespread and lasting fame.

3. Similarly, the components of sacred temples must have dimensions which are precisely commensurable with the totality of their greatest dimensions derived from the sum of their single elements.

Again, the central point of the human body is naturally the navel. So that if a man were laid out on his back with his hands and feet spread out and compasses are set at his navel as the centre and a circle described from that point, the circumference would intersect with all his fingers and toes. In the same way, just as the figure of a circle can be traced out on the human body, so too the figure of a square can be elicited from it. For if we measure from the soles of the feet to the crown of the head and that dimension is compared with the distance between the outstretched hands, we find that the breadth is the same as the height, as in the case of plane surfaces which are perfectly square.

4. Therefore, if nature has composed the human body so that its individual limbs correspond proportionately to the whole figure, it would seem that the ancients had good reason to decide that, when constructing their buildings, the individual components should be exactly commensurable with the configuration of the whole structure. Therefore, while they handed down rules for all kinds of buildings, they did so especially for the temples of the gods, constructions of which the merits and defects usually last for ever.

5. Furthermore, they derived the system of mensuration clearly essential for all buildings from members of the body, such as the finger, palm, foot and cubit:[4] they distributed them in a perfect number, called τέλεον [*teleon*] by the Greeks; the ancients established the number defined as 'ten' as the perfect number, since they derived it from the number of fingers on the hand.[5] But if ten, derived from the digits of both hands, is perfect according to nature, Plato also maintained that the same number was perfect for the reason that it derives from single

units that the Greeks called μονάδες [*monades*; units]:[6] but as soon as they become eleven or twelve and so become supernumerary, they cannot remain perfect until they arrive at the next decade; for the [first four] single units are the components of that number.[7]

6. But mathematicians, maintaining a different view, have declared that the number known as 'six' is perfect because it incorporates subdivisions which amount to that number according to their calculations. So, one is a sixth [of six], two is a third, three is a half, and four is two-thirds, which they call δίμοιρον [*dimoiron*]; five is five-sixths, which they call πεντάμοιρον [*pentamoiron*], and the perfect number is six. When the number six increases by the addition of a unit, it is called ἔφεκτον [*ephekton*; 6 + 1]; when the number becomes eight with the addition of a third, it produces a number called ἐπίτριτος [*epitritos*], meaning the number plus a third [6 + 6/3 = 8]; when half is added, the result is nine, producing the number plus a half, called ἡμιόλιος [*hemiolios*; 6 + 6/2 = 9]; the addition of two [thirds] more produces ten, or the number plus two-thirds, called ἐπιδίμοιρον [*epidimoiron*; 6 + 4]; the number eleven is produced by the addition of five, making the number plus five-sixths, called ἐπίπεμπτον [*epipempton*; 6 + 5]; twelve, which comprises the basic number multiplied by two, is called διπλάσιον [*diplasion*].

7. Further, they fixed six as the perfect number because a man's foot is a sixth of his height.[8] They also realized that the cubit consists of six palms and twenty-four fingers. For this reason, too, it seems that the Greek cities, using the example of the cubit of six palms, subdivided the drachma, which they used as the unit of coin, into six stamped bronze coins, like our *asses* [pounds], which they call obols; and, to correspond to the twenty-four fingers, they further subdivided the drachma into twenty-four quarter-obols, which some call *dichalca* and others *trichalca*.

8. But at first my countrymen selected the ancient number and divided the *denarius* into ten bronze pieces, which is why to this day the name devised for the unit retains the idea of a tenth. And they called the quarter-*denarius*, which comprises

two and a half *asses*, the *sestertius*. Afterwards, however, when they realized that both six and ten were perfect numbers, they combined them into one unit, thus creating the absolutely perfect number sixteen: they found authority for this in the foot. For if two palms are taken from a cubit, a foot of four palms is left: but the palm has four fingers: this is why the foot includes sixteen fingers and the bronze *denarius* the same number of *asses*.

9. Therefore: if it is agreed that the numerical system was derived from human members, and that there should be a commensurable relationship based on accepted units between those members taken separately and the form of the body as a whole, it remains for us to demonstrate the greatest respect for those who, when building temples for the immortal gods, arranged the elements of the buildings in such a way that, thanks to the proportions and modularity, the arrangement of the individual elements and of the whole corresponded to each other.

CHAPTER II

Types of Temples

1. But in the case of temples there are certain principles on which their external appearance depends. First, there is the temple *in antis*, which the Greeks call ναὸς ἐν παραστάσιν [*naos en parastasin*]; then the prostyle, amphiprostyle, peripteral, pseudodipteral, dipteral and hypaethral temples. Their configurations are developed on these principles [Fig. 5].

2. A temple will be *in antis* when it has the *antae* of the walls surrounding the *cella* at the front and there are two columns in the middle between the *antae*; and above, a pediment constructed using the modular system which will be described in this book. An example of this type will be found at the three temples of Fortune, in the one nearest to the Colline gate.

3. The prostyle temple includes all the same components as

the temple *in antis*, except that it includes two columns at the corners aligned with the *antae*, and above the architraves[9] the components are the same as in the temple *in antis*, but with single sections [of architrave] to left and right down the sides: an example of this is the Temple of Jupiter and Faunus on the Island in the Tiber.

4. The amphiprostyle type has all the same components as the prostyle, but in addition it includes, on the back façade [*posticum*], the same arrangement of columns and pediment as at the front.

5. A temple will be peripteral if it has six columns on both the front [*frons*] and back and eleven down the sides including those at the corners. But these columns should be located so that a space equal to the breadth of an intercolumniation is left between the walls around the *cella* and the inside faces of the rows of columns, and it should have a corridor around the *cella* of the temple, as in the cases of the Temple of Jupiter Stator by Hermodorus in the Portico of Metellus, and the Temple of Honour and Valour constructed without a colonnade at the back by Mucius near the Monuments to Marius [Plates 6, 7].

6. The pseudodipteral temple, by contrast, should be laid out so that there are eight columns at both the front and back façades and fifteen down the sides including the corner columns: but the walls of the *cella* should correspond to the four middle columns at front and back: in that way there will be a space of two intercolumniations plus one column-diameter all round between the walls of the *cella* and the inner faces of the rows of columns. There is no example of this in Rome, but at Magnesia there is the Temple of Diana by Hermogenes and also that of Apollo at Alabanda built by Menesthes.

7. The dipteral temple is also octastyle at both the *pronaus* and back façades, but there are two rows of columns around the building, as at the Doric Temple of Quirinus and the Ionic Temple of Diana at Ephesus laid out by Chersiphron [cf. Plates 3, 4, pp. 15, 16].

8. The hypaethral temple is decastyle at the *pronaus* and at the opisthodomus.[10] All its other components are the same as those of the dipteral temple, but inside it has two storeys of

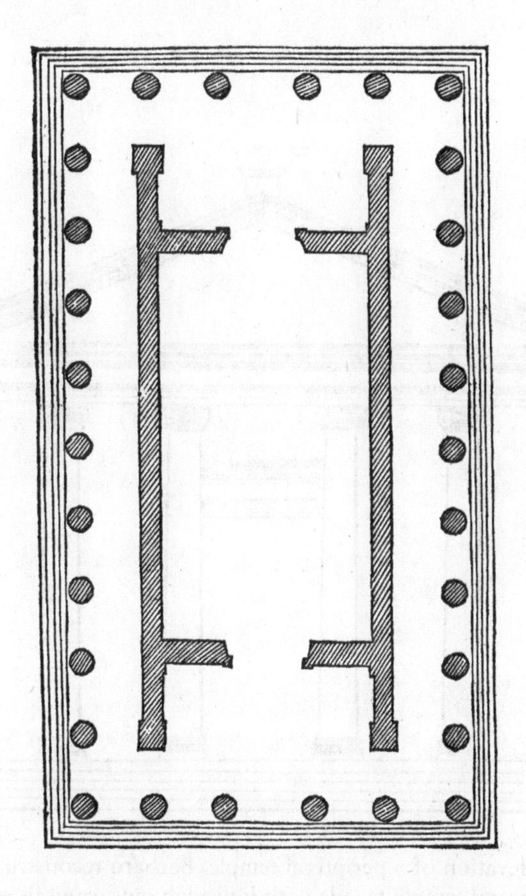

Plate 6. Ground-plan of a peripteral temple. Vitruvius probably had in mind a *pronaus distyle in antis* and *cella* without the *opisthodomus* shown here by Palladio, as well as a podium on three sides. Palladio adheres to Vitruvius' text in that he shows a peripteral temple with 6 × 11 columns and a colonnade separated from the *cella* walls by one intercolumniation. He could not have known anything about the examples cited by Vitruvius, Hermodorus' Temple of Jupiter Stator (probably Ionic) or Mucius' Temple of Honour and Valour (Ionic and without *posticum*), and instead the plan may owe something to the southern Doric temple in the Forum Holitorium at S. Nicola in Carcere, recently illustrated by Antonio Labaco (1552); the latter has 6 × 11 columns with a wider central intercolumniation on the façade.

Plate 7. Elevation of a peripteral temple. Barbaro reconstructs Vitruvius' peripteral temple façade with Ionic columns, capitals somewhat like those of the Temple of Saturn (Palladio, 1997, IV.127, p. 339), the celebrated Vitruvian bases lacking the lower torus and a pulvinate frieze, leaving a wider intercolumniation at the centres of the façades, not specifically mentioned by Vitruvius at this point.

columns set away from the walls [of the *cella*] creating ambulatories all round like those in the porticoes of colonnaded courtyards [*peristylia*]. The space in the centre is open to the sky and has no roof. There are entrances with folding doors at either end [of the *cella*] in the *pronaus* and opisthodomus. There is no example of this type in Rome, but there is the octastyle temple in Athens, that is, the Temple of Olympian Zeus [Plate 8].

CHAPTER III

Columns and Intercolumniations

1. There are five categories [*species*] of temple, for which these are the terms: pycnostyle, with columns set close together; systyle, with columns slightly further apart; diastyle, with columns further apart still: araeostyle, with columns further apart than they should be; eustyle, with the correct distribution of intercolumniations [Fig. 6].

2. Thus the pycnostyle temple is that in which a diameter and a half of a column can be inserted in the intercolumniation: such as, for example, the Temple of the Deified Julius, that of Venus in the Forum of Caesar, and any others designed like them.[11] Again, the systyle temple is one in which the diameters of two columns can be placed in the intercolumniation, and the breadth of the plinths [*plinthides*] of the bases [*spirae*] should equal the distance between them, as for example, in the Temple of Fortuna Equestris near the Stone Theatre,[12] and all others built on the same principles.

3. Both these kinds [pycnostyle and systyle] present practical difficulties because mothers ascending the stairs to make supplications cannot pass through the intercolumniations arm in arm but must go single file:[13] and again the view towards the folding doors is obstructed and the cult statues themselves are hidden because the columns are so close together: also, walks around the temple are impeded by the narrowness of the spacings.

Plate 8. Section of decastyle, dipteral, hypaethral temple on a podium (*orthographia* at right, *sciografia* at left). Vitruvius' Ionic dipteral temple with steps on all four sides here becomes Composite, an order not described by Vitruvius, with a podium on three sides. The spacing is slightly less than pycnostyle and includes a wider intercolumniation in the centre of the façade. The splendid elevation includes, on the exterior of the *cella* at right, the drafted ashlar and horizontal string course with little waves copied by Palladio from the *cella* of the Temple of Mars Ultor (cf. Palladio, 1997, IV.22, p. 232); the pulvinate garlanded frieze in the *cella* presumably adapted from the base of Trajan's column; the *cella* is partially roofed, but only over the side aisles, as

though it were an enormous *peristylium* (cf. Vitruvius, 3.2.8: 'like those in the porticoes of colonnaded courtyards'); the combination of giant Composite exterior columns, plus a *cella* comprising Ionic columns below and Corinthian columns above, resembles Palladio's reconstruction of the Temple of Serapis (Palladio, 1997, IV.43–6, pp. 255, 258), as does the sloping roof over the aisles of the *cella*. The upper columns in the *cella* are three-quarters the height of those below, as Vitruvius advises (5.1.3), and the door, which must match the lower order inside the *cella*, is Vitruvian and Ionic. Appropriately, a statue of Zeus, with eagle and thunderbolt, can be seen on the altar.

4. The arrangement will be diastyle when we can insert three column-diameters in an intercolumniation, as is the case of the Temple of Apollo and Diana. This scheme presents the difficulty that the architraves fracture because of the breadth of the spans.

5. Stone or marble architraves cannot be used for araeostyle temples, but whole wooden beams must be set on the columns: also, these temples look as though they are squatting, top-heavy, low and broad, and their pediments are decorated with terracotta statues or gilt bronze in the Tuscan fashion: for example, the Temple of Ceres in the Circus Maximus and Pompey's Temple of Hercules; so too the one on the Capitol.

6. Now we must give an explanation of the principles of the eustyle temple, which is much the most praiseworthy and incorporates a system of design which is excellent with respect to function, appearance and stability. Between the columns, the intercolumniations should be two and a quarter column-diameters wide and the central intercolumniations at the front and back, three column-diameters. For this way the effect of the design will be elegant, access to the entrance unimpeded and the walkway around the *cella* impressive.

7. This is how one may put these principles into practice: if a tetrastyle temple is to be built, let the front of the site which has been chosen be divided into eleven and a half units, ignoring the plinths [*crepidines*] and the lateral projection of the bases; if it is to be hexastyle, it should be divided into eighteen units; if it must be built as octastyle, it should be divided into twenty-four and a half units. Whether the temple is to be tetrastyle, hexastyle or octastyle, let one of these units be chosen, and this will be the module [*modulus*]: one module will equal the diameter of the columns. Each intercolumniation, apart from those in the middle, will be two and a quarter modules. The middle intercolumniations at the front and back will each be three modules. The columns themselves will be nine and a half modules high. So, as a result of these subdivisions, the intercolumniations and the heights of the columns will have the correct proportions. 8. We have no example of this type in Rome, but at Teos in Asia there is the hexastyle Temple of Father Liber.

Hermogenes established these modular systems, and it was he who first invented the octastyle temple[14] or rather the principles of the pseudodipteral temple. He removed the inner rows of thirty-four columns appropriate to the modular system of the dipteral temple and with this procedure saved on labour and expense. He ingeniously created a much wider space in the intermediate area for the walkway around the *cella* which did not detract in the slightest from the external appearance of the temple, but in fact maintained the dignity of the whole structure without creating any sense of loss for the superfluous columns.

9. The idea of the surrounding colonnade [*pteroma*], that is, the placement of columns around a temple, was devised particularly so that its appearance would be imposing because of the sense of depth created by the intercolumniations, and also so that, if torrential rain were to take a mass of people by surprise and force them to shelter inside, they would have a lot of free space in the temple and around the *cella* in which to wait. These are the advantages incorporated in the plans of pseudodipteral temples, from which it is clear that Hermogenes created spectacular buildings with impressive and penetrating ingenuity and has left written sources from which posterity could derive the theory underlying these disciplines.

10. In araeostyle temples, the columns should be built so that their diameters are an eighth of their height: and in diastyle temples, the height of the column should be divided into eight and a half units, of which one should be allocated to the diameter of the column. In the systyle temple, the height should be divided into nine and a half units, and one should be given to the diameter of the column. Again, in the pycnostyle temple, the height should be divided into ten units, of which one should be used for the diameter of the column. But in the eustyle temple, the height of the column should be divided into nine and a half units, as in the systyle, of which one should be designated as the diameter at the bottom of the shaft [*scapus*]. In this way the system of intercolumniations will be developed in accordance with the module.

11. When the distances between columns increase, the diameters of the column-shafts should be enlarged proportionately.

For if, in the case of an araeostyle temple, only a ninth or tenth [of the column height] is given to the diameter, the column will look thin and insubstantial, since the air seems to consume and reduce the breadth of the shafts because of the width of the intercolumniations. By contrast, if, in the case of pycnostyle temples, an eighth of the height is allocated to the diameter, it will make the columns look swollen and inelegant because of their closeness and the reduced intercolumniations. So we must follow the modular system appropriate to each kind of building. Again, corner columns should be made thicker than the rest by a fiftieth of their own diameter, because they are strongly silhouetted against the air and appear more slender to observers. Accordingly, we must compensate for these misleading optical effects with calculations based on theory.

12. However, it seems that the diminution of the neck [*hypotrachelium*] at the top of the column should be made so that if the column ranges in height from the smallest size up to fifteen feet, the diameter at the bottom should be divided into six units, of which five should be allocated to the diameter at the top: again, if the column is to be between fifteen and twenty feet, the bottom of the shaft should be divided into six and a half units, and five and a half should be allocated to the upper diameter; in the case of columns ranging from twenty to thirty feet, the bottom of the shaft should be divided into seven units, of which six should be given to the contraction at the top; the column which will measure between thirty and forty feet should be divided at the bottom into seven and a half units, of which there will be six and a half at the top according to the rules of contraction; columns from forty to fifty feet should be divided into eight units, which should be contracted to seven at the top of the shaft under the capital. In the case of still taller columns, the contractions should be determined according to the same principle of proportion.

13. But the upper diameters of columns should be enlarged to compensate for the increasing distances for the glance of the eye as it looks up. For our sight searches for beauty, and if we do not satisfy its desire for gratification by increasing proportions with additions derived from modules in order to correct false impressions with an appropriate adjustment, the building will present an awkward and clumsy sight to onlookers. With respect to the enlargement applied to the middle of columns, which the Greeks call ἔντασις [*entasis*], an explanatory diagram will be added at the end of the book showing how to make it appropriate and pleasing.

CHAPTER IV

Building Temples

1. The foundations of these buildings should be dug out of solid ground, if it can be found, and as deep down into such solid ground as seems proportionate to the size of the building; the whole of the platform for the foundations should be built with the most robust masonry possible. Above ground, walls should be built thicker by a half than the columns which will be placed on them so that the substructures will be stronger than the superstructure: these walls are called stereobates because they take the load. The projections of the columns' bases should not be wider than the substructure. Similarly above the walls [of the stereobate] the width [of the stylobate] is to be maintained in the same way, but the spaces between the walls [of the stereobate] should be vaulted or made stable by ramming down hard the earth infill between them so that the walls are held apart.

2. If solid ground cannot be found, and the site consists of loose earth right down to the bottom, or is marshy, then it must be dug out, completely emptied and packed with piles of alder, olive or charred oak. The piles must be rammed together as tightly as possible with pile-drivers and the spaces between them filled with charcoal: finally the foundations should be filled in with the strongest possible masonry. After the foundations have been built the stylobates [*stylobatae*] must be put in place absolutely horizontally.

3. The columns should be distributed on top of the stylobates as described above: close together in the case of pycnostyle; in the cases of systyle, diastyle or eustyle in the way they have been described and worked out above. In araeostyle temples one is free to set the intervals as one likes. But in peripteral temples the columns should be placed so that there are twice as many intercolumniations down the sides as there are on the front; in that way the length of the building will be twice its breadth. For those who have simply doubled the number of columns have clearly gone wrong because the length is evidently greater than it should be by one intercolumniation.[15]

4. The steps on the front must be arranged so that they always comprise an odd number: for since one climbs the first step leading with the right foot, then one's right foot should arrive at the top level of the temple first. I think that the height of these steps should be worked out so they are no more than five-sixths of a foot in height nor less than three-quarters; in that way climbing the steps will not be tiring. It seems that the depth of the treads should not be made less than a foot and a half or greater than two. And if there are to be steps all around the temple, then they should all be built the same size.

5. But if a podium is to be built on three sides of a temple, it should be constructed so that the plinths, base mouldings, the shafts of the pedestals [*trunci*], the cornices and the final moulding [*lysis*] of the podium conform to the stylobate itself which will be under the bases of the columns. The level of the stylobate must be adjusted so that it curves upwards in the middle thanks to the use of little projections of varying height:[16] for if it is laid absolutely horizontally, it will look concave to the eye. But with regard to the question of how the little projections can be made correctly for this purpose, a drawing and explanation will be presented at the end of the book.

CHAPTER V

Proportions of the Ionic Order

1. Once these operations have been completed, the bases must be located in the correct positions and should be made in conformity with the modular system so that their height, including the plinth [*plinthum*], will equal half the column-diameter, and their projection, called ἐκφορά [*ekphora*] in Greek, half of that: thus the base will be one and a half times the diameter of the column in length and breadth.

2. If the base is to be Attic, its height must be subdivided so that the upper part will be a third of the column-diameter, and the rest should be left for the plinth. Excluding the plinth, the rest should be subdivided into four units; a quarter should be allocated to the upper torus and the other three divided equally in two units[17] so that one will be allocated to the lower torus and the other to the *scotia*, which the Greeks call the τροχίλος [*trochilos*] with its fillets [*quadrae*].

3. But if Ionic bases are to be built, their modular system should be established so that the breadth of the base[18] will equal the diameter of the column plus three-eighths. Its height will be the same as that of the Attic base, and so too that of the plinth. Apart from the plinth, the rest of the base, which will be a third of the column-diameter, should be divided into seven units: three for the torus at the top; the other four should be divided equally, one part forming the upper *scotia* with its astragals [*astragali*] and the upper fillet [*supercilium*], the other part being left for the lower *scotia*: but the lower *scotia* will look bigger [than the upper *scotia*] because it will project as far as the edge of the plinth. The astragals are to be made an eighth of the scotia. The projection of the base will be three-sixteenths of the column diameter [Plate 9].

4. Once the bases have been finished and put in place, the middle columns of the *pronaus* and at the back façade are to be set up vertically on the axes of the bases; but the corner columns and those which will be aligned with them to right and left down the sides of the building should be built so that their inner sides facing the walls of the *cella* have vertical surfaces, but the outer surfaces should be contracted in the manner described above. In this way the visual impact of the designs of temples will be perfected by contraction calculated using this method.

5. Once the shafts of the columns have been set up, the arrangement of the capitals, if they are to be pulvinate,[19] will conform to this modular system: the abacus should be equal in length and breadth to the lowest diameter of the column-shaft plus an eighteenth, and the height, with the volutes, should be half that. But the faces of the volutes must be set inwards in relation to the edge of the abacus by an eighteenth and a half. Then the height of the capital must be divided into nine and a half units, and, following the abacus, lines called *catheti* are to be let down in the four sectors with the volutes from the fillet [*quadra*] at the edge of the abacus. Then, one and a half of the nine and a half units should be reserved for the height of the abacus, and the other eight should be allotted to the volutes.

6. Then another vertical line should be set one and a half units further in [horizontally along the abacus] than the line let down following the edge of the abacus. Next, these lines should be divided up in such a way that four and a half units are left

Plate 9. Ionic capital and base (*spira*). Base: **A** Plinth (*plinthus*). **B** Scotia (*trochilus*). **1 1** Astragals (*astragali*). **C** Torus. **D** Fillet or *listello*. **f** *Apophysis* and its construction (**abc**). Many Renaissance architectural writers had strong objections to the Vitruvian Ionic base, which lacks a lower torus: they denounced it as ugly and, since they could not find ancient examples of it, never used it; in favour were Sansovino, Trissino, Bertani, Alessi, Vignola and Pellegrino, who all built with it. Capital: **o** [letter not marked] Plan of the capital. **Ah** Abacus. **n** Rim of the volute (*axis*). **m** Channel of the volute (*canalis*). **l** Echinus (*cymatium*). **p** Eye of the volute (*oculus*). **e** *Apophysis* at top of column. Frustratingly Vitruvius' text fails us at the very moment when we want to know how the diminishing arcs of the scrolls of the volutes were drawn. A very attractive solution had been developed by G. P. Salviati in 1552 (cf. Palladio, 1997, I.34, p. 38).

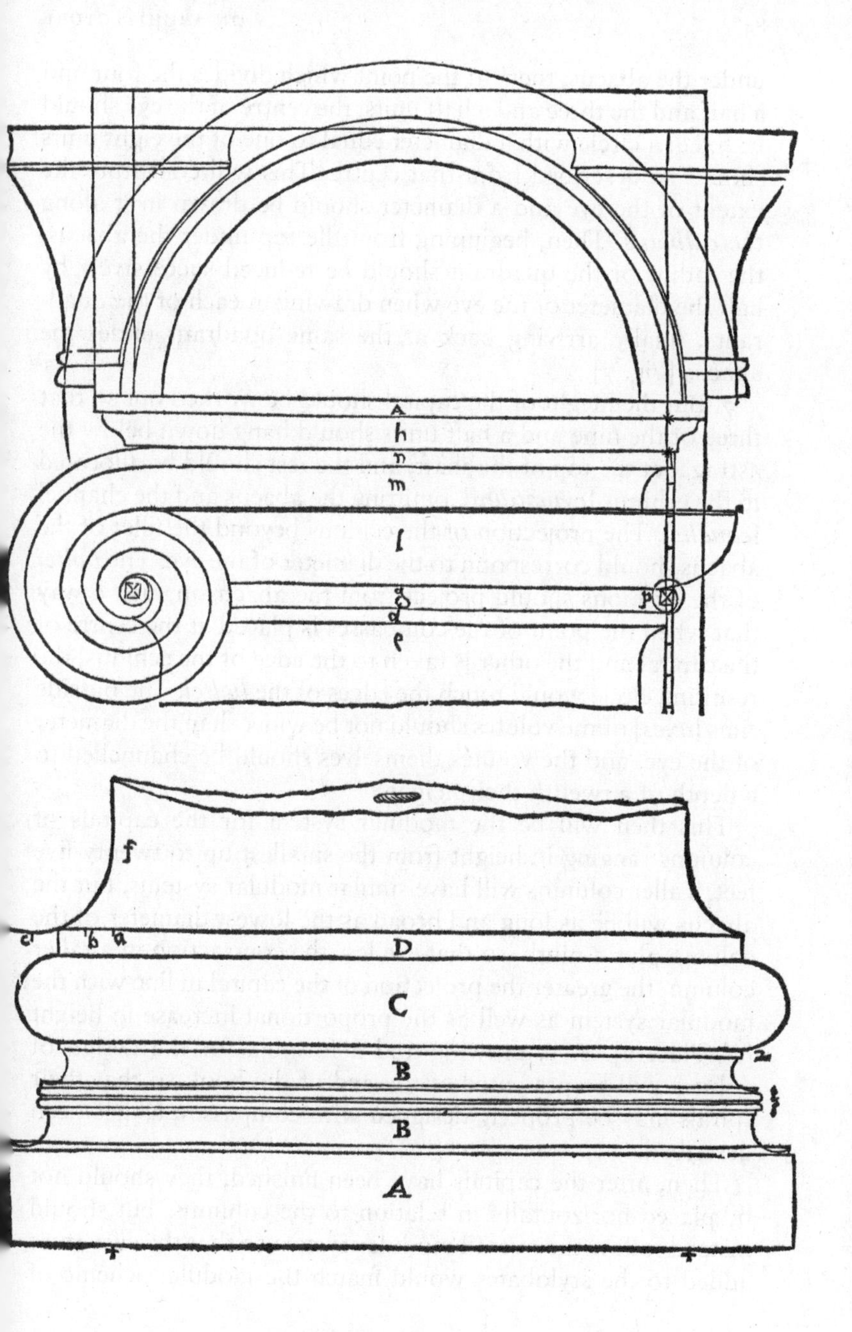
A
h
n
m
l
g
d
e
p
f
c
b
a
D
C
B
B
A

under the abacus; then, at the point which divides the four and a half and the three and a half units, the centre of the eye should be fixed; a circle with a diameter equal to one of the eight units should be described from that centre. This will constitute the extent of the eye and a diameter should be drawn in it along the *cathetus*. Then, beginning from the top under the abacus, the radius of the quadrant should be reduced successively by half the diameter of the eye when drawing in each of the quadrants, finally arriving back at the same quadrant under the abacus [Fig. 7].

7. But the height of the capital should be worked out so that three of the nine and a half units should hang down below the astragal at the top of the shaft, and the rest should be allocated to the echinus [*cymatium*], omitting the abacus and the channel [*canalis*]. The projection of the echinus beyond the fillet of the abacus should correspond to the diameter of the eye. The *baltei* of the cushions should project from the abacus in such a way that when the point of the compasses is placed at the centre of the capital and the other is taken to the edge of the echinus, the resulting circle would touch the edges of the *baltei*. The outside rims [*axes*] of the volutes should not be wider than the diameter of the eye, and the volutes themselves should be channelled to a depth of a twelfth their height.

This then will be the modular system for the capitals of columns ranging in height from the smallest up to twenty-five feet. Taller columns will have similar modular systems, but the abacus will be as long and broad as the lowest diameter of the column plus a ninth, so that the less the contraction in a taller column, the greater the projection of the capital in line with the modular system as well as the proportional increase in height [cf. Plate 9]. 8. A drawing and a formula for the design of volutes will be presented at the end of the book so that their spirals may be properly designed with compasses, as has been described.

Then, after the capitals have been finished, they should not be placed horizontally in relation to the columns, but should instead follow the overall modular system so that the curvature added to the stylobates would match the modular scheme of the architraves in the upper part of the building. The principle to be adopted for architraves should be this: if the columns are to be no less than twelve up to fifteen feet tall, the height of the architrave should be half the diameter of the column at the bottom of the shaft; if the columns range from fifteen to twenty feet tall, the height of the column should be divided into thirteen units, of which one should be allocated to the height of the architrave; and if the columns range from twenty to twenty-five feet high, their height should be divided into twelve and a half units, and one of them should be allocated to the height of the architrave; finally, if the columns are from twenty-five to thirty feet tall, their height should be divided into twelve units, with one comprising the height of the architrave. So the heights of the architraves must be derived from the heights of the columns[20] according to a proportional system calculated in the same way.

9. For the further up the gaze of the eye has to climb, the less easily can it penetrate the density of the air; and so it falters when the height is great, and having lost strength, transmits to the senses an unreliable estimate of the dimensions of the modules. For this reason, one must always incorporate in the calculations an increase in the size of the components worked out according to the modular system, so that whether buildings are constructed on very high sites or are themselves built on a grand scale, they will maintain the proportions appropriate to their size. The breadth of the soffit of the architrave to be placed on the capital should match the diameter of the top of the column below the capital, and the breadth of the top of the architrave should match the diameter of the column at the bottom of the shaft.

10. The moulding [*cymatium*] at the top of the architrave should be made a seventh of the height of the whole architrave, and its projection the same. Apart from the moulding, the rest of the architrave is to be divided into twelve units, of which the lowest fascia should comprise three, the second four and the upper five. Then the frieze [*zophorus*] above the architrave will be less tall than the architrave by a quarter; but if it is appropriate to carve reliefs in the frieze, then it should be a quarter taller than the architrave so that the sculpture may be more

impressive. The moulding [at the top of the frieze] will be a seventh of the height of the frieze, and its projection the same as its height.

11. Above the frieze, the dentil must be built as high as the middle fascia of the architrave, with a projection equalling its height. This connecting element [*intersectio*], in Greek μετοχή [*metoche*; joining],[21] is to be subdivided in such a way that the front of each dentil should be half as wide as it is high, but the width of the cavities in the connecting element should be two-thirds of the width of the dentil: the moulding should be a sixth of the height of the dentil. The cornice with its moulding, excluding the *sima*, should be as tall as the middle fascia of the architrave, and the projection of the cornice, including the dentilation, should equal the height from the frieze to the top of the gola of the cornice: and, in general, all projections appear more attractive when they are equal to their own height.

12. But the height of the tympanum in the pediment [*fastigium*] should be established like this: the whole of the front of the cornice, from the extremities of the moulding, should be divided into nine units, of which one should be set on the axis of the apex of the tympanum, while the architraves and the neckings of the columns should be perpendicular to it. The cornices above the tympanum, excluding the *simae*, are to be built in the same way as those beneath it. Above the cornices [of the tympanum], the *simae*, called ἐπαιετίδες [*epaietides*] in Greek, should be made an eighth taller than the cornices. The acroteria at the corners should be as tall as the apex of the tympanum, and those in the middle an eighth taller than those at the corners [Plate 10].

13. All the elements which are to be built above the capitals of the columns, that is, the architraves, friezes, cornices, tympana, pediments and acroteria should be inclined forward by a twelfth of their own height, because if two lines are extended from our eye when we stand in front of façades, with one extending to the bottom of the building and the other to the top, the one which arrives at the top will be longer. Therefore, the longer the line of sight that extends to the top, the more the top of the

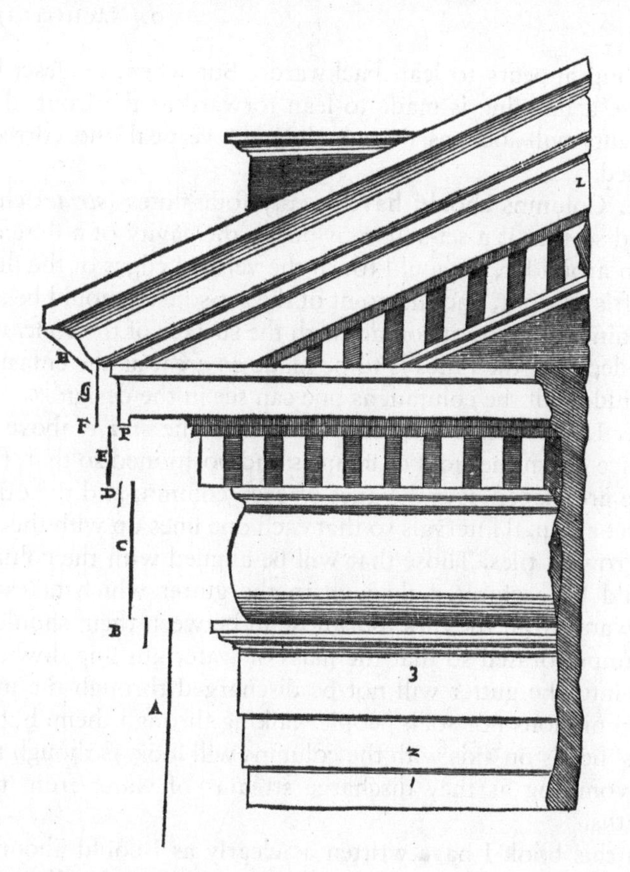

Plate 10. Architrave, frieze and cornice of the Ionic order. Palladio inserts dentils in the tympanum, ignoring Vitruvius' prohibition that principal and common rafters (*cantherii*, *asseres*) could not be shown projecting from it (4.2.5). A Architrave (*epistylium*). 1 First fascia. 2 Second fascia. 3 Third fascia. B Moulding (*cymatium*) of architrave. C Frieze (*zophorus*), which Palladio makes pulvinate, which is not unreasonable given that Vitruvius describes Ionic capitals thus (3.5.5). D Moulding of frieze. E Dentilation (*denticulus*). O Barbaro applies the term 'intersection' (*intersectio*) to the spaces between dentils (also called *metopae*), but it seems more likely that the term refers to the strip of dentilation as a whole. F Moulding of the dentils. G Cornice (*corona*). L Pediment (*fastigium*). K Tympanum. I Acroteria (*acroterium*). H *Sima* in the form of a *gola recta*.

building appears to lean backwards. But when, as described above, a building is made to lean forward at the front, these elements will look as though they are vertical and correctly aligned.

14. Columns should have twenty-four flutes [*striae*] channelled so that if a set-square is put in the cavity of a flute and taken around it, it should touch the vertical edges of the flutes to left and right, and the point of the set-square should be able to maintain constant contact with the surface of the concavity. The depth of the flutes is to be made as great as the entasis at the middle of the column, as one can see in the design.

15. Lions' heads should be carved in the *simae* above the cornice down the sides of temples and positioned so that, first, some are placed over the axes of each column, and the others are set at equal intervals so that each one lines up with the axis of a row of tiles. Those that will be aligned with the columns should be perforated through to the gutter which takes the rainwater from the tiles, but those in between them should be left unperforated so that the mass of water gushing down the tiles into the gutter will not be discharged through the intercolumniations nor soak people walking through them: but the lions' heads on axis with the columns will look as though they are vomiting as they discharge streams of water from their mouths.

In this book I have written as clearly as I could about the organization of Ionic temples; in the next book I will explain the proportions of Doric and Corinthian temples.

BOOK IV

Introduction

1. When I noticed, Supreme Ruler, that many authors have left us teachings and volumes of commentaries on architecture which are not systematic but merely inchoate, like scattered fragments, it occurred to me some time ago that it would be worthwhile and extraordinarily useful to organize the whole discipline into an orderly form and explain the requisite characteristics of each subject in separate books. So, Caesar, in my first book I described for you the duties of the architect and the subjects in which he should be knowledgeable; in the second, I discussed the supplies of material with which buildings are constructed. In the third, which deals with the organization of temples and the variety of their types, I illustrated what and how many categories they should have and the plans which should be used for each order, using the conventions of the Ionic order, one of the three orders which should incorporate the most precisely measured modules in their systems of proportion. Now, in this book, I will discuss the rules established for Doric, Corinthian and the others and will explain their differences and characteristics.

CHAPTER I

The Origins of the Three Orders and the Corinthian Capital

1. Corinthian columns, apart from the capitals, incorporate the same modular system as Ionic in all respects, but the height of the capitals makes them proportionately taller and more graceful, because, while the height of the Ionic capital is a third of the diameter of the column, the Corinthian capital is as tall as the whole diameter of the shaft. Therefore, because two-thirds of the diameter of Corinthian shafts are added [to the height of their capitals], this makes the columns' appearance more graceful because of their height.

2. The other components placed on top of Corinthian columns are built using either the Doric modular system or Ionic conventions, because the Corinthian order itself did not have its own conventions for cornices and the other mouldings [of the entablature], but either they are arranged in the Doric manner, following the system of triglyphs with mutules in the cornices and *guttae* in the architraves, or they are decorated following Ionic conventions with friezes adorned with reliefs arranged with dentils and cornices.

3. Thus, a third architectural order used in buildings was evolved from the other two by the insertion of a [different] capital; for the names of the three orders, Doric, Ionic and Corinthian, were derived from the designs of their columns, of which the first to evolve, long ago, was the Doric.

For Dorus, son of Hellen and the nymph Orseis, ruled Achaea[1] and all the Peloponnese, and he built a temple of Juno in the ancient city of Argos, a sanctuary which happened to have the characteristics of this order, and then others of the same order in other Achaean cities, at a time when the system of proportions had not yet been created.

4. But after the Athenians, acting on the oracular responses of Delphic Apollo and by general agreement of the whole of Hellas, led thirteen colonies simultaneously to Asia, they installed leaders in each colony and gave overall command to Ion, son of Xuthus and Creusa, whom Apollo at Delphi had, in fact, declared to be his own son in his oracular responses. Ion led the colonies to Asia and occupied the territory of Caria, and founded there the great cities of Ephesus, Miletus and Myus, which was swallowed up by the water long ago, and its religious cults and voting rights transferred to the Milesians by the Ionians; also Priene, Samos, Teos, Colophon, Chios, Erythrae, Phocaea, Clazomenae, Lebedos and Melite – this was the Melite which, on account of the arrogance of its citizens, was destroyed by the other cities as a consequence of a war declared by general agreement, and afterwards, through the mediation of King Attalus and Arsinoe, the city of the Smyrnaeans was accepted in its place as one of the Ionian cities.

5. After the Carians and Lelegans had been driven away, the inhabitants of these cities named that part of the world Ionia after their leader Ion, and, setting up temples for the immortal gods there, began to build sanctuaries; first they built a temple to Apollo Panionios like the ones they had seen in Achaea; they called it Doric because they had originally seen temples of this type in the Dorian cities.

6. When they wanted to set up the columns in the temple, they had no proportional system appropriate to them, and so, trying to find out what procedures could be adopted to ensure that the columns would be capable of bearing the loads and that the beauty of their appearance would be assured, they measured the sole of a man's foot and applied it to his height. When they discovered that a man's foot is a sixth of his height, they applied that unit to the column and allocated six times the diameter they had established for the bottom of the shaft to its height, including the capital: that is why the Doric column began to exhibit the proportions, strength and grace of the male body in buildings.

7. Some time later they built a Temple of Diana; searching for a look for the new order, they used the same plans, adapting them to feminine gracefulness, and first made the diameter of the column an eighth of its height so that it would appear taller. At the bottom they substituted a base for the shoe, and

on the capital they placed volutes at right and left like graceful curls hanging down from the hair; they decorated the fronts with convex mouldings and runs of fruit arranged like hair, and sent flutes down the whole trunk [*truncus*] like the folds in the robes traditionally worn by married women. And that is how they developed two different types of column: one which looked naked, undecorated and virile, the other characterized by feminine delicacy, decoration and modularity. **8.** In fact, later builders, becoming more sophisticated with regard to elegance and subtlety of judgement and, delighting in more graceful modules, established seven diameters as the height for the Doric column and nine for the Ionic: the latter, which the Ionians built first, is therefore called the Ionic.

The third order, in fact, called Corinthian, imitates the elegance of a virgin, because virgins, who are endowed with more graceful limbs because of their tender age, achieve more elegant effects in their ornament.

9. The initial discovery of this type of capital is said to have been made like this: a virgin and citizen of Corinth, already of marriageable age, died as a result of illness. After her funeral the nurse collected and arranged in a basket her loving cups, which had great sentimental value for the girl when she was alive; she carried them to the tomb and put them on top and, so that the cups would last much longer in the open, covered them with a tile. This basket happened to have been placed on top of an acanthus root. In time the acanthus root, pressed down in the middle by the weight, sent out leaves in the spring, as well as stalks, which, growing up the sides of the basket, and restricted at the corners by the weight of the tile, were forced to form the spirals of volutes at the ends. **10.** Then Callimachus, whom the Athenians called κατατηξίτεχνος [*katatexitechnos*] because of the refinement and delicacy of his skill in working marble, walked by the tomb and saw the basket with the tender young leaves growing around it: delighted by the style and novelty of the form, he built some columns at Corinth following this example and developed a modular system, as a result of which he evolved the rules for the completion of buildings in the Corinthian order.

11. The modular system of this capital should be established in such a way that the lower diameter of the column should equal the height of the capital with the abacus. The breadth of the abacus is to be proportioned so that the length of diagonals taken from corner to corner should be twice the height of the capital. In that way the faces of the abacus will have fronts of the right breadth on all sides. The faces should curve inwards from the points of the angles of the abacus by a ninth of the breadth of the face. At the bottom, the capital should be as wide as the top of the column, disregarding the *apothesis*[2] and the astragal. The height of the abacus should be a seventh that of the capital.

12. Omitting the height of the abacus, the rest should be divided into three units, of which one should be given to the lower leaves. The second row of leaves should occupy the middle level. The stalks should be of the same height: other leaves grow from them and project obliquely so as to support the volutes that spring from the stalks and run to the points at the corners: smaller spirals should be carved in the space defined by the central curve of the abacus[3] under the flowers. The flowers on the four sides [of the abacus] should be made as large as the height of the abacus: with this modular system Corinthian capitals will achieve their definitive form [Plate 11].

There are, however, other kinds of capitals built on top of the same columns [Doric, Ionic and Corinthian] and described with various terms, but we cannot identify their particular modular systems or their columns as though they were another order; instead we can see that their technical vocabulary is taken over, and adapted from the Corinthian, the pulvinate [Ionic] and the Doric, of which the modular systems have been transferred to match the delicate forms of new architectural sculpture.

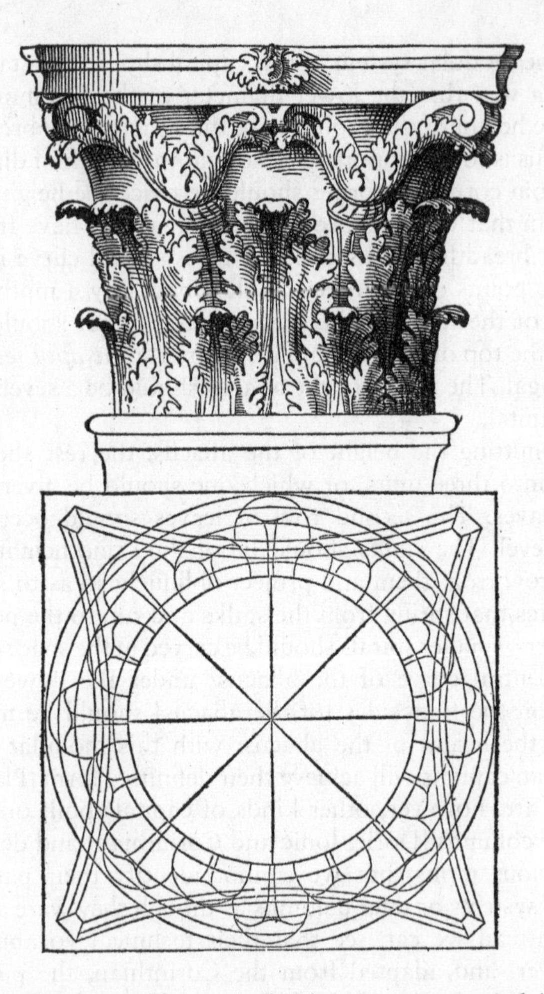

Plate 11. The Corinthian capital. The Vitruvian capital is related to the base column-diameter. The upper diameter of the column is six-sevenths, and the total height of the capital at seven-sevenths: one-seventh for the abacus, and two-sevenths for each of the three levels of the bell (the first layer of leaves, the second layer of leaves and the layer with the volutes). But Barbaro and Palladio, while maintaining all the elements mentioned by Vitruvius, adjust the proportions to produce a capital taller than Vitruvius' in relation to its breadth by making its diameter at the bottom slightly less than that of the top of the column shaft.

CHAPTER II

The Origins of the Orders: From Wood to Stone

1. But since the origins and invention of the orders of columns have been described above, I do not think it irrelevant to speak about their entablatures[4] using the same criteria, showing how they came about and on the basis of what principles and original components they were invented.

The timberwork placed above all buildings is described with various terms; the variety of the terminology corresponds to the variety of uses to which the carpentry is put. For beams [*trabes*] are built on top of columns, pilasters [*parastaticae*] and *antae*, and there are joists [*tigna*] and planks [*axes*] in wooden flooring. Under the roofs, if the span is very large, there are cross-beams [*transtra*] and struts [*capreoli*]; if the span is moderate, there is only a ridge-piece [*columen*] with the principal rafters [*cantherii*] extending to the outer edges of the eaves. Above the principal rafters are the purlins [*templa*], and then above these, and under the roof-tiles, are the common rafters [*asseres*], extending so far that the walls are sheltered by their projection [Plate 12].

2. And so every element has its own location, type and order of position. Starting from these components and this kind of carpentry, builders adapted them for the relief work of the stone and marble structures of sacred buildings, convinced that such inventions should be copied. And so ancient carpenters, building in some place or other, put in place joists projecting from the interior walls to the outside, then built masonry between the joists, and decorated the cornices and gables above them with carpentry of great elegance; then they cut off the projections of the joists flush with the vertical planes of the walls; but when this looked clumsy to them, they fixed wooden boards, shaped in the same way that triglyphs are now made, on the faces of the cut-off joists, and painted them with blue wax so that the cut-off ends, now covered, would not be unpleasant to look at: so it was that in Doric buildings the

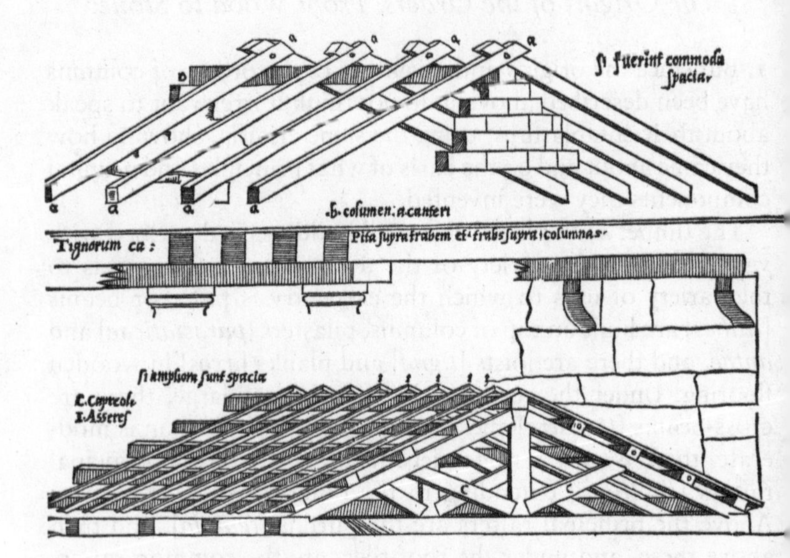

Plate 12. Woodwork in roofs. The illustration shows the beams (*trabes*) above the columns, then, laid at right angles to them, the ends of the joists (*tigna*) which, according to Vitruvian theory, became the triglyphs in stone architecture. The roof includes only the principal rafters (*cantherii*) and the ridge-piece (*columen*) supported by masonry. **Woodwork in roofs with wider spans** (below). Above the large horizontal beams (*transtra*) is the roof with the ridge-piece and principal rafters surmounted by the purlins (*templa*) laid crossways, then the common rafters (*asseres*) on top parallel with the principal rafters; the king-post is braced to the principal rafters with sloping struts (*capreoli*) and the principal and common rafters bound together by the ridge-piece. Fra Giocondo (1511) was the source for the essential ideas behind this illustration.

separation of the joists faced by an arrangement of triglyphs began to provide space for the metopes between them.

3. Later on, other carpenters working on other buildings extended the projecting principal rafters so that they were perpendicular to the triglyphs, and then cut off their projections. The result of this was that just as triglyphs were developed from the arrangement of joists, so too the concept of mutules under cornices was evolved from the projection of the principal rafters. This is why the mutules in stone and marble structures are usually made of sloping relief-work because they imitate the principal rafters; for it is essential that they are built with an incline because of the rainfall. Therefore, in general, the scheme of triglyphs and mutules in Doric buildings derives from this imitation [of timber buildings].

4. It cannot be true, as some have maintained erroneously, that triglyphs represent windows, because they are located at the corners of the frieze and over the axes of columns, positions where the insertion of windows is absolutely impossible: for the binding elements at the corners of buildings would fall apart if openings for windows were to be made there. Again, if we suppose that there were apertures for illumination where the triglyphs are now placed, on the same logic, the dentils in Ionic buildings would appear to have replaced windows, since the spaces between both dentils and triglyphs are called metopes. The Greeks call the housings of joists and of common rafters ὀπαί [*opai*; holes] while our people call these cavities 'pigeon-holes'. Hence, they call the space between joists, that is, between two *opai*, μετόπη [metope].

5. Just as the system of triglyphs and mutules was developed in Doric buildings, so too the creation of dentils has its own rationale in Ionic structures: just as mutules retain the appearance of the projection of the principal rafters, so too, in Ionic buildings, the dentils imitate the projection of common rafters. Consequently in Greek buildings nobody puts dentils under mutules, as it is impossible that common rafters come under principal rafters. Therefore, if that which in reality must be placed above principal rafters and purlins [i.e. common rafters] were, in imitations, to be placed underneath them, then it would

demonstrate fallacious working principles. Moreover, the fact that the ancients did not approve of or inaugurate the practice of putting mutules or dentils[5] in pediments, but plain cornices, is due to the fact that neither principal nor common rafters are built into the fronts of pediments, nor can they project from them, but are placed so as to slope down to enable the run-off of rain. Thus they thought that what is impossible in reality could not be based on sound principles if it was replicated in imitations of that reality.

6. For they converted all these elements for the perfection of their buildings with a precise sense of propriety and sound practice derived from nature, and gave their seal of approval to those things which, explained in rational argument, have the force of truth. So starting from beginnings like this, they left for us the modular systems and proportions which they had established for each order. Following in their footsteps I have spoken above about the conventions of Ionic and Corinthian architecture, and now I will explain briefly the Doric system and its overall appearance.

CHAPTER III

The Modularity of Doric Temples

1. Some ancient architects denied that temples should be built with the Doric order because false and unpleasant modular systems were generated in such buildings: this is what Tarchesius said, but also Pytheos, and especially Hermogenes; Hermogenes, in fact, after collecting a supply of marble for the construction of a Doric temple, changed the project and, using the same materials, built an Ionic temple to Father Liber instead:[6] not because the general appearance, order or shape is unattractive or lacks dignity, but because during construction the distribution of triglyphs and coffers[7] is restricting and incommensurable.

2. For it is essential that triglyphs should be placed on the central axes of columns and that the metopes between them should be as broad as they are tall: conversely, the triglyphs over the angle-columns are set at the corners [of the frieze] and not over the central axes of the columns. So the metopes placed next to the angle triglyphs do not turn out square, but are broader by half the breadth of a triglyph. But those who wish to make the metopes equal contract the end intercolumniations by half the breadth of a triglyph. This solution, however, is defective whether it results from alterations to the lengths of the metopes or contractions of the intercolumniations. This seems to be the reason why the ancients avoided the Doric modular system in their sacred temples.

3. However, as the sequence of our argument requires, we shall explain this system here as we have learnt about it from our masters, so that anyone wishing to proceed following these rules in this way will have at his disposal a full explanation of the proportions with which he can bring sacred temples in the Doric style to completion correctly and without errors. The façade of a Doric temple where the columns are placed[8] should be divided into twenty-seven units if it is to be tetrastyle, and forty-two units if it is to be hexastyle. One of these parts will be the module, called ἐμβάτης [*embates*] in Greek, and the subdivisions of the whole building are based on calculations derived from the establishment of the module.[9]

4. The diameter of the columns will be two modules and their height fourteen, including the capital. The height of the capital will be one module, the breadth two and a sixth. The height of the capital should be divided into three units, of which one should be allocated to the plinth [= abacus] with its moulding, the second to the echinus with its fillets [*anuli*], and the third to the necking. The column should be contracted using the procedure described in Book III on Ionic temples. The height of the architrave with its *taenia* and *guttae* should be one module, the *taenia* a seventh, and the run of *guttae*, lying under the *taenia* and perpendicular to the triglyphs, should hang down with the *regula* for a sixth of a module. The breadth of the soffit of the architrave should match the diameter of the necking at the top of the column. The triglyphs, one and a half modules

high and a module wide at the front, must be placed above the architrave with their respective metopes, and distributed so that they are located over the central axes of the angle-columns and of the columns between them, with two triglyphs over each intercolumniation and three over the middle intercolumniations of the *pronaus* and back façade.[10] With the central openings widened like this, access for those approaching the cult statues of the gods will be unimpeded.

5. The width of the triglyphs should be divided into six units, of which five should be marked off with a ruler in the middle, and two halves to right and left. One unit in the centre should be marked off as the 'thigh' [*femur*], called μηρός [*meros*] in Greek. On each side of it small channels [*canaliculi*] fitting the corner of a set-square should be excavated; and in parallel, other such 'thighs' should be created, one to the right and one to the left; at the lateral edges small half-channels facing in opposite directions should be made. After the triglyphs have been positioned in this way, the metopes between them should be as tall as they are wide, and again, at the angles [of the frieze], half-metopes, half a module wide, should be inserted. In this way all defects, whether in the metopes, the intercolumniations or the coffers, would be corrected because all the spacings would be made uniform.

6. The capitals [*capitula*] of the triglyphs are to be made a sixth of a module high. The cornice should be placed above the capitals of the triglyphs, projecting two-thirds of a module and supplied with a Doric moulding below, and another one on top. Again, the cornice with its moulding [*cymatium*] will be half a module

Plate 13. The Doric order: capital, base and part of the entablature. Base. **A** Plinth. **B** Lower torus. **C** *Scotia*. **D** Upper torus. **E** Fillet or *listello*. **F** *Apophysis*. Capital. **G** Moulding. **H** Abacus. **I** Echinus. **K** Fillets, *listelli*. **L** Frieze of capital, collar (*hypotrachelium*). **M** Tondino. **N** *Apophysis*. **O** Architrave. **P** *Guttae*. **q** Small block to which the *guttae* are attached (*regula*). Palladio mislabels this element here, which is the *taenia*; the *regula* is the little block immediately below from which the *guttae* hang. **R** 'Thigh' (*femur*, *meros*) of triglyph. **S** Patera. Following ancient examples ('nell'antico si trova'), Barbaro increases the breadth of the abacus.

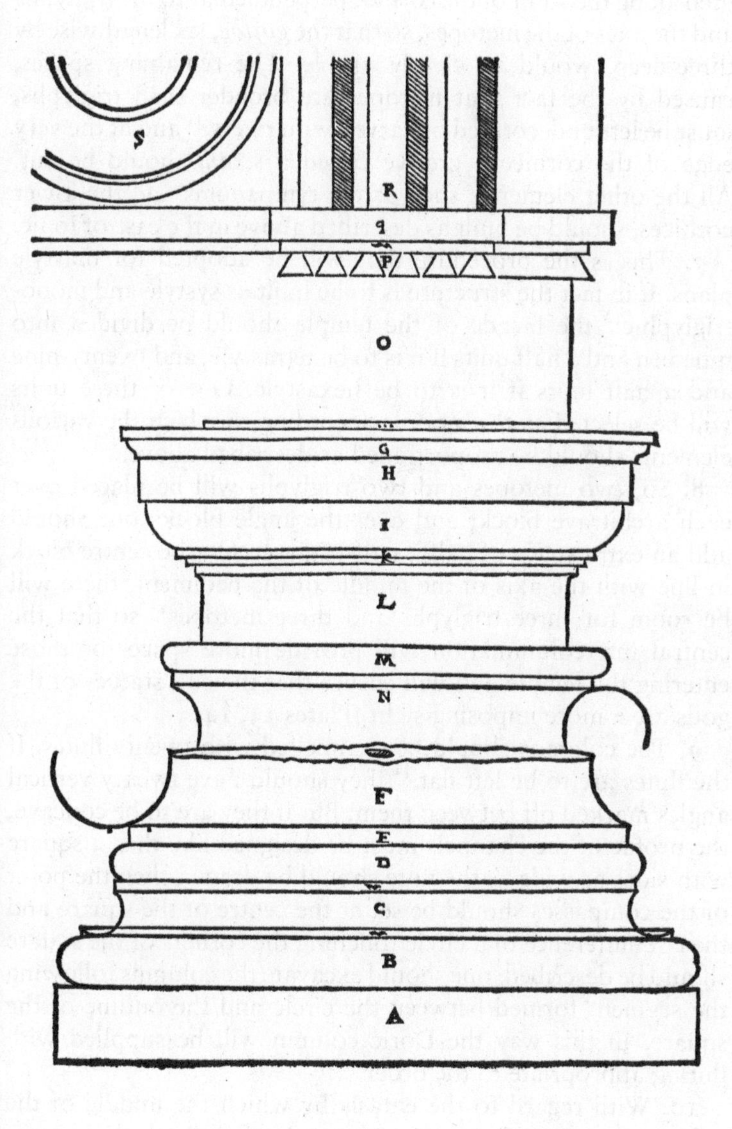
S
R
q
P
O
G
H
I
L
M
N
F
E
D
C
B
A

high. The runs of *viae* and the groups of *guttae* are to be distributed along the soffit of the cornice perpendicular to the triglyphs and the axes of the metopes, so that the *guttae*, six lengthwise by three deep, would be clearly visible. The remaining spaces, caused by the fact that metopes are broader than triglyphs, must be left undecorated or carved with rivers,[11] and at the very edge of the cornice a groove called a *scotia* should be cut. All the other elements, such as the tympanums and the lower cornices, should be built as described above in the case of Ionic.

7. This is the procedure that will be adopted for diastyle plans. If in fact the structure is to be built as systyle and monotriglyphic[12] the façade of the temple should be divided into nineteen and a half units if it is to be tetrastyle, and twenty-nine and a half units if it is to be hexastyle. One of these units will be selected as the module according to which the various elements should be proportioned as described above.

8. So, two metopes and two triglyphs will be placed over each architrave block; and over the angle blocks one should add an extra space of half a half-triglyph. On the centre block in line with the axis of the middle of the pediment, there will be room for three triglyphs and three metopes[13] so that the central intercolumniation will provide more space for those entering the temple and will ensure that the cult statues of the gods are a more imposing sight [Plates 13, 14].

9. The columns should be channelled with twenty flutes. If the flutes are to be left flat,[14] they should have twenty vertical angles marked off between them. But if they are to be concave, the profile of the channels must be designed like this: a square with sides as wide as the flute should be drawn, then the point of the compasses should be set at the centre of the square and the circumference of a circle touching the corners of the square should be described; one should excavate the columns following the segment formed between the circle and the outline of the square. In this way the Doric column will be supplied with fluting appropriate to the order.

10. With regard to the entasis by which the middle of the column is increased, what I wrote in detail in Book III on Ionic columns should be applied to Doric columns.

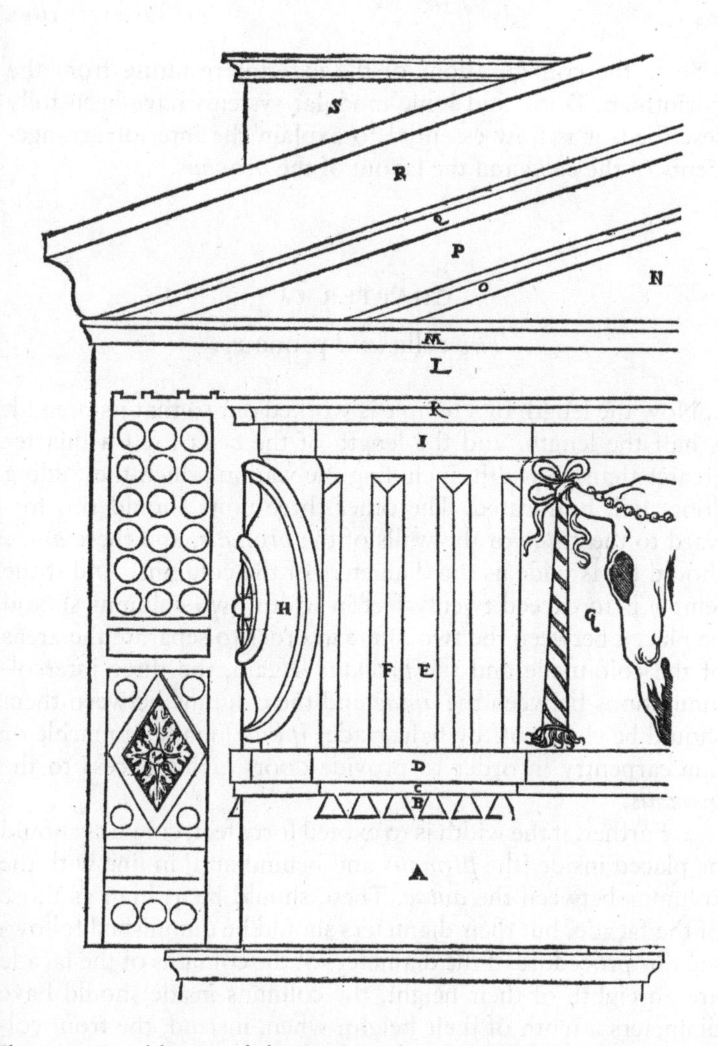

Plate 14. Entablature of the Doric order. A Architrave (*epistylium*). B *Guttae*. C Small block to which the *guttae* are attached (*regula*). D Continuous moulding marking top of architrave (*taenia*). E 'Thigh' (*femur*, *meros*) of triglyph. F Flute, channel (*canaliculus*). G Metope. H Half-metope. I Capital (*capitulum*) of triglyph. K Moulding of the triglyph. L Cornice. M Doric moulding of cornice. N Tympanum. O, P, Q Elements of pediment corresponding to cornice. R *Gola recta*. S Acroterion.

Since the configurations of the exterior resulting from the Corinthian, Doric and Ionic modular systems have been fully described, it is now essential to explain the interior arrangements of the *cella* and the layout of the *pronaus*.

CHAPTER IV

The cella *and* pronaus

1. Now the length of a temple is worked out so that its breadth is half the length, and the length of the *cella* itself a quarter greater than its width, including the wall in which the folding doors will be located. The other three units should run forward to the *antae* of the walls of the *pronaus*, and these *antae* should be as wide as the diameters of the columns. And if the temple is to exceed twenty feet in width, two columns should be placed between the two *antae* in order to separate the areas of the colonnade and the *pronaus*. Again, the three intercolumniations between the *antae* and the columns between them should be closed off by balustrades [*plutei*] made of marble or fine carpentry in order to provide doors giving access to the *pronaus*.

2. Further, if the width is to exceed forty feet, columns should be placed inside [the *pronaus* and behind and] in line with the columns between the *antae*. These should be as high as those of the façade, but their diameters should be diminished following this procedure: if the diameters of the columns of the façade are an eighth of their height, the columns inside should have diameters a tenth of their height: when, instead, the front columns have diameters of a ninth or a tenth of their height, [the inside columns must be adjusted] proportionately. For in an enclosed space columns that have been made more slender would not be noticeable. But if the columns should seem too slender, they should be given twenty-eight or thirty-two flutes when the exterior columns have twenty or twenty-four flutes. Thus, an increase in the number of flutes will compensate for

what is removed from the body of the shaft, making it less noticeable, and so the diameters of the inner columns will be compensated for using a different method.[15]

3. The reason for this effect is that the eye, touching on a greater number of reliefs[16] set closer together, has to traverse a wider range of vision. For if the circumferences of two columns of equal diameter, one fluted and the other not, were to be measured with string, and the string goes round the concavities of the flutes and their vertical edges, the length of string taken around them will not be equal, even though the columns share the same diameter, because the circuit around the edges of the flutes and their concavities makes the length of the string longer. If, then, this seems to be the case, it is not inappropriate to set up columns with more slender proportions on restricted sites or in enclosed spaces when the adjustment of the number of flutes comes to our aid.

4. The thickness of the walls of the *cella* itself should be established in proportion to its size in the sense that the *antae* should be equal in breadth to the diameters of the columns. And if the walls are to be built of masonry, they should be constructed of very small stones: but if they are to be built of slabs of squared stone or marble, it seems that they should be made of carefully cut blocks of uniform size, because the central points of the blocks [in one course] holding down the vertical joints [of the blocks of the course below] will make the whole structure more stable, and again, the rusticated panels delimiting the vertical and bed-joints [*coagmenta*, *cubilia*] make its appearance more picturesque.[17]

CHAPTER V

The Orientation of the Temple

1. The directions in which the sacred temples of the immortal gods must face should be established in such a way that, if nothing precludes it and the architect has a free choice, the

temple and the statue to be placed in the *cella* should face the western area of the sky, so that those approaching the altar to make offerings or sacrifices would look in the direction of the eastern zone of the sky at the cult statue in the temple; in this way those undertaking vows would look at the temple and the eastern sky, and the cult statues themselves would seem to rise up to watch those making their prayers and sacrificing; this is evidently why all the altars of the gods must face east.

2. But if the nature of the site prevents this, then the arrangements for orientation must be changed so that the largest possible extent of the city-walls can be seen from the sanctuaries of the gods. Again, if the sacred temple is to be built on the banks of a river, for example, along the Nile in Egypt, then it should appear to look out towards the banks of the river. Similarly, if the sacred temples of the gods are to lie along public roads, they should be orientated so that passers-by can see them and make their salutations in front of them.

CHAPTER VI

The Doorways of Temples

1. These are the rules for doors [*ostia*] and their frames [*antepagmenta*] in temples: first, one must decide of which order they will be, since the orders of doorways [*thyromata*] are Doric, Ionic and Attic.

The modular systems for Doric doors are distinguished by the following characteristics: the top of the cornice, which should be placed above the frame, must be level with the tops of the capitals of the columns put in place in the *pronaus*. But the aperture [*hypaethrum*] of the doorway should be established in such a way that the height of the temple from pavement to coffered ceiling should be divided into three and a half units: two and a half of these should be allocated to the height of the opening of the folding doors [*valvae*]. This should then be divided into twelve units, of which five and a half should form the breadth of the aperture at the bottom. If the aperture ranges between very small and sixteen feet in height, the width at the top should be reduced by a third of the width of the frame; if the aperture varies between sixteen and twenty-five feet in height, the breadth at the top should be contracted by a quarter of the frame; if the aperture ranges between twenty-five and thirty feet, the top should be contracted by an eighth of the frame. Other, even taller apertures should, it seems, have vertical sides.

2. These frames should be reduced at the top by a fourteenth of their width: the height of the lintel [*supercilium*] will be equal to the breadth of the frames at the top. The moulding should be made a sixth of the frame, and the projection should equal its height; one should carve the moulding, with its astragal, in Lesbian style. The frieze [*hyperthyrum*] which should be placed above the moulding of the lintel should be of the same height as the lintel, and on it one must carve a Doric moulding, a Lesbian astragal and a *sima* cut with the chisel. The cornice and its moulding should be undecorated and its projection will equal its height. The projections of the lintel placed above the frame at left and right should be made in such a way that their rims [*crepidines*] project and are jointed obliquely into the moulding [Plate 15].

3. If the doors are to be built in the Ionic order, it seems clear that the height of the aperture should be established in the same way as for Doric. The width should be worked out by dividing the height into two and a half units, one of which provides the breadth of the aperture at the bottom. The diminutions should be arranged in the same way as for Doric. The width of the frame at the front should be a fourteenth of the height of the aperture and the moulding a sixth of this width. The rest, apart from the moulding, is divided into twelve units: the first fascia [*corsa*] with its astragal should comprise three of these units, the second, four, the third, five, and these fascias should run uniformly all round the frame with their astragals.

4. The friezes of Ionic doorways should be constructed in the same way and with the same modules as in Doric doorways; the projecting brackets [*ancones*] carved at left and right,

A
P
O
N
K
I
F
D
H
V
Q
M
Y
Z
S
T
S
X
E
B
R
C
G

also called *parotides* [earpieces, hence consoles], should hang down to the level of the bottom of the lintel, excluding their leaves. On the fronts they should be two-thirds the width of the door-frame, but at the bottom they should be less wide than at the top by a quarter.

Panelled doors [*fores*] should be constructed so that the hinge-stiles [*scapi cardinales*] are a twelfth of the width of the whole aperture. The panels [*tympana*] between the two stiles should each occupy three of the twelve units.

5. The cross-bars [*impages*] should be arranged so that once the height has been divided into five units, two should be allocated to the upper section and three to the lower; the middle bars should be placed above the centre of the door, and with regard to the rest, some should be fixed at the top and others at the bottom. The height of a cross-bar should be a third of the breadth of a panel, and its moulding a sixth of the height of the cross-bar. The widths of the inner stiles should be half that of the cross-bar, and the cover-joint [*replum*] two-thirds of it. The stiles running alongside and in front of the frame should be half the width of the cross-bar. But if the door is to have folding panels, the height will remain the same, but the breadth should be increased by the width of one leaf: if the door is to have four panels, its height should be increased again [Plate 16].

6. Attic doorways are built with the same procedures as for Doric doorways. But additionally, the fascias under the

Plate 15. Doric doorway. Palladio increases the height of the cornice above the door greatly because he does not make the top of the door arrive at the height of the adjacent capitals as Vitruvius instructs, and presents a door with only one panel. **A, B** Height from pavement to coffers. **C, D** Height of opening of door. **C, E** Width of opening of door. **D, F** Width of aperture at top. **C, G** Width of door-frame (*antepagmentum*) at bottom. **D, H** Width of door-frame at top. **I** Lintel (*supercilium*). **K** Moulding (*cymatium*) round door-frame. **M** Door-frame. **N** Frieze (*hyperthyrum*). **O** Doric moulding and Lesbian astragal of frieze. **P** Cornice (*corona*) with moulding, as high as that of the abacus of the capital. **Q, R** Height of hinge-stile (*scapus cardinalis*). **S** Panel (*tympanum*). **T** Cross-bar (*impages*).

Plate 16. Ionic doorway. Top left. **f** Scrolls, consoles (*ancones*, *parotides*). **d** Frieze. **c** Leaf below scroll. **e** Moulding of the door-frame. Lower left. **g** Moulding (*cymatium*). **h** Cover-joint (*replum*). **I, E** Stile. Right. **D** Cornice. **G** Frieze (*hyperthyrum*). **H** Moulding. **I** First fascia of door-frame. **K** Second fascia of door-frame. **L** Third fascia of door-frame. Centre. **M** Panels (*tympana*). **N** Cross-bars (*impages*). **O** Stiles of door. **C** Frieze. **A, B** Height from pavement to the beams of the ceiling. Palladio, appropriately, uses the Vitruvian Ionic base; the cornice of the door does not rise to the height of the capitals of the columns as it should.

mouldings are taken round the frames and should be arranged so that they occupy two-sevenths of the jamb, excluding the moulding. The doors themselves are not covered with relief-work and do not have two panels, but are folding doors with leaves opening outwards.

I have explained to the best of my understanding the rules that should be employed for the design of sacred temples when building in Doric, Ionic and Corinthian in accordance with approved practices. Now I shall talk about how projects for Tuscan temples should be organized.

CHAPTER VII

Tuscan Temples

1. If the site on which a temple is to be built is six units long, one should be removed, and the result allocated to the breadth. But the length should be divided into two units and the one on the inside should be allocated for spaces for the *cellae*, and that nearer to the façade should be left for the layout of the columns.

2. Again, the width should be divided into ten units, of which three on the right and three on the left should be given to the smaller *cellae*, or to corridors[18] if there are to be any there, and the other four should be assigned to the temple in the middle. The space in front of the *cellae* in the *pronaus* should be marked out with columns so that angle-columns are placed in front of the *antae* in line with the exterior walls;[19] the middle two columns should be placed in line with the walls between the *antae* and the temple in the centre;[20] and the other columns should be laid out midway between the *antae* and the front columns along the same axes.[21]

The diameters of the columns at the bottom should be a seventh of their height, the height a third of the width of the temple: the top of the column should be contracted by a quarter relative to the diameter at the bottom.

3. The height of their bases should be half the diameter of the bottom of the column. Their bases should have circular plinths, half as tall as the height of the base, and the torus above with the *apophysis* should be as tall as the plinth. The height of the capital should be half the diameter of the column. The breadth of the abacus must be the same as that of the diameter at the bottom of the column. The height of the capital should be divided into three units, of which one should be given to the plinth constituting the abacus, the second to the echinus and the third to the neck with the *apophysis*.

4. Architrave beams tied together should be laid on top of the columns to the height in modules required by the size of the structure. And these compound-beams should be arranged so that they have the same width that the necking at the top of the column will have; they should be locked together with mortises and dovetail joints so that a space of two digits is left in the joints: for when the beams touch each other and do not allow the circulation of air currents through them, they heat up and decay rapidly.

5. Above the beams and walls, the projections of the mutules should extend outwards by a quarter of the height of the column; again, protective facings [*antepagmenta*] should be nailed on their fronts, and above that should be placed the tympanum of the pediment made of masonry or wood: above the pediment, the ridge-piece, the principal rafters and purlins should be placed so that the eaves [*stillicidia*] correspond to a third of the entire roof [Plate 17; Fig. 8].

CHAPTER VIII

Circular Temples and Other Varieties

1. Circular temples are built as well; some are constructed in monopteral form [*monopteroe*], with columns but without a *cella*, while others are defined as peripteral [*peripteroe*]. Those built without a *cella* have a platform [*tribunal*] and a flight of stairs a third of the diameter in height. The columns on top of

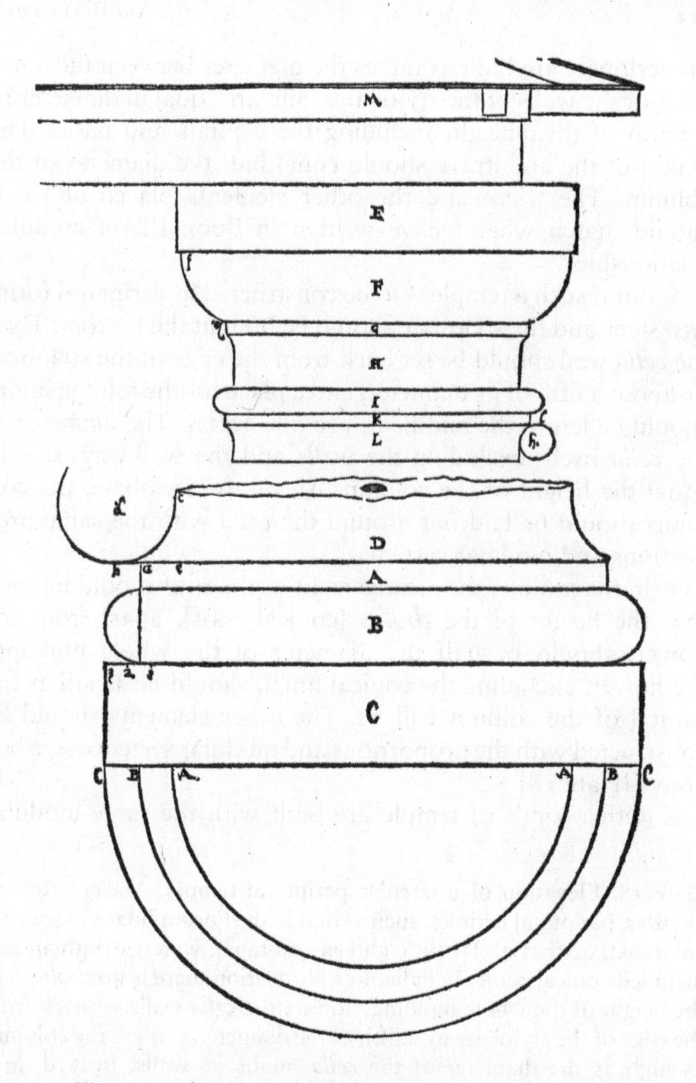

Plate 17. Tuscan order: capital and base. A Plan of the circumference of the column above the *apophysis*. B Plan of the circumference of the torus, excluding its convex sides, also B in the elevation. C Plan of circular plinth, also C in elevation. D Curvature of *apophysis*. E Abacus (*plinthus* = *abacus*). F Echinus. G Fillet, *listello*. H Collar of capital (*hypotrachelium*). I Tondino. K Fillet, *listello*. L *Apophysis*.

the stylobate are built as tall as the diameter between the outer faces of the walls of the stylobates, and are equal in diameter to a tenth of their height including the capitals and bases. The height of the architrave should equal half the diameter of the column. The frieze and the other elements placed above it should match what I have written in Book III on modular relationships.

2. But if such a temple is to be constructed in peripteral form, two steps and the stylobate should be built at the bottom. Then the *cella* wall should be set back from the edge of the stylobate by about a fifth of its diameter, and a place for the folding doors should be left in the middle to provide access. The diameter of the *cella* itself, excluding the walls and the walkway, should equal the height of the column. Above the stylobate, the columns should be laid out around the *cella* with the same proportions and modular system.

3. In the middle, the arrangement of the roof should be such that the height of the *tholus* [conical roof], apart from the flower, should be half the diameter of the whole building: the flower, excluding the conical finial, should be as tall as the capital of the column will be. The other elements should be constructed with the proportions and modular system described above [Plate 18].

4. Other kinds of temple are built with the same modular

Plate 18. Elevation of a circular peripteral temple. Ancient Roman circular, peripteral temples, such as that in the Forum Boarium (twenty columns), or that at Tivoli (eighteen columns), were Corinthian and included conical roofs. In Palladio's illustration there is no *tholus* half the height of the whole building, nor are the *cella* walls set back from the edge of the stylobate by a fifth of the diameter, nor are the columns as high as the diameter of the *cella* minus its walls. Instead, in a splendid Renaissance fantasy, but with twenty widely spaced columns on pedestals, Palladio inserts a tambour, deriving from Bramante's celebrated Tempietto at S. Pietro in Montorio, resting on the *cella* walls and a stepped cupola *à la* Pantheon. In the lantern, of which Barbaro says he saw examples on coins of Nero, he shows in plan half of the flower mentioned by Vitruvius, and used originally by Michele Sanmichele at the top of the lantern of the Pellegrini chapel in Verona.

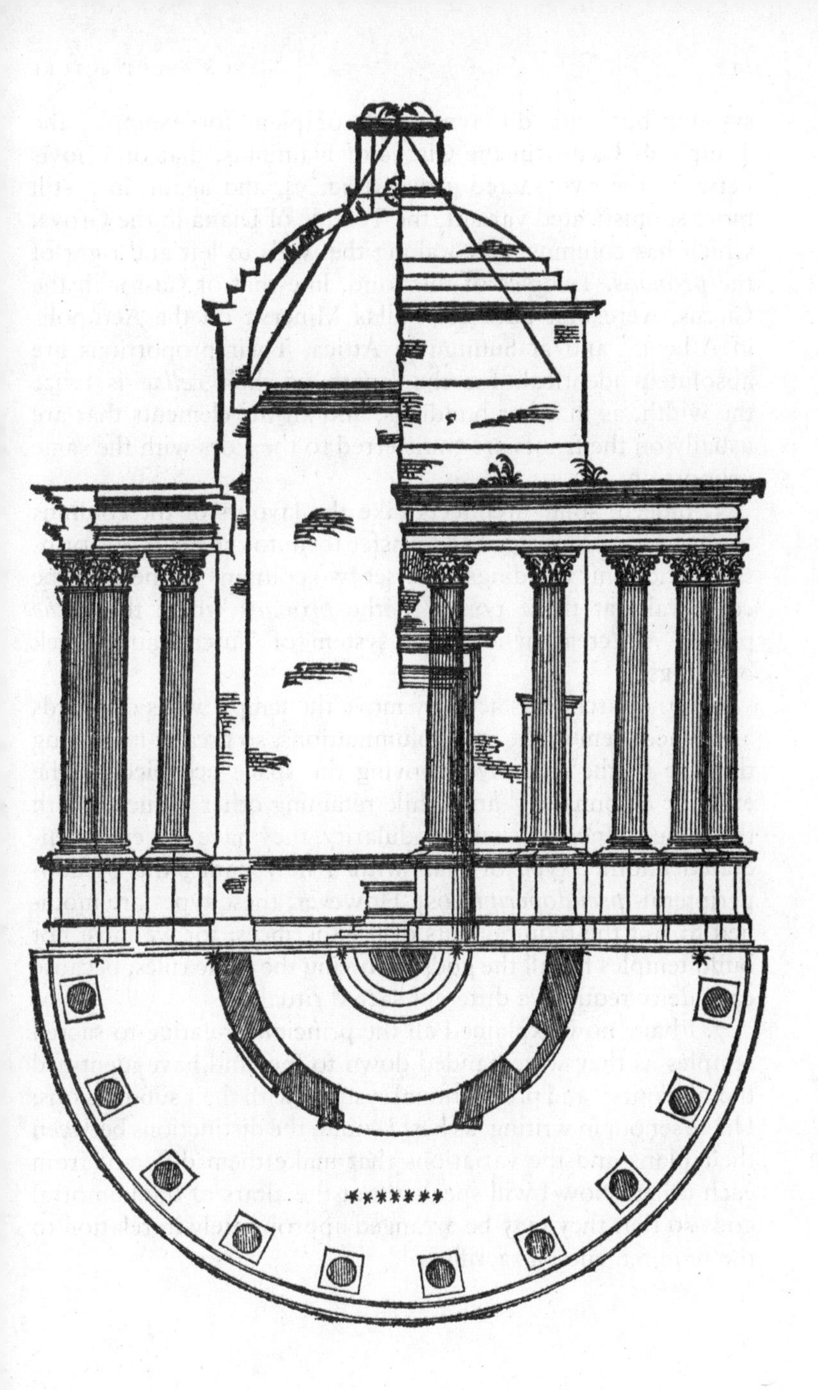

systems but with different kinds of plan: for example, the Temple of Castor in the Circus of Flaminius, that of Veiovis between the two sacred groves [Fig. 9], and again, in a still more sophisticated variant, the Temple of Diana in the Grove, which has columns added along the walls to left and right of the *pronaus*. Temples of this kind, like that of Castor in the Circus, were first built for Pallas Minerva on the Acropolis in Athens[22] and at Sunium in Attica. Their proportions are absolutely identical: for the length of their *cellae* is twice the width, as in other buildings, and all the elements that are usually on the fronts are transferred to the sides with the same proportions.

5. Indeed, some architects take the layouts of the columns from the Tuscan order and transfer them to the plans of Corinthian and Ionic buildings; they set two columns in line with the *cella* walls at those points in the *pronaus* where the *antae* project, so creating a mixed system of Tuscan and Greek buildings.

6. Other architects actually move the temple walls outwards and place them in the intercolumniations, so greatly increasing the size of the *cella* by removing the space occupied by the exterior colonnade;[23] and while retaining other elements with the same proportions and modularity, they have evidently generated another type of plan with a new name, the pseudoperipteral [*pseudoperipteros*]. However, these types are modified to suit the requirements of the sacrifices; for we must not build temples for all the gods following the same rules, because each deity requires a different sacred ritual.

7. I have now explained all the principles relating to sacred temples as they were handed down to me, and have identified their layouts[24] and proportional systems with their subdivisions; I have set out in writing, as best I could, the distinctions between their plans and the variations that make them different from each other. Now I will speak about the altars of the immortal gods so that they may be arranged appropriately in relation to the requirements of sacrifices

CHAPTER IX

Altars

1. Altars should face east and should always be placed lower than the cult statues housed in the temple, so that people looking up at the divinity when praying and sacrificing should find themselves at different levels, as is appropriate for each god. The levels of the altars should be set so that they are built as high as possible for Jupiter and the gods of the heavens; they should be set low for Vesta, the Earth and the Sea. In this way, the architect's designs for altars will be developed in an appropriate form. Having described the projects for sacred temples in this book, we shall provide explanations of the planning of public buildings in the next.

BOOK V

Introduction

1. Supreme Ruler, those who have set out their intellectual theories and teachings in more ample volumes than these have endowed their writings with prodigious and magisterial authority. If only the same applied to my own field of study so that the authoritativeness of these teachings would also be enhanced by various amplifications; but this is not as easy as is thought. For one does not write about architecture as one writes history and poetry. Historical narratives grip the reader by themselves, for they offer expectations of a variety of novelties; while the metres and feet of poetry, the elegant arrangement of words and of the emotional interactions between the various characters, as well as the recital of verses, transport the minds of readers onwards and lead them to the end of the work without boring them.

2. But this cannot happen in the case of architectural treatises since the technical terms which had to be devised to cater for the particular requirements of the art produce obscurities of meaning because of the unaccustomed vocabulary. Therefore, since the material is not of itself accessible and the relative terminology is not familiar in common language, it follows that the writings of teachers who ramble too far, unless they are concise and expressed in clear and limpid propositions, will create misunderstandings in the minds of the readers because of the confusion caused by the density and mass of their discourse: accordingly, when I mention obscure terms and the commensurability of the components of buildings, my account will be terse so that they may be committed to memory, because in that way people's minds can grasp such information more readily.

3. This applies all the more since I have noticed that our citizens are particularly engrossed in public affairs and private business, and so I decided to write briefly so that my readers will be able to absorb my suggestions rapidly in their brief moments of leisure.

Again, Pythagoras and those who followed his beliefs thought it appropriate to write down their principles in books using a system of cubes: they decided on a cube with two hundred and sixteen verses and thought that there should be no more than three cubes in any one treatise. **4.** Now a cube is a body with flat, square surfaces of which all sides are equal in breadth. When it has been thrown it remains completely immobile as long as it is not touched, no matter which face it rests on, like the dice which players throw on a board. The Pythagoreans seem to have adopted the analogy of the cube because this number of lines would create an indelible record of itself in whichever mind it lands, like a dice. The Greek comic poets, too, subdivided their dramatic works by inserting songs by the chorus: so, by using the principle of the cube to create interludes, they lightened the actors' recitations.

5. Since these rules based on natural law were observed by our ancestors, and I am very much aware that I have to write on matters that are unusual and obscure to many people, I decided to write them up in short books so that they may penetrate the reader's consciousness more readily: in that way they will be easy to understand. And I have organized the various subjects in such a way that those looking for them do not have to put them together from different places but may find treatments of the various topics in a single comprehensive work and in its single books. So, Caesar, I explained the theory behind sacred temples in Books III and IV, and in this one I will discuss the organization of public spaces. First I will discuss how the forum should be laid out, because the administration of public and private affairs by the magistrates takes place in it.

CHAPTER I

The Forum and the Basilica

1. The Greeks set out their forums on a square plan with very spacious double colonnades; they decorate them with close-set columns and stone or marble architraves, above which they build walkways on wooden floors. But the same format must not be followed in the cities of Italy, because there the custom of giving gladiatorial shows in the forum has been handed down by our ancestors. 2. Accordingly, the intercolumniations around the structures built for shows should be laid out much wider; the offices of the bankers should be distributed around the porticoes and the balconies on the wooden floors above, which would then be well placed in terms of practical use and public revenue.

But the dimensions of the forum should correspond to the number of inhabitants lest the space should be too small for practical purposes or seem deserted because of a lack of people. Its breadth should be determined in such a way that, having divided the length into three units, two of these are required for its breadth. For in this way its plan will be oblong and the layout well adapted for the purpose of putting on spectacles.

3. The upper columns should be built a quarter less tall than those below because load-bearing elements must be more robust than those above, and further, because we really should imitate the nature of things that grow, for example, trees with smooth trunks such as fir, cypress and pine, which are all thicker just above the roots, and then, as they develop, reach upwards and grow with a uniform natural diminution up to the top. Therefore if the nature of growing organisms demands it, it is right to build in such a way that the elements above should be less massive in height and breadth than those below [Plate 19].

4. The sites of basilicas should be connected with forums in the warmest areas possible so that in the winter businessmen can gather in them without being bothered by bad weather:

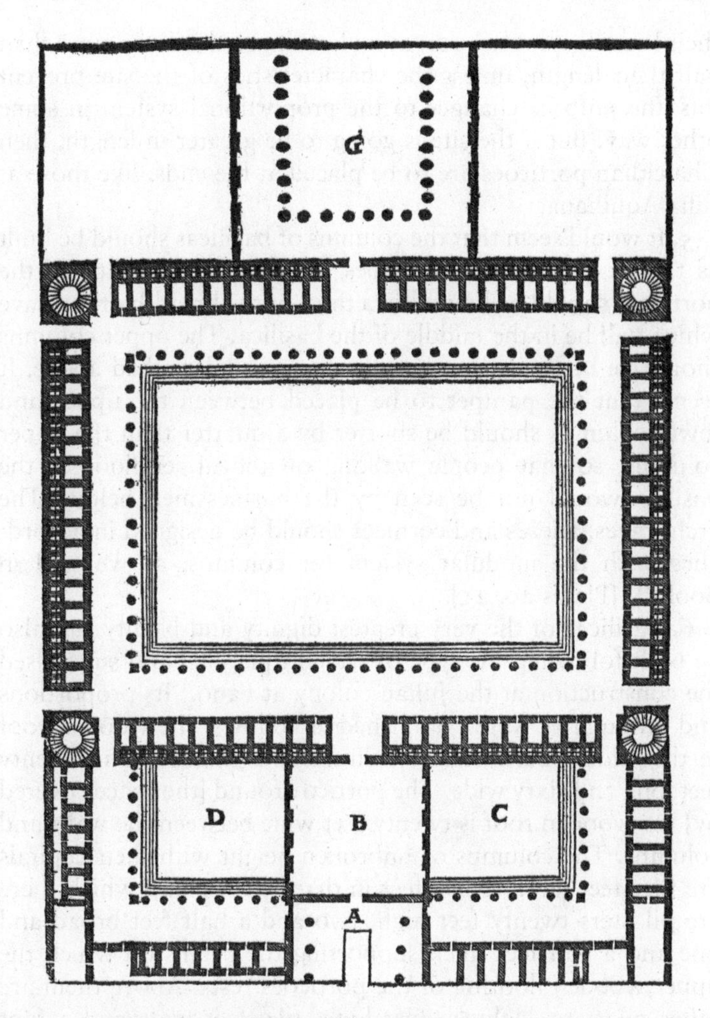

Plate 19. Forum and other buildings. Vitruvius' description of the complex including the basilica, forum and *curia* was unspecific enough to permit Renaissance fantasies. A Wings of palace. B Senate house (*curia*), square in plan as Vitruvius recommends. C Square in front of the jail. D Square in front of mint (*aerarium*). G Basilica. In the middle, the plan of the forum, roughly 2 × 3 as Vitruvius recommends, with the shops and columns around it raised on a platform with four steps.

their breadth should be set at no less than a third nor more than half their length, unless the characteristics of the site prevent this and enforce changes to the proportional system in some other way. But if the site is going to be greater in length, then Chalcidian porticoes are to be placed at the ends, like those at Julia Aquiliana.

5. It would seem that the columns of basilicas should be built as tall as the internal porticoes are broad; the width of the porticoes should be limited to a third of the breadth of the nave which will be in the middle of the basilica. The upper columns should be built smaller than the lower, as described above. It seems that the parapet to be placed between the upper and lower columns should be shorter by a quarter than the upper columns, so that people walking on the upper floor of the basilica would not be seen by the businessmen below. The architraves, friezes and cornices should be designed in accordance with the modular system for columns, as we said in Book III [Plates 20, 21].

6. Basilicas of the very greatest dignity and beauty can also be built following the type of which I planned and supervised the construction at the Julian colony at Fano.[1] Its proportions and modularity were established as follows: the wooden roof in the middle between the columns is a hundred and twenty feet long and sixty wide. The portico around [the space covered by] the wooden roof is twenty feet wide between the walls and columns. The columns of unbroken height with their capitals are fifty feet tall and five feet in diameter, behind which there are pilasters twenty feet high, two and a half feet broad and one and a half feet thick supporting the beams on which the upper wooden flooring of the porticoes rests. Above them are other pilasters, eighteen feet high, two feet wide and a foot thick which, in their turn, carry the beams supporting the principal rafters and the roofs of the porticoes, which have been positioned at a lower level than the main wooden roof.

7. The rest of the spaces between the beams placed on the pilasters and columns have been left for the light that pours in through the intercolumniations. There are four columns at right and left along the breadth of the [space covered by the] main

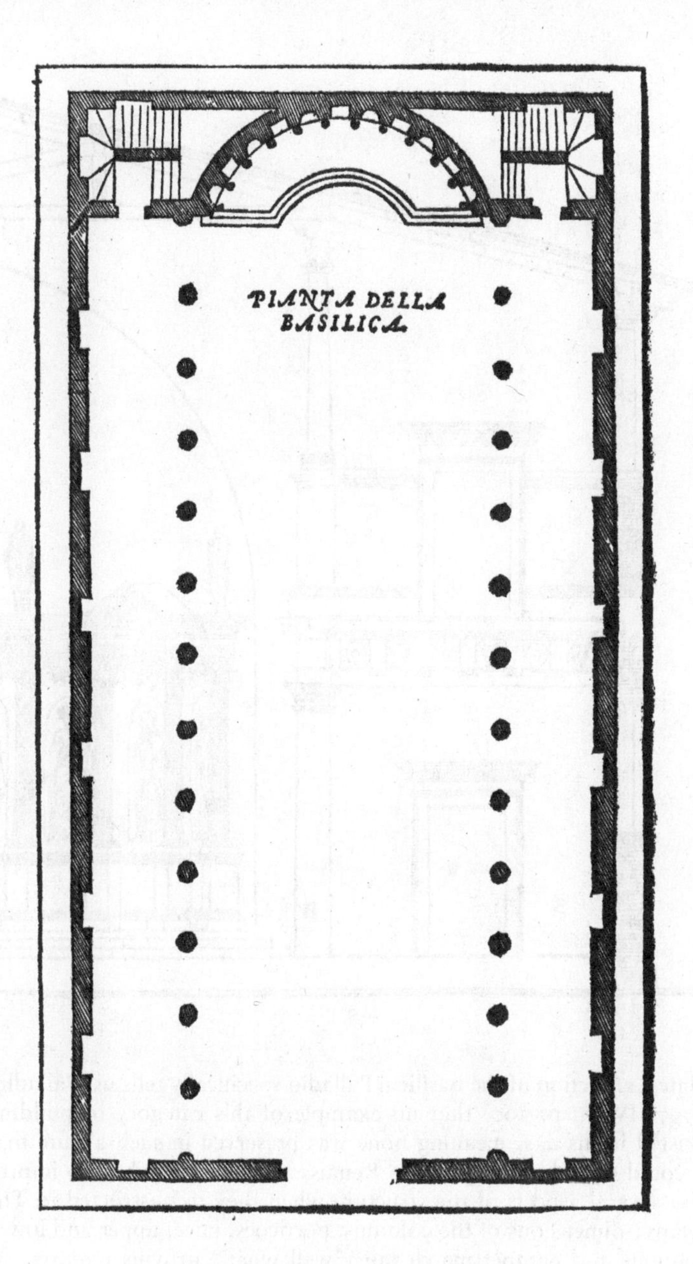

Plate 20. Plan of the basilica.

Plate 21. Section of the basilica. Palladio specifically tells us (Palladio, 1997, IV.38, p. 200) that no example of this category of building existed in his day, meaning none was preserved in such a state that it could reliably be copied, so Renaissance architects had to improvise several aspects of the structure when they reconstructed it. The relative dimensions of the columns, porticoes, nave, upper and lower columns and parapet match fairly well what Vitruvius requires except that in plan the nave is slightly too wide and in Plate 20 the proportions are 2 : 1, not 3 : 2. The lower interior order has been

made Doric by Palladio with his favourite metopes and *bucrania*, and the doors are Doric; he omits the bases of the Doric columns, as at the Theatre of Marcellus; in the upper order, the columns are Ionic, to judge by the bases, but he does not provide Ionic doorways. The magnificent curved wall behind the tribune with split tympana on pedestals and running entablatures tying the tabernacles together is similar to that of Palladio in the choir of S. Giorgio Maggiore in Venice and in his drawing of the Temple of Serapis (Palladio, 1997, IV.43, 46, pp. 255, 258).

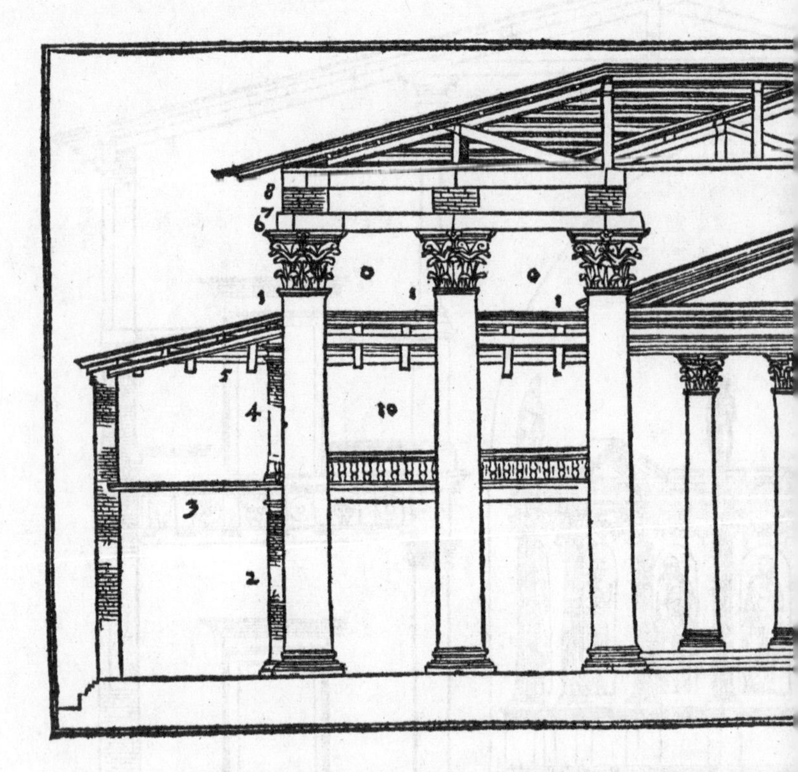

Plate 22. The Basilica at Fano. Internal section (left) and outside view, seen from the side opposite the *aedes* (temple) of Augustus, which he proposes was tetrastyle *in antis*, whereas, if the *pronaus* had columns, there were probably only two, that is, the ones that Vitruvius says were not included in the colonnade on the long side near the temple. Here, the columns are Corinthian and the spacing eustyle (2¼ column-diameters), matters on which Vitruvius is not specific. The roofs here are hipped whereas the famous description of them as having double gables (*duplex fastigiorum dispositio*) suggests either the presence of

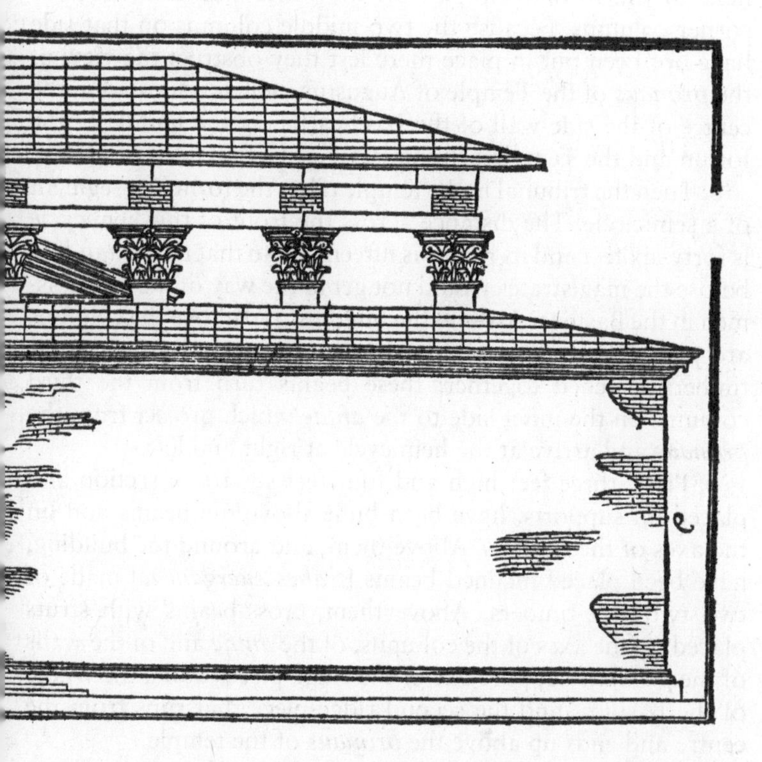

two gables or of two sets of two gables, surely the latter; that is, one ridge-piece with gables at either end across the length of the building, the other, in some way, across the breadth towards the Temple of Augustus. 1 Unbroken columns of fifty feet. 2 Twenty-foot backing pilasters. 3 First floor of portico. 4 Upper backing pilasters of eighteen feet. 5 Beams supporting roof of portico. 6 Corinthian capitals with wooden beams acting as architrave. 7 Three-foot piers used instead of frieze. 8 Compound-beams acting as cornices. 9 Wall round basilica. 10 Parapet of first floor of portico. o Windows.

wooden roof, including those at the corners; there are eight columns, including the same corner columns, on the long side next to the forum: and six on the other side including the corner columns, because the two middle columns on that side have not been put in place there lest they obstruct the view of the *pronaus* of the Temple of Augustus, which is located in the centre of the side-wall of the Basilica facing the middle of the forum and the Temple of Jupiter.

8. Then the tribunal in the temple takes the form of a segment of a semicircle. The distance across the front of this hemicycle is forty-six feet and its radius is fifteen feet so that those standing before the magistrates would not get in the way of the business-men in the Basilica. Around the building and above the columns are placed beams [as architraves] comprising three two-foot timbers fastened together: these beams turn from the third columns on the inner side to the *antae* which project from the *pronaus* and arrive at the hemicycle at right and left.

9. Piers, three feet high and four feet square in section and placed on supports, have been built above the beams and on the axes of the capitals. Above them, and around the building, have been placed inclined beams [*trabes euerganeae*] made of two two-foot timbers. Above them, cross-beams with struts placed on the axes of the columns, of the *antae* and of the walls of the *pronaus* support one of the ridge-pieces along the whole of the Basilica, and the second ridge-piece that runs from the centre and ends up above the *pronaus* of the temple.

10. Thus the arrangement with double pediments lends an elegant appearance to the roof outside and to the upper wooden roof inside. Again, the elimination of the mouldings above the architraves, of the row of parapets and of the upper columns removes tedious difficulties and reduces the overall expense considerably; by contrast, the columns themselves of uninterrupted height leading up to the beams of the wooden roof seem to add the magnificence of conspicuous expenditure and dignity to the building [Plate 22; Figs. 10, 11].

CHAPTER II

The Treasury, Prison and the Senate House

1. The treasury, the prison and the senate house must be connected to the forum but in such a way that their scale and modularity match those of the forum. The senate house especially should be constructed primarily to match the dignity of the municipality or citizenry. And if it is to be square, the height should be established by taking the breadth and adding a half: but if it is to be rectangular, the length and breadth should be added together and half of that sum given to the height of the coffered roof.

2. Besides that, the interior walls should be circled halfway up with cornices made of woodwork or stucco [*opus albarium*]. If such cornices are not present, the voices of the disputants will be carried right up to the top and will not be intelligible to the audience. But when the walls are entirely circled with cornices, the voice from below, blocked before being carried upwards and dissipating itself, will be intelligible to the ears.

CHAPTER III

The Theatre

1. Once the forum has been laid out, one must then select the healthiest possible site for the theatre for the spectacles of the games on the festival days of the immortal gods, following what I have written on healthiness in connection with the siting of walled cities in Book I. This is because people sitting with their wives and children for some time through the games are captivated with pleasure, and their bodies, remaining immobile with enjoyment, present open pores into which wafts of air penetrate, which would infuse their bodies with noxious vapours if they come from marshy or other unhealthy areas.

And so if the site of the theatre is chosen very carefully, problems like this will be avoided.

2. We must also be careful that it is not exposed to the south. For when the sun fills the curvature of the theatre, the air, enclosed in the auditorium without the possibility of circulating, heats up and glows, then burns, dries out and exhausts the fluids in the spectators' bodies. This is why one must at all costs avoid areas which have disadvantages like this, and select healthy ones.

3. The procedure for building the foundations will be easier if they are in hilly areas; but if practical necessity requires that they should be built on a plain or a marshy site, reinforcements and substructures must be built in accordance with what was written on the foundations of sacred temples in Book III. Above the foundation walls, tiers of seats [*gradationes*] should be built up from the substructure with materials of stone and marble.

4. The transverse aisles [*praecinctiones*] should evidently be constructed according to a predetermined unit derived from the heights of theatres, and they should not be higher than the width of their own passageways. For if they are higher [at the back], they will throw the voice back and drive it away from the upper section, preventing the word-endings from arriving in intelligible form at the ears of those sitting in their seats above the transverse aisles. All told, the theatre must be organized so that when a piece of string is extended from the lowest to the highest step, it would touch all their top edges and angles; in this way the voice would not be obstructed.

5. It is appropriate to arrange a large number of wide entrances: the upper ones should be made discontinuous in relation to the lower ones, and should be built straight and without curves from all sectors so that when crowds leave the shows they would not be hemmed in but would have separate and unobstructed exits from all parts of the theatre.

Again, one must take great care that the site is not 'deaf', but that the voice can resonate in it with great clarity: this can be achieved if the site is chosen in an area where the voice is not impeded by reverberation.

6. For the voice is like a flowing breath of air and is detected on contact[2] by our sense of hearing. It moves itself in an infinity of concentric circles, just as when a stone is dropped in standing water, innumerable expanding circles of ripples develop and spread out from the centre as widely as they can, unless interrupted by the narrowness of the pond or some other obstacle which prevents the patterns of ripples from reaching the end of their course. And so when they are interrupted by obstacles, the first ripples turn back and disturb the pattern of those that follow them.

7. In the same way the voice diffuses itself in concentric circles, but while the circles in water move horizontally on a flat surface, the voice not only expands outwards horizontally but also rises in regular stages. Therefore, as with the pattern of ripples in water, so too with the voice: when no obstruction blocks the first wave, it does not interfere with the second nor with those following, but they all reach the ears of the spectators in the lowest and highest seats without rebounding.

8. Therefore the ancient architects, taking their lead from nature, designed the tiers of seats in theatres on the basis of their investigations into the rising of the voice, and tried, with the help of mathematicians' principles and musical theory, to devise ways in which any voice uttered onstage would arrive more clearly and pleasantly at the ears of the spectators. For just as organs constructed with bronze sheets or with ἠχεῖα [*echeia*; resonators] made of horn can be made to achieve the clarity of string-instruments, similarly the ancients devised methods of amplifying the voice in theatres using the science of harmonics.

CHAPTER IV

Harmonics

1. Harmonics is an obscure and difficult branch of musical literature, particularly for those who do not know the Greek literature on the subject. If we wish to explain it, it is in fact

essential that we use Greek terms because a number of them do not have equivalents in Latin. I will therefore interpret it as clearly as I can using the writings of Aristoxenus[3] and, adding his diagram at the end, will indicate the positions of the sounds so that anybody paying careful attention can understand them more readily.

2. For when the voice is altered by variations of pitch, it sometimes becomes high, sometimes low, and it changes in two ways, of which one has a continuous result, the other episodic. The continuous voice does not become stationary within any particular limits or in any particular place, and does not generate perceptible limits; but the intervals in between are perceptible as, for example, when we say 'sol', 'lux', 'flos', 'vox' [sun, light, flower, voice] in conversation. For now one cannot perceive where the voice begins or where it tails off, but our ears notice that the voice passes from high to low and vice versa. The opposite applies to the voice which moves in intervals: for when it is modulated in pitch, the voice in motion positions itself within the limits of a certain sound, and then within the limits of another, and doing this frequently back and forth, appears to our senses to be continuous, as when we produce a variation in modulation by altering our voice in songs. So, when the voice is altered by intervals, it is clear when it stops and starts within the clear boundaries of the notes, but the intermediate points, though clear in themselves, are obscured by the intervals.

3. There are three types of modulation: first, that which the Greeks call ἁρμονία [*harmonia*]; second, the χρῶμα [*chroma*]; third, the διάτονος [*diatonos*]. Harmonic modulation is artificially constructed, and, for that reason, singing with it produces above all a severe and impressive sense of authority. The chromatic, with its subtle virtuosity and frequent transitions, provides a sweeter pleasure. The distance between the intervals of the diatonic is easier to perceive because it is natural. The arrangement of the tetrachords is different in these three types, because in the harmonic, the tetrachords consist of two tones and two *dieses*. A *diesis* is a quarter of a tone, and so two *dieses* are allocated to a semitone. Two semitones are arranged in

succession in the chromatic, and the third interval consists of three semitones. There are two consecutive tones in the diatonic, and the third semitone completes the tetrachord. So, in the three modes, the tetrachords are composed equally of two tones and a semitone, but when they are examined separately and within the limits of each mode, the arrangement of their intervals is different.

4. Therefore, with respect to the voice, nature has divided it up using the intervals of tones, semitones and tetrachords and has set limits to it with measures according to the extent of the intervals, and established their characteristics according to distinct modes; also, the craftsmen who build organs prepare to perfection instruments which make the correct sounds by taking advantage of these modes established by nature.

5. In each mode there are eighteen notes, called φθόγγοι [*phthongoi*] in Greek, of which eight are constant and fixed in the three modes, and the other ten are variable when they are all modulated together. But the fixed notes are those which are placed between the variable sounds, and they comprise the unit of the tetrachord, and stay fixed in their boundaries with respect to the differences of the modes. They are defined thus: *proslambanomenos* [very low, the added note at the bottom of the scale], *hypate hypaton* [the lowest of the low], *hypate meson* [the lowest of the middle notes], *mese* [middle], *nete synemmenon* [the last of the joined notes], *paramese* [near the middle], *nete diezeugmenon* [the last of the disconnected notes] and *nete hyperbolaeon* [the last of the high notes]. The movable notes are those which, positioned amongst the fixed ones in the tetrachord, change from place to place in the different modes. The terms for them are: *parhypate hypaton* [near the lowest of the low], *lichanos hypaton* [finger-struck low note], *parhypate meson* [near the lowest of the middle], *lichanos meson* [finger-struck middle note], *trite synemmenon* [third of the joined notes], *paranete synemmenon* [near the last of the joined notes], *trite diezeugmenon* [third of the disconnected notes], *paranete diezeugmenon* [near the last of the disconnected notes], *trite hyperbolaeon* [third of the high notes] and *paranete hyperbolaeon* [near the last of the high notes].

6. Because they are movable, these notes acquire other qualities, for they have intervals and increasing distances between them. So the *parhypate*, which is half a semitone distant from the *hypate* in the harmonic mode, has a semitone interval when transferred to the chromatic mode. What is called *lichanos* in the harmonic mode is a semitone distant from the *hypate*; when moved to the chromatic mode, it advances two semitones; and in the diatonic mode it is three semitones distant from the *hypate*. Thus the ten sounds produce three different kinds of modulation on account of their switches between the three modes.

7. There are five tetrachords: the first and lowest is called ὕπατον [*hypaton*] in Greek; second is the middle one, called μέσον [*meson*]; the third, the joined one, called συνημμένον [*synemmenon*]; the fourth, the separated one, is called διεζευγμένον [*diezeugmenon*]; the fifth, the highest pitched, is called ὑπερβολαῖον [*hyperbolaeon*] in Greek. There are six concords, called in Greek συμφωνίαι [*symphoniai*], which men can naturally modulate: the *diatessaron* [the fourth], the *diapente* [the fifth], the *diapason* [the octave], the *disdiatessaron* [octave and a quarter], then the *disdiapente* [octave and a fifth] and the *disdiapason* [the double octave].

8. So their names are taken from a numerical system, for when the voice stops at one particular sound and then, modulating itself and moving away, arrives at the fourth position, it is called *diatessaron*; when it arrives at the fifth it is called *diapente*, at the sixth, *diapason*, at eight and a half, *diapason* and *diatessaron*, at nine and a half, *diapason diapente*, and at the twelfth, *disdiapason*.

9. For there can be no concords between two intervals when the sounds of strings or of a singing voice have been produced, nor in the third, sixth or seventh; but, as I wrote above, the concords deriving from the nature of the voice have limits of a *diatessaron*, a *diapente* and so on up to the *disdiapason*, and because of this, concords are created by the conjunction of sounds, which are called φθόγγοι [*phthongoi*] in Greek.

CHAPTER V

Placement of the Sounding Vessels in the Theatre

1. So, bronze vases should be made in proportion to the size of the theatre on the basis of these investigations carried out on mathematical principles, and they should be cast so that when struck, they are capable of emitting, collectively, sounds including a *diatessaron*, *diapente* and so on up to *disdiapason*. Then, after constructing chambers between the seats of the theatre, the vases should be placed in them following musical principles so that they do not touch any wall-surface, there is clear space around them and nothing on top of them; they should be placed upside down and the sides that face the stage should have wedges pushed under them no less than half a foot high; in front of these chambers apertures two feet wide and half a foot tall should be left in the bed-joints [*cubilia*] of the steps below [Fig. 12].

2. The arrangement of these vases with respect to their locations should be the following. If the theatre is not going to be very large, a horizontal level halfway up should be marked out, and thirteen vaulted chambers separated by twelve equal distances should be built round it so that with regard to the resonators mentioned above, those which give the note *nete hyperbolaeon* should be located in the first compartments at the extremities at right and left; second [in from the extremities], those which give notes *diatessaron* to *nete diezeugmenon*; third, those giving *diatessaron* to *paramese*; fourth, *nete synemmenon*; fifth, *diatessaron* to *mese*; sixth, *diatessaron* to *hypate meson* and, in the centre, one only that gives notes from *diatessaron* to *hypate hypaton* [Fig. 13].

3. With this arrangement, the voice, emitted from the stage as from a centre, diffuses itself and strikes the concavities of the individual vases, and produces increased clarity and, from this concord, a harmonious sound in unison with itself. But if the theatre is to be much larger, then the height should be divided into four bands so that three transverse levels of chambers can be laid out: one for the harmonic, another for

the chromatic and the third for the diatonic modes. And the lowest level, which will be the first, should be constructed following the harmonic mode as though in a smaller theatre, in the way I have described above.

4. In the middle band, the vases which give the sound of the chromatic *hyperbolaeon* should be placed first, at the extreme ends; second, those giving the *diatessaron* in the chromatic *diezeugmenon*; third, those that give the chromatic *synemmenon*; fourth, those giving the *diatessaron* in the chromatic *meson*; fifth, those giving the *diatessaron* in the chromatic *hypaton*; sixth, those that give the *paramese*, for this provides a concord both of the *diapente* in the chromatic *hyperbolaeon* and of the *diatessaron* in the chromatic *meson*.

5. Nothing is to be placed in the middle because there is no other type of note in the chromatic mode that can generate a concord of sound. In the zone of chambers at the ends of the highest section should be placed vases cast so as to give the sound of the diatonic *hyperbolaeon*; in the second set of chambers, those which give the *diatessaron* in the diatonic *diezeugmenon*; third, those giving the diatonic *synemmenon*; fourth, the *diatessaron* in the diatonic *meson*; fifth, the *diatessaron* in the diatonic *hypaton*; sixth, the *diatessaron* in the *proslambanomenos*; in the middle, the *mese*, because this gives a concord of sounds of both the *diapason* in the *proslambanomenos* and of the *diapente* in the diatonic *hypaton*.

6. Anyone wishing to put these principles into practice with ease should consult the diagram at the end of the book drawn in accordance with musical theory; this was left for us by Aristoxenus, who devised it with immense diligence and labour after distinguishing the modulations according to their modes. If someone follows the rules laid out in the diagram attentively, he could create theatres which cater much more easily for the natural characteristics of the voice and the pleasure of the audience.

7. Perhaps someone will say that plenty of theatres are built in Rome every year in which no consideration at all has been taken of these rules; but he would be wrong about this because all public theatres made of wood include a number of timber

partitions which necessarily resonate. In fact, one can also see this in the case of the *cithara* players, who, wishing to sing in a higher key, turn towards the folding doors of the stage and so enhance their vocal resonance with their help. But when theatres are built of hard materials such as masonry, stone or marble, which cannot resonate, then these rules must be put into practice using resonating vases.

8. But if one asks in which theatre these vases have been incorporated, we cannot point to any in Rome, but we can in the provinces of Italy and in many Greek cities: and we also have as a witness Lucius Mummius, who, after the destruction of the theatre in Corinth, brought such bronze vessels to Rome and dedicated them as part of his booty in the Temple of the Moon.[4] And besides, many skilful architects who built theatres in cities of moderate size chose, for lack of finance, earthenware vessels resonating in the same way, and produced excellent results by arranging them with the same technique.

CHAPTER VI

The Plan of the Theatre

1. The plan of the theatre itself is to be laid out so that, whatever the extent of the perimeter of the lowest part,[5] a centre point should be established in it and a circumference described around it; four equilateral triangles should be inscribed in it which should touch the circumference of the circle at equal distances: astrologers also use such triangles to make their calculations regarding the twelve celestial signs on the basis of the musical consonance of the stars. The façade of the stage-set [*scaenae frons*] should be set in the area where the side of the triangle with its base nearest to the stage-set cuts off a segment of the circle: and a line parallel with that section should be drawn through the centre of the circle, marking the separation of the platform [*pulpitum*] of the stage [*proscenium*] from the area of the orchestra [*orchestra*].

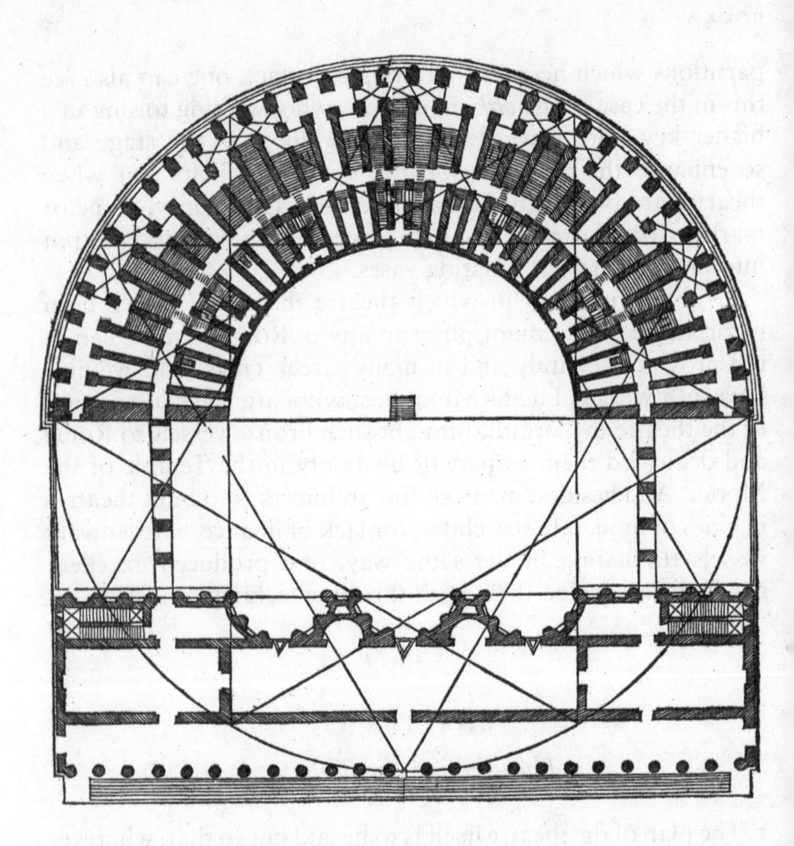

Plate 23. Plan of Roman theatre. Vitruvius meant that the four equilateral triangles are to be constructed in the orchestra, as at the Theatre of Marcellus, and as had been understood by only a couple of architectural designers before Barbaro. But this wonderful plan famously misunderstands the Vitruvian scheme, as do many other Renaissance artists (Fra Giocondo, 1511, Cesariano 1521 (see n. 14 of Introduction, above), etc.). Palladio, adapting a drawing he had already made reconstructing the Berga Theatre at Vicenza, uses triangles and concentric circles to establish the main elements of the whole structure, not just of the orchestra, as he does at the Teatro Olimpico at Vicenza. The result is that the distance between the stage-set and the centre of the orchestra, hence the depth of the platform (*pulpitum*), is far too great; he also added the lateral structures supported on four piers on either side. The triangular *periaktoi* are placed in the apertures for the doors.

2. In this way the platform of the stage will be made deeper than that of the Greeks because all [our] actors perform on stage: in the orchestra, instead, are the places reserved for the seats of the senators. The height of the platform should not be more than five feet so that those sitting in the orchestra can see the gestures of all the actors. The wedge-shaped sections of the spectators' seats [*cunei*] in the auditorium [*theatrum*] should be divided in such a way that the apexes of the triangles which run around the circumference of the circle indicate the alignments of the ascending gangways [*ascensus*] and the steps [*scalae*] between the sections of seating as far up as the first transverse aisle [*praecinctio*], but above, the upper sections of seating should be divided in the middle with alternating passageways [*itinera*: in relation to those between the blocks of seating below].

3. The apexes indicating the directions of the flights of steps will be seven in number in the lowest part of the theatre,[6] and the other five will govern the arrangement of the stage-set:[7] the apex in the middle should correspond to the royal door [*valvae regiae*], while the apexes at right and left will indicate the location of the doors of the guests' quarters [*hospitalia*]; and the two at the ends will point in the direction of the passages in the wings [*versurae*]. With regard to the steps [*gradus*] in the spectators' part where the benches [*subsellia*] are to be organized, they should be no less than a foot and a palm in height, nor more than a foot and six digits; their depth should be fixed at no more than two and a half feet and no less than two.

4. The roof of the portico which will be placed on the top row of the steps should be arranged[8] so as to be horizontal with the top of the stage-set in order that the voice, as it diffuses itself, will arrive simultaneously at the highest steps and the roof. For if the roof of the portico is not equal in height [to the top of the stage-set], in that it is lower, the voice will be dissipated at the height at which it arrives first.

5. A sixth of the diameter of the orchestra between the lowest seats should be divided off; at the corners[9] on either side [of the orchestra] where the entrances are, the innermost seats[10] must

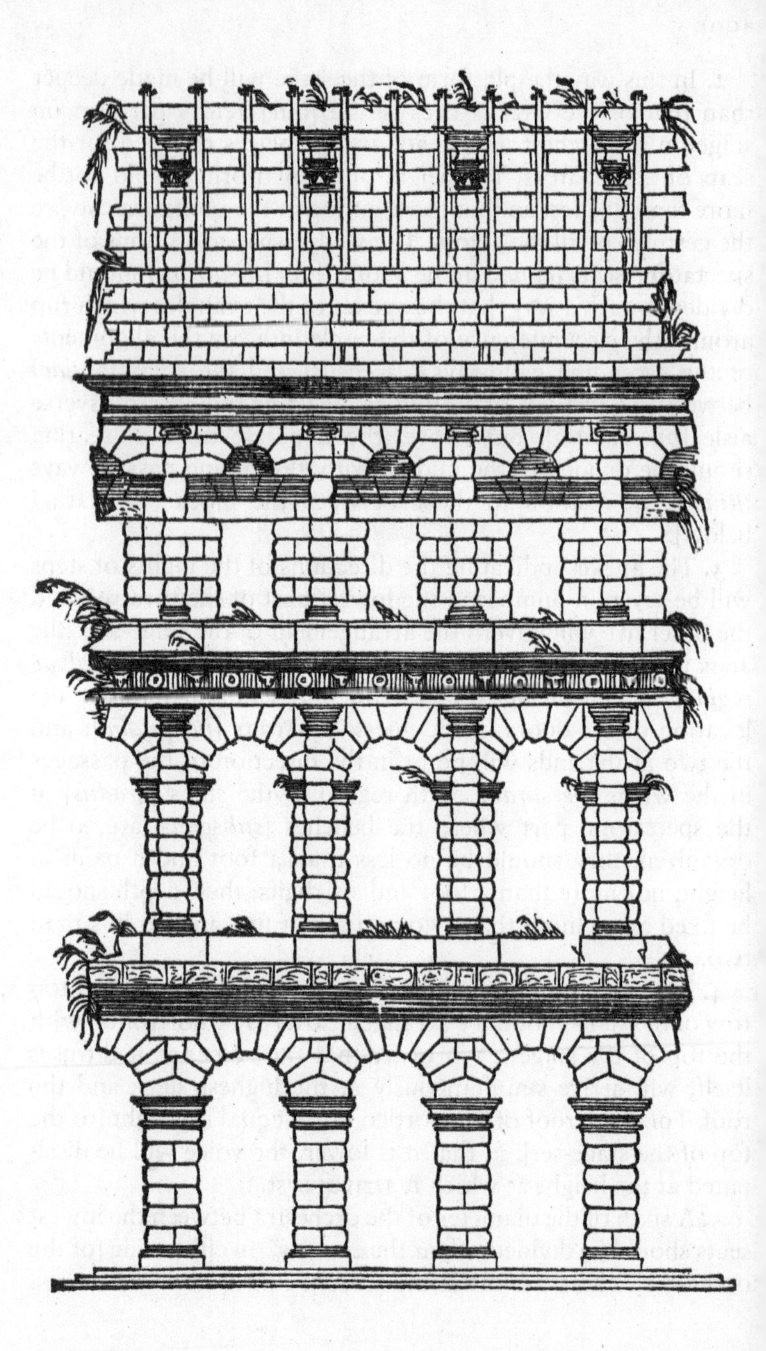

be cut perpendicularly to a height equalling that sixth,[11] and the springing of the arches of the passages should be set where the cut has been made; in this way their vaults will be sufficiently high.

6. The length of the stage-set [*scaena*] should be double the diameter of the orchestra. The height of the podium,[12] with its cornice and moulding, should be a twelfth of the diameter of the orchestra from the level of the platform of the stage. Above the podium, the columns with their capitals and bases should be a quarter of the same diameter high, and the architraves and their mouldings should be a fifth of the height of the columns. The parapet above, with its moulding [*unda*] and cornice, should be half the height of the parapet below. The columns above this parapet should be shorter than those below by a quarter, and the architraves and mouldings a fifth of their height. Again, if there is to be a third storey on top of the stage-set, the parapet at the top should be half the height of the one in the middle, the columns at the top a quarter less tall than those in the middle, and similarly the architraves with their cornices should be a fifth of the height of these columns.

7. These modular systems, however, cannot cater for the requirements and desired effects in all theatres, but the architect

Plate 24. Exterior of Roman theatre. Vitruvius says nothing about the decoration of the exterior of theatres. Palladio's elevation is entirely rusticated with the sequence Tuscan, Doric, Ionic, Corinthian, a mini-lexicon of antique forms seen through Renaissance eyes; the careful accentuation of the vertical axes in the rustication of the piers in the lowest order was used frequently in the Cinquecento (Bramante, Giulio Romano, Sansovino, Palladio); the bulgy blocks comprising the piers are reminiscent of the Porta Maggiore or the Claudianeum in Rome or possibly the amphitheatres at Pola and Aosta; the apertures in the third level combining flat and semicircular arches (a well-known antique form) were popularized by Bramante at the Palazzo Caprini, and copied innumerable times in the Cinquecento. Palladio understood perfectly well the function of the brackets of the fourth level of the Colosseum as supports for poles from which the great awnings were pulled across the interior of the theatre but he resists the temptation to copy the frieze and cornice from the same building.

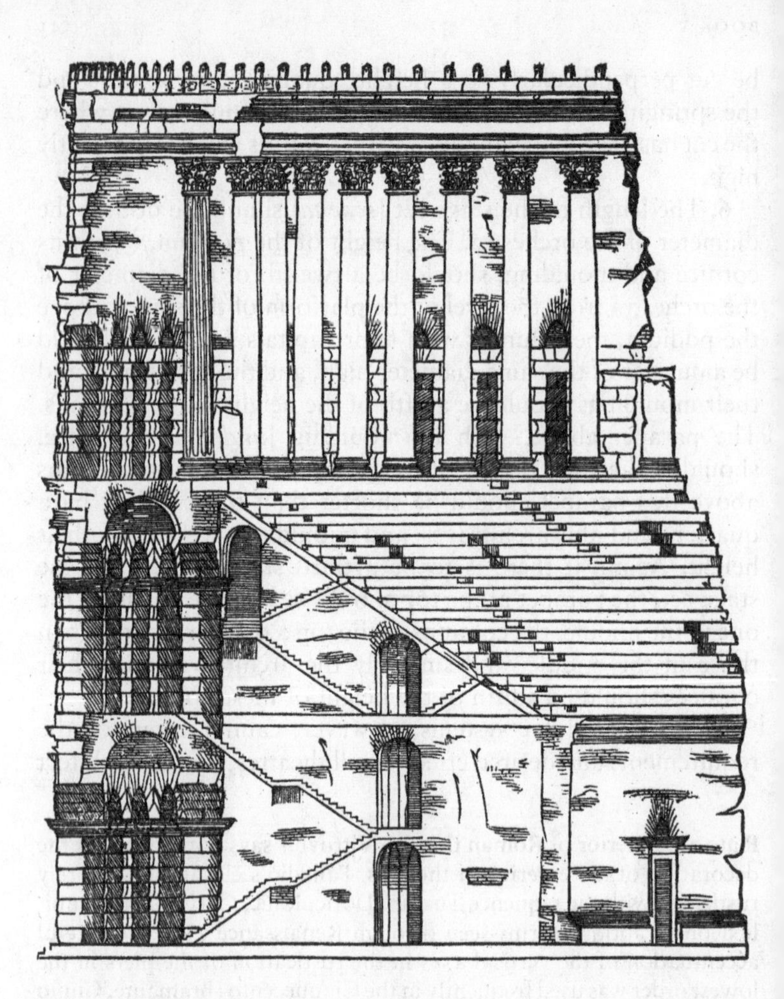

Plate 25. Section of interior of Roman theatre. The height of Palladio's vast colonnade, matching the top two storeys of the exterior, would have been far too great. The small objects at the top of the building are bollards through which the ropes of the awnings were pulled, according to Palladio; some can still be seen embedded in the ground, in fact, near the Colosseum, in their original locations.

must be alert to the proportions he should follow to conform to the modular system and those which should be modified to suit the character of the site or the size of the building. For there are some components which, for practical purposes, must be made the same size in both small and large theatres, such as the steps, the transverse passageways [*diazomata*], the parapets, the passages, the ramps of stairs, the actors' platform, the boxes for the magistrates [*tribunalia*] and anything else that crops up which necessarily requires some divergence from modularity lest the functioning of the theatre should be interfered with. Again, if some shortage of materials occurs during construction, for example, of marble, wood or of any other supplies, it will not be inappropriate to add or subtract something so long as this is not done too rashly, but with common sense. This will happen if the architect has practical experience and also is not devoid of mental flexibility and technical skill.

8. The stage-sets themselves follow well-defined rules such that the central folding door has the decoration appropriate to a royal palace, the doors to right and left are those of the guests and beyond them are the areas provided for the set-decorations, which the Greeks call περίακτοι [*periaktoi*; turning on a pivot], so called because in these areas there are revolving triangular devices, each provided with three differently decorated faces; when there is about to be some dramatic change of scene or the appearance of the gods with sudden claps of thunder, they are rotated to change the decorative field they present on their surfaces. Next to these spaces are the extended wings that provide entrances onto the stage, one from the forum, the other from outside the city.

9. There are three types of stage-set, one called the tragic, the second, the comic, and the third, the satiric. Their decorations differ in style and content because tragic scenes are represented by columns, pediments, statues and other regal objects; by contrast, comic scenery represents the houses of private citizens with balconies and overhangs with windows arranged like those in ordinary buildings; by contrast, satiric scenery is decorated with trees, caves, mountains and other rustic features designed to resemble the countryside [Plates 23–6; Figs. 14, 15].

Plate 26. Stage-set of Roman theatre. The illustration shows the right half of the stage, with the royal door at the left. Palladio makes no attempt to illustrate the *periaktoi* in the wings. The sequence of elements in the two storeys, podium (pedestal), columns (Corinthian, then Composite), entablature with pulvinate, garlanded wreath like that of the torus of the base of Trajan's Column, a *pluteus* plus another order follows Vitruvius' recommendation well enough, but the proportions

shows that Palladio is not following him exactly (for example, the lower podium is too high in relation to the other elements and the second storey of columns is not three-quarters of the lower columns in height, etc.). The low structure, evidently comprising vaults on four piers arriving at the bay to the right of the guests' door, is that shown in the plan at right angles to the stage, and was apparently conceived by Palladio as a low walkway giving access to the back of the stage.

CHAPTER VII
Greek Theatres

1. Not all the elements of Greek theatres should be built following the same principles: since, first, in the Latin theatre, the points of four triangles touch the circumference of the lowest part, whereas in the Greek theatre, the corners of three squares touch the line of the circumference and the stage is fixed where the side of the square closest to the stage cuts off a segment of the circle. A line parallel to this should be drawn tangent with the lowest point of the circumference, along which the façade of the stage-set is established; another line is then drawn through the centre of the orchestra and parallel with the stage and points should be marked where it intersects the lines of the circumference at the furthest extensions of the semicircle to right and left;[13] and having fixed the compass on the point at the right, an arc with a radius based on that distance[14] is extended down to the left hand side of the stage. Then, having fixed a centre at the left end, another arc with a radius based on that distance[15] is extended down to the right hand side of the stage.

2. In this way the Greeks, basing their plan on three centres, obtain a more spacious orchestra, a stage-set placed further back and a shallower stage, which they call the λογεῖον [*logeion*], because the tragic and comic actors perform on the stage, while other artists perform their roles moving around in the orchestra. Because of this the actors are given different names in Greek: *scaenici* [actors on stage] and *thymelici* [the chorus in the orchestra]. The height of the stage [*logeion*] should not be less than ten feet nor more than twelve. The ramps of stairs between the blocks of seats and the seats themselves should be aligned with the corners[16] of the squares as far as the first transverse aisle; above the transverse aisles, intermediate flights should again be laid out between the stairs, and at the top, the number of intermediate flights should always be increased to match a greater number of cross-aisles [Plate 27; Fig. 16].[17]

CHAPTER VIII

The Acoustics of the Theatre

1. Since all these matters have now been explained with the greatest care and attention, we must take great pains that a site should be selected in which the voice falls on the ear lightly and is not driven back so that, by rebounding, it conveys unintelligible sounds to the ears. For there are some sites which naturally impede the propagation of the voice, for example, dissonant locations, which are called κατηχοῦντες [*katechountes*; pushing sound back] in Greek; circumsonant locations, which they call περιηχοῦντες [*periechountes*; repelling sounds all round], and again, resonant locations, which they call ἀντηχοῦντες [*antechountes*; creating echoes]; and consonant locations, which they call συνηχοῦντες [*synechountes*; amplifying sound]. Dissonant locations are those in which the first sound uttered hits solid objects above when carried up high, and, being driven back, rebounds downwards and blocks the rise of the following sound. **2.** Circumsonant locations, by contrast, are those in which the voice is forced to dissipate itself by circling around, and, leaving only the middles of words intelligible, fades away there incomprehensibly without sounding the case-endings. Again, resonant locations are those in which the words, having rebounded after striking a hard surface, create echoes and make the most recently uttered case-endings sound double. And consonant locations are those in which the voice is helped from below, and, increasing in volume as it rises, arrives at the ears with clear and distinct words. So, if great care is taken in the selection of sites, the audibility of the voice will be improved with respect to its use in theatres by taking these precautions. But the configurations of their plans can be identified by these

Plate 27. Plan of Greek theatre. Palladio understood correctly that the three squares with which the Greek theatre was constructed, according to Vitruvius, applied to the orchestra; but presumably he was so taken by the Braga Theatre at Vicenza that he decided not to apply this insight to the triangles of the Roman theatre. Not knowing any examples of Greek theatres he again wanted to make a huge stage-set with another circle concentric to the first, not mentioned by Vitruvius, which he also used to establish the line of the periphery of the auditorium and the building behind the stage-set. He also misunderstood the statement by Vitruvius that the theatre was established using three circles; this leads him to describe two whole circles centred at points **D** and **E** using half the diameter of the orchestra, not arcs reaching the stage-wall (*proscaenium*) using the whole diameter as is required by Vitruvius' text; he then takes these circles to the line of the façade of the stage-set, uselessly, because the line of the stage-set has already been established by the two corners of the squares at the bottom of the orchestra. To judge by the plan, it may be that he also wanted to built a stage-set with giant order detached columns, with the minor columns marked on the plan between them as those of the first order.

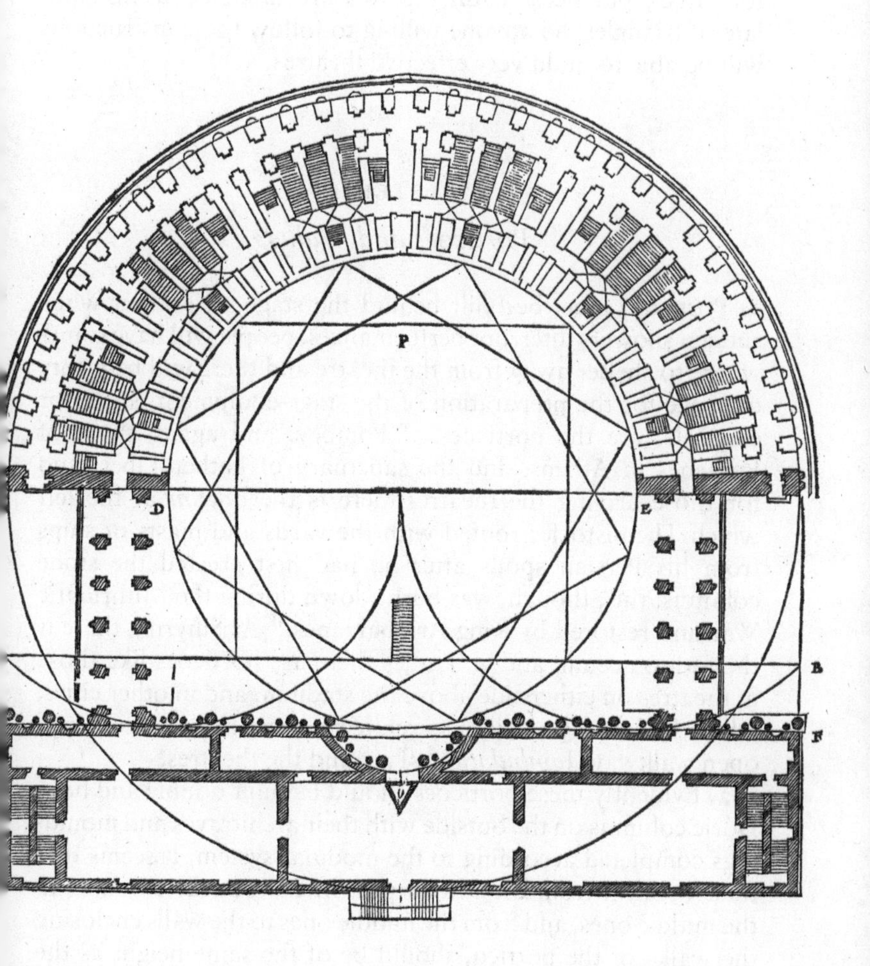
P
D
E
B
F

differences: if they are designed using squares, then they cater for Greek practices; Latin theatres are designed using equilateral triangles. So anyone willing to follow these instructions will be able to build very effective theatres.

CHAPTER IX

Porticoes and Walks

1. Porticoes must be built behind the stage-set so that when sudden showers interrupt performances, people will have somewhere to shelter away from the theatre and there will be plenty of space for the preparation of the stage-equipment; such, for example, are the porticoes of Pompey, and again, those of Eumenes in Athens, and the sanctuary of Father Liber; and for those leaving the theatre, there is the *odeum* at the left which Themistocles roofed with the yards and masts of ships from his Persian spoils after he had first erected the stone columns; this, though, was burnt down during the Mithridatic War and restored by King Ariobarzanes.[18] At Smyrna, there is the Stratoniceum, and at Tralles there are porticoes like those in theatres on either side above the stadium, and in other cities, which had very punctilious architects, there are porticoes and open walkways [*ambulationes*] around the theatres.

2. Evidently these porticoes should be built double and have Doric columns on the outside with their architraves and mouldings completed according to the modular system. It seems that their breadth from the lowest part of the outside columns to the middle ones, and from the middle ones to the walls enclosing the walks of the portico, should be of the same height as the exterior columns. Instead, the columns in the middle should be taller than the outer ones by a fifth, but designed using either the Ionic or Corinthian order.

3. The proportions and modularity of the columns will not be calculated with the same methods as those I have described for sacred temples because the temples of the gods should have their own kind of dignity, which is different from the elegance of porticoes and other such buildings. Thus, if the columns are to be built with the Doric order, their heights, including the capitals, should be subdivided into fifteen units; one of these should be chosen as the module, and the articulation of the whole building is to be developed on the basis of that module. First, the diameter of the column should be two modules; the intercolumniation five and a half; the height of the column, apart from the capital, fourteen; and the capital, one module in height and two and a sixth in breadth. The modular dimensions of the rest of the building should be developed as has been described in Book IV on sacred temples.

4. But if the columns are to be Ionic, the shaft, excluding the base and capital, should be divided into eight and a half units, of which one should be allocated to the diameter of the column: the base with its plinth should be set at half the diameter; the dimensions of the capital should be worked out as shown in Book III. If the column is to be Corinthian, the proportions should be worked out so that the shaft and base conform to those of the Ionic order, but the capital matches what was described in Book IV. And if a convex curve is to be added to the stylobates by means of small projections of unequal height [*scamilli impares*] the procedure should follow the diagram[19] supplied above in Book III. The architraves, cornices and all other elements should be adjusted to match the proportions of the columns following what has been written in the preceding books.

5. In fact the areas between the porticoes which will be open to the sky should, it seems, be enlivened with greenery; open-air walks are very healthy, particularly for the eyes, since the refined and rarefied air emanating from plants flows into the body because it is in motion, and sharpens our vision; and by removing the heavy discharge from the eyes, leaves our sight acute and images sharp. Moreover, when the body warms up with the exercise of walking, the air, drawing the moisture out of our limbs, reduces superfluity and diminishes the intolerable excess of it in the body by consuming it.

6. We can see that this is true from the fact that when there

are springs of water under cover or even a plentiful supply of marsh-water underground, no clouds of vapour rise up from them; but in sites without cover or those open to the sky, when the warmth of the rising sun touches the world, it drives out masses of vapours from such humid locations, and rolling them up together, takes them up high. Accordingly if, as seems clear, the more dangerous fluids are drawn out of our bodies by the air in locations open to the sky in the same way that they are evidently drawn from the earth by clouds, I do not think there can be any doubt that very broad and elaborately decorated walks, uncovered and open to the sky, should be built in cities.

7. One must proceed like this to ensure that they remain dry and free from mud. They should be excavated and cleared out as deeply as possible. Masonry drains should be built at right and left, and small pipes sloping on an incline into the drains should be inserted in the walls facing the walkway. Once these have been built, the site should be filled with charcoal, and then the walks should be covered with coarse sand and levelled off. In this way, overflowing water will be collected by the naturally porous charcoal and by the arrangement of pipes running into the drains, and so the walks will be kept dry and free from moisture.

8. Furthermore, it was the tradition that storerooms for materials indispensable for the citizens were constructed in these buildings. For during sieges supplies of wood are more difficult to come by than those of any other materials; salt, for example, can easily be brought in before a siege, corn can readily be collected by the state or by private citizens, and, if it runs short, can be compensated for with vegetables, meat or pulses; and water can be provided by digging wells or collecting it from the roof-tiles when there are sudden downpours. But a supply of wood, which is absolutely essential for cooking food, is difficult and arduous to collect because it takes a long time to transport and is consumed in bulk.

9. So, these walks are opened up at such times, and quantities of wood are allocated to each individual in each tribe. In this way uncovered areas like these ensure that the walks have two great advantages: first, they ensure healthiness in peacetime and

second, salvation in wartime. Therefore, laying out walks on these principles not just behind the stage-sets of theatres but also in the sanctuaries of all the immortal gods will be of great advantage to cities.

Now, since we seem to have discussed these themes enough, our explanation of the planning of baths will follow.

CHAPTER X

The Baths

1. First, a site as warm as possible must be chosen, that is, one which faces away from the north and north-east winds. In particular, hot [*caldaria*] and warm baths [*tepidaria*] should be lit from the south-west; but, if the character of the site prevents this, they should be lit from the south, because customarily the most popular time for bathing is from midday to evening. Again, we must take care that the hot bathrooms for women and men are adjacent to each other and located in the same areas; for in this way one can ensure that they share the use of the water-tanks [*vasaria*] and the under-floor furnace [*hypocausis*]. Three bronze tanks should be placed above the under-floor furnace, one for hot, another for warm and a third for cold water; they should be located in such a way that the amount of hot water [*sic*] flowing from the warm into the hot tank will be replaced by an equal amount flowing from the cold into the warm tank; and the half-cylinders of the baths[20] should be heated by the same under-floor furnace.

2. The suspended floors of the hot bathrooms should be constructed in such a way that first, the ground below them is laid with tiles a foot and a half square inclined towards the underground furnace, so that a ball released into it would not stay inside but would roll of its own accord to the mouth of the furnace; that way the flames will spread more easily under the suspended floor. Piers comprising bricks of two-thirds of a foot should be built on the ground, and positioned so that two-foot tiles can be placed on top of them; the piers should be two feet high and laid with clay mixed with hair; two-foot tiles should be placed above them to support the suspended pavement.

3. The vaults will certainly be more durable if they are built of masonry; but if they are made of wooden beams, a surface of tiles [*opus figlinum*] should be applied to their soffits. They should be built as follows: iron bars or arches should be cast and suspended from the wooden beams with iron hooks placed as close together as possible: these bars or arches should be arranged so that tiles without flanges can be set between any two of them and will be supported by them: so the whole of the vault would be built in such a way that it is supported on iron. The joints on the outer surfaces of the vaults should be coated with clay mixed with hair, and conversely, the soffit facing the pavement should first be reveted with ground-up terracotta mixed with lime and then finished off with stucco or plaster [*opus tectorium*]. In hot baths such vaults will function better if they are made double because then the moisture of the vapour will not be able to rot the woodwork of the frame, but will circulate between the two vaults.

4. But clearly the dimensions of the baths should be related to the number of people who use them. They should be arranged like this: whatever the length, the breadth should be a third less, excluding the niches for the basin [*labrum*] and the bath [*alveus*]. Obviously the basin should be put in a well-lit spot lest those who stand around it obscure it with their shadows. The niches for the basins should be made spacious enough that when the first-comers have taken their places, the others have enough room to stand around and wait their turn. But the depth of the bath between the wall and the parapet should not be less than six feet, so that the lower step and the socle [*pulvinus*] can take up two feet.

5. The Spartan hot-room [*laconicum*] and the sweating-rooms [*sudationes*] should be adjacent to the warm bathroom and should be as broad as they are high up to the base of the curve of the hemispherical dome. An aperture should be left in the centre of the hemisphere with a bronze disc suspended from

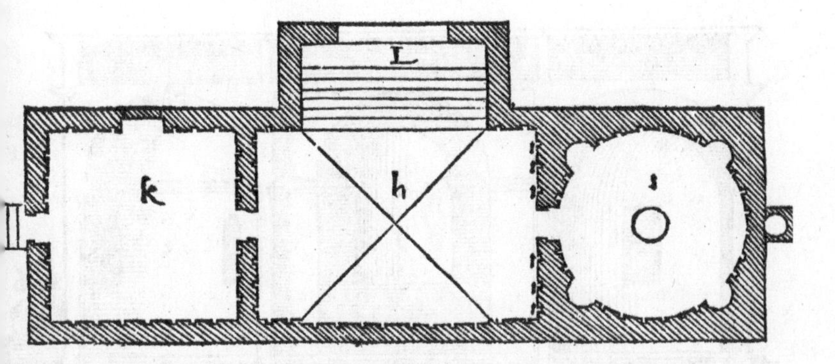

Plate 28. Plan of baths. The overall proportions are 1 × 3, not 2 × 3 as Vitruvius recommends. I Spartan hot-room (*laconicum*). H Warm bath (*tepidarium*). K Cold bath (*frigidarium*). L Basin (*labrum*) with steps leading up to the platform on which it stands, referring to Vitruvius' *pulvinus*; but Palladio replaces the bath (*alveus*) with the basin. Palladio not unreasonably provides the rectangular *tepidarium* with a cross-vault of which he saw many examples in Roman baths of much greater scale.

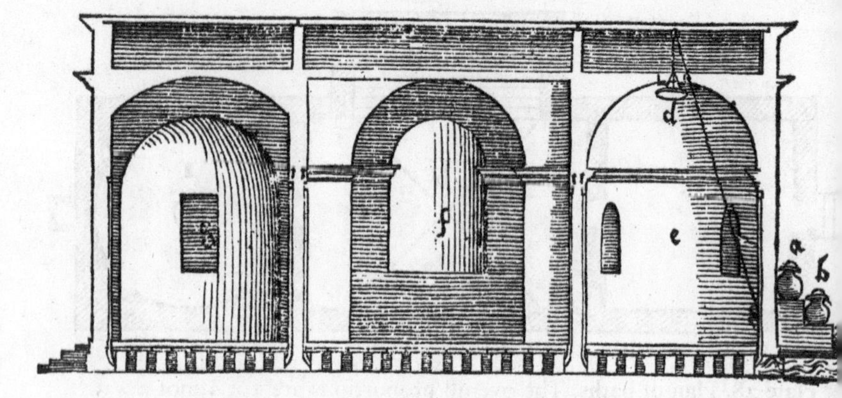

Plate 29. Section of baths with hypocaust below and furnace at right (*hypocausis*). **a** Cold water tank. **b** Warm water tank. **c** Hot water tank. **e** Spartan hot-room (*laconicum*). Vitruvius says the height of the *laconicum* up to the springing of the vault should equal its breadth, and that it should be circular in plan, as Palladio draws it here. **d** Bronze disc. **f** Warm bath (*tepidarium*) with cross-vault. **g** Cold bath (*frigidarium*). The charming illustration of the three water-tanks at the right is taken from Fra Giocondo (1511). Curiously, Palladio thinks that the hypocaust (a term Vitruvius does not use) should extend under the cold bathroom.

it on chains, by which the temperature of the sweating-room can be adjusted by raising and lowering it. Evidently the Spartan hot-room should be made circular so that the intense heat of the fire and vapour can diffuse itself evenly around the circumference from the centre [Plates 28, 29].

CHAPTER XI

The Greek Gymnastic Complex [palaestra]

1. Now, despite the fact that the construction of gymnastic complexes is not an Italian custom, I think it appropriate to discuss them and show how they are organized by the Greeks, since the method of construction has been handed down to us. In gymnastic complexes, the colonnaded courtyards should be made square or rectangular so that they have a walkway with a circuit of two *stadia*, which the Greeks call δίαυλος [*diaulos*; double circuit]: three of these porticoes should be built with single rows of columns, and the fourth, which faces south, should be double, so that showers cannot penetrate its interior when there are windy storms.

2. In the three porticoes, spacious *exedrae*[21] should be built provided with seats on which philosophers, orators and others who delight in erudition can sit and dispute. But in the double portico, the following spaces should be built: the young men's hall [*ephebeum*] should be in the middle; this, in fact, is a very spacious *exedra* with seats, which should be a third longer than it is wide. At the right should be the room with the punchbags [*coryceum*] and next to that, the dusting-room [*conisterium*]; beyond the dusting-room, and at the corner of the portico, the cold bath [*frigida lavatio*], which the Greeks call the λουτρόν [*loutron*]. At the left of the young men's hall is the oiling-room [*elaeothesium*], and next to that, the cold bathroom [*frigidarium*], beyond which there is access to the warm bathroom [*propnigeum*] at the corner of the portico. Next, on the inside and near the cold bathroom the vaulted sweating-room should

be placed, twice as long as it is broad: at the corners of the sweating-room there should be the Spartan hot-room on one side, built as described above, and the hot bathroom [*lavatio calda*] on the other. The colonnaded courtyards in the gymnastic complex should be laid out as described above in all respects.

3. But on the outside [of the *palaestra*], three porticoes should be laid out, one for those leaving the colonnaded courtyard, and one at the right and another at the left, each a stade long: the one that is to face north should be made double and very wide: the other should be made single so that there will be borders, like pathways, no less than ten feet wide in the areas along the walls and columns: their middle sections should be dug out so that the steps going down from the borders to the level area will be a foot and a half: the level area should be not less than twelve feet wide. In this way clothed people walking round on the borders will not be disturbed by the anointed athletes doing their exercises.[22]

Plate 30. The Greek gymnastic complex (*palaestra*). Derived in part from Fra Giocondo (1511), Palladio's plan is pretty accurate in relation to Vitruvius. Orientation inverted. Where modern scholars disagree about the reconstruction, they distribute the outside elements, the *xysta*, stadium and other porticoes either at the south or the north, and not at the side as here. The main colonnaded courtyard includes a double portico facing south; along the three single porticoes are the *exedrae* of varying shape, rectangular, some with semicircular niches, and cross-vaulted with and without corner columns. The range of buildings at the north of this courtyard includes: **A** Young men's hall (*ephebeum*), with cross-vault. **B** Hall with the punchbags (*coryceum*). **C** Dusting-room (*conisterium*). **D** Cold baths (*frigida lavatio*, *loutron*). **E** Oiling-room (*elaeothesium*). **F** Cold bathroom (*frigidarium*). **G** *Propigneum*: Barbaro does not say what he thinks the word means in this context: here it is translated as a warm bathroom. **H** Vaulted sweating-room (*sudatio*). **I** Spartan hot-room (*laconicum*). **K** Hot bathroom (*lavatio calda*). Exterior porticoes: **M** Double portico facing north. **N** Portico in which athletes exercise. **O** Plantations between porticoes. **P** Uncovered walkways where athletes exercised in the summer, called *paradromides* by the Greeks, *xysta* by the Romans. **Q** Stadium where people stood to watch the athletes. 1, 1, 1, 1 in centre of main courtyard, places to rest (*stationes*).

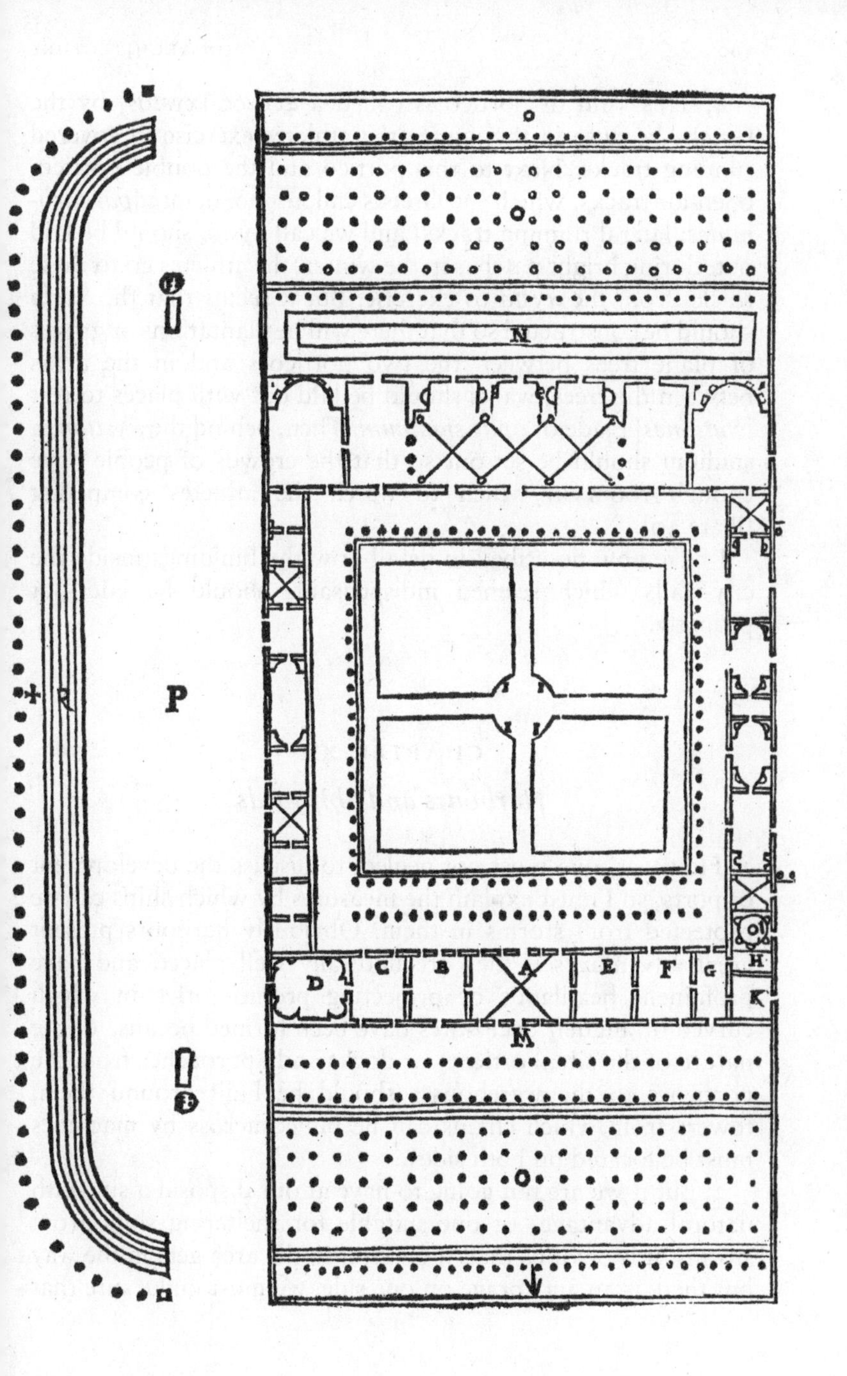
N
P
R
C
B
A
E
F
G
H
K
D
M
O

4. This kind of portico is called a ξυστός [*xystos*] by the Greeks because in the winter the athletes exercise in covered running tracks. Next to this portico and the double portico, open-air tracks, which the Greeks call παραδρομίδες [*paradromides*; lateral running tracks] and we call *xysta*, should be laid out; during bright patches in the winter, the athletes go to these tracks from the *xysta* to exercise. But it seems that the *xysta* should be constructed so that there will be plantations or groves of plane trees between the two porticoes and in the areas between the trees, walks should be laid out with places to rest [*stationes*] made of *opus signinum*. Then, behind the *xystum*, a stadium should be set out so that the crowds of people have enough room in which to watch the athletes competing [Plate 30].

I have now described in detail how the buildings inside the city-walls which seemed indispensable should be laid out properly.

CHAPTER XII

Harbours and Shipyards

1. However, one must not neglect to discuss the development of ports, so I must explain the measures by which ships can be protected from storms in them. Obviously harbours present great advantages if they are naturally well placed and have prominent headlands or projecting promontories in which curved or angular enclosures have been formed because of the nature of the site. Porticoes or docks and approaches from the porticoes to the warehouses should be built around them; towers from which chains can be pulled across by machines must be located on both sides.

2. But if we are not going to have at our disposal a site with natural advantages or one suitable for sheltering ships from storms, it is obvious that, if no river in the area gets in the way but there is an anchorage on one side, we must make sure that

piers of masonry or embankments are built on the other side, so forming enclosed harbours. The masonry to be built under water should evidently be built using the powder [pozzolana] brought in from the areas stretching from Cumae to the promontory of Minerva: it should be mixed in the mortar [with lime] so that it has a proportion of two to one.

3. Then, in the site to be decided on, cofferdams made of oak piles bound together with chains should be sunk into the water and fixed securely; then the floor on the inside below water level must be levelled off and cleared out by men working on small cross-beams;[23] and there the concrete from the mortar-trough, premixed with lime prepared as described above, must be piled up; finally, the space between the coffers should be filled in with masonry. For this is the natural advantage presented by the sites described above.

But if the props cannot hold the cofferdams together because of tides or the swells of the open sea, then an immensely strong platform should be constructed of masonry[24] beginning from the dry land itself or from a breakwater; less than half of this platform should be built with a horizontal surface[25] while the rest facing the [opposite] shore should slope forward.[26]

4. Then at the water's edge and down the sides of the platform, containing-walls about one and a half feet thick should be constructed around the platform and level with the flat surface mentioned above; then the slope at the front should be covered with sand and brought up to the level of the container-wall and the flat surface of the platform. Then a pier of the appropriate size should be constructed on the level surface of the platform which, once built, should be left to dry out for no less than two months. After this the container-wall holding in the sand should be demolished: and so the sand, undermined by the waves, will make the pier collapse into the sea. Using this procedure whenever necessary, one can proceed further out to sea.

5. But in places where this powder does not occur naturally, these methods must be employed: double cofferdams held together with planks and chains should be built in the location to be designated and clay in baskets made of swamp-rushes should

be packed between the props. When it has been compacted down well and is as solid as possible, then, having put water-screws, waterwheels and drums in position, the site surrounded by this enclosure should be evacuated and drained and the foundations dug out inside the enclosure. If the floors of the foundations consist of earth, they must be cleared out and drained until one reaches a layer of solid ground wider than the wall to be built above it; then it should be filled in with rubble, lime and sand.

6. But if the site is soft, then the foundations should be packed with piles of charred alder or olive wood, then filled with charcoal in the way described above for the foundations of theatres and walls. Finally, a wall should be built of square stone slabs with the longest possible joints so that, above all, the centres of the blocks [in one row] would be held in place by the vertical joints [of the row above].[27] Then the cavity inside the wall should be filled in with concrete or masonry. In that way the wall will be so strong that a tower can be built on it.

7. When all these structures have been completed, the docks should be built so that, as a rule, they face north. For, because of the heat, exposure to the south produces rot, woodworm, shipworm and other types of destructive insects which it keeps alive by nourishing them: such buildings should be constructed with an absolute minimum of wood because of the danger of fire. No limit on their dimensions need be fixed, but they should be built for the biggest types of ship so that even if very large ones are brought to safety they will have plenty of space.

In this book I have written about everything essential for the proper functioning of public places in cities that occurred to me, and how they should be planned and executed: in the next book, I shall discuss the functions of private buildings and their modular systems.

BOOK VI

Introduction

1. It is said that when Aristippus, the Socratic philosopher, was cast up in a shipwreck on the shore of Rhodes, he noticed geometric figures drawn there and exclaimed to his companions: 'Let's hope for the best, since I see signs of men's presence.'[1] He immediately made for the city of Rhodes and went straight to the gymnasium: there he discussed philosophy and was so well rewarded with gifts for it that he was able not only to provide for himself but also to supply his companions with clothing and the other necessities of life. But when they wanted to return to their own country and asked him what he wanted them to report to those back home, he told them to say that it would be a good idea to provide their children with equipment and provisions for travel that could float ashore with them even after a shipwreck.

2. For our real safeguards in life are those which neither the treacherous storms of fortune, changes of political regimes nor the devastation of war can damage. Theophrastus, in particular, developing this thought further, put it this way when he exhorted men to educate themselves rather than put their trust in money:[2] of all people, only the educated man is neither a stranger when in unfamiliar countries, nor deprived of friends even when he has lost his family and intimates, but is a citizen of every city and can look dispassionately and fearlessly at the troublesome vicissitudes of fortune; but the man who believes that he is not defended by the safeguards of culture but by those of fortune is tormented by an unstable and insecure life as he makes his way along its treacherous paths.

3. Certainly, Epicurus says in much the same way that fortune bestows little on the wise, but that the greatest and most indispensable of her gifts are governed by the reasoning-processes of the mind and of the intellect. And many other philosophers have said the same thing; and similarly the poets who wrote the ancient comedies in Greek expressed identical sentiments in their verses on stage, for example, Eucrates, Chionides, Aristophanes and along with them, Alexis in particular, who says the Athenians are to be applauded because, while the laws of all the Greeks require that parents should be looked after by their children, those of the Athenians do not apply to all parents, but only to those who have educated their children in the arts. For all the gifts that Fortune bestows can easily be snatched away by her, but scientific knowledge allied to spiritual values never fails us, and remains with us constantly till the end of our lives.

4. I am therefore infinitely grateful and eternally thankful to my parents because they entirely approved of this Athenian law and took pains to educate me in an art which cannot be practised correctly without the discipline of literature and a comprehensive knowledge of all fields of study. So, having mastered a rich array of disciplines thanks to the solicitousness of my parents and the instruction of my teachers, and, delighting myself with linguistic and technical subjects and with the writings about them in commentaries, I equipped my mind with these intellectual assets, of which the principal fruit is that there is no necessity to own more and more and that the richest possession of all is to wish for nothing whatsoever. But it could be that some people regard these arguments as ineffectual and therefore think that the wise are those who are wealthy, which is why the majority of people pursue this goal and, relying on boundless self-confidence, even achieve celebrity along with their wealth.

5. But I, Caesar, have not applied my mind to the acquisition of money through my art, and have regarded it as right to pursue a life of modest means with a reputation intact rather than wealth with infamy. Accordingly I have never become particularly well known; but once these volumes are published,

I hope I will be known even to future generations. And it is no wonder that I am unknown to most people; other architects solicit and court the powerful to practise architecture, but I was taught by my masters that one should ensure that one was asked rather than did the asking, because one flushes with embarrassment when asking for something dubious. In fact it is those who grant favours who are courted, not those who receive them. For what suspicions would we imagine a man would harbour when he is continuously asked to commit money from his own financial resources for the benefit of some petitioner – if not that the operation is to be carried out for the profit and gain of that person?

6. And so our ancestors entrusted work to architects who were approved of on the grounds of their birth, in the first place, and then made enquiries as to whether they had been honestly educated, judging that work should be entrusted to men of honourable modesty rather than those of arrogant assertiveness. And these same craftsmen taught nobody except their own children or kin and brought them up as honest men to whom the funds for such great projects could indeed be committed on trust without hesitation.

But instead, when I see that a knowledge of such a fundamental discipline is boasted of by the uneducated and unskilled, and by those who are not only ignorant of architecture but know absolutely nothing of construction, I cannot but praise the heads of families who are so confident in what they have read about architecture that they build for themselves, reckoning that if they must trust someone unskilled, then they themselves are more entitled to spend sums of money to satisfy their own, rather than other people's desires.

7. So it is that, with the exception of architecture, nobody attempts to practise any other art at home, such as shoemaking, dyeing or any of the easier ones, because those who proclaim themselves to be architects are so-called not because of any genuine skill – but falsely. For these reasons I decided that I should write a comprehensive treatise on architecture and its principles with scrupulous care in the belief that in the future it would not be an unwelcome gift for all peoples. Therefore, since I wrote at length about the development of public buildings in Book V, in this one I will explain the principles and commensurability implicit in the modular systems of private houses.

CHAPTER I

Climate and Houses

1. Houses, then, will be correctly planned if, first, we take careful notice of the regions and latitudes of the world in which they are to be built. For it clearly makes sense that one type of building should be built in Egypt, another in Spain, some other type in Pontus and a different one again in Rome, and so on, depending on the different characteristics of the countries and regions. The reason is that one part of the earth is oppressed by the trajectory of the sun, another is a long way from it and yet another, lying along the middle zone, is temperate. Therefore, since the position of the heavens relative to the mass of the earth is naturally governed by the inclination of the circle of the zodiac and the course of the sun, producing very different results, it is obvious that the siting of houses must be organized similarly with reference to the characteristics of the regions and variations of climate.

2. It is clear that houses in the north should be roofed over, as closed as possible and provided with few apertures; and they should be orientated towards the warm zones. By contrast, houses in southern regions under the impact of the sun should be provided with more apertures and face north and north-east because those regions are oppressed by heat. In that way, that which nature damages most must be corrected by art. Again, we must make adjustments in the same way in other regions, depending on how the sky is positioned in relation to the latitude.

3. Now these characteristics should be identified and considered in relation to natural phenomena, but they are also observable in the limbs and bodies of different peoples. For the sun maintains our bodies at a reasonable temperature in those areas where it radiates heat in moderate amounts; it burns others when passing close by them and upsets the proper balance of moisture by scorching them. On the other hand, moisture is not drawn out of people's tissue in chilly regions because they are a long way from the south, but the saturated air from the sky diffuses itself in people's bodies, creating more robust physiques and making the pitch of their voices lower. This is the reason why people in the north develop enormous physiques and grow up with fair complexions, straight red hair, grey-blue eyes and a great deal of blood caused by the abundance of moisture and the coldness of the atmosphere.

4. By contrast, those who are closest to the southern part of the world's axis and lie under the trajectory of the sun grow up with shorter bodies, dark complexions, curly hair, black eyes, strong legs and less blood because of the impact of the sun. So their lack of blood makes them more scared of facing weapons, but they can resist heat and fevers without fear because their limbs have been developed by heat. Equally, bodies which develop in the north are made more timorous and weak by fever, but face up to weapons fearlessly because of their rich supply of blood.

5. In the same way, pitches of voice vary from people to people and have distinct characteristics because the limits of east and west around the earth's equator, by which the upper and lower parts of the world are divided, seem to present a naturally levelled circumference which astronomers accordingly call the ὁρίζων [*horizon*]. Therefore, keeping this idea firmly in mind as a given, one should draw a line from the edge of the northern region to that which lies above the southern axis, and from that, another line rising obliquely to the highest pole behind the stars of the Great Bear; as a result we should realize that, without room for doubt, the world has a triangular configuration, like the musical instrument the Greeks call the σαμβύκη [*sambyke*; Fig. 17].

6. And so the people in southern regions in the area nearest to the lower pole relative to the line of the axis produce thin and very shrill vocal sounds, like the string nearest to the corner of a musical instrument, because of the reduced altitude with respect to the heavens above them. The other strings after this one produce sounds that are less shrill in pitch in the nations as far as Greece, which is placed in the middle, and so, growing regularly from the middle to the extreme north under the highest points of the sky, the sounds uttered by people are naturally expressed in deeper tones. So it seems that, because of its inclination, the whole system of the world was created as perfectly as possible in relation to musical harmony by the appropriate temperature of the sun.

7. Therefore, the nations that lie in the middle between the pole of the southern section of the axis and that of the northern section produce a sound like the intermediate notes on a musical scale when they talk; and the people encountered by those proceeding north have their vocal emissions reduced by moisture to the deep and the very deep because the distances from the sky are greater, and they are naturally compelled to use lower tones. In the same way, for those proceeding from the middle to the south, the people produce voices with the thinness of sound corresponding to the highest pitch.

8. One can understand that it is true that sounds are made lower by naturally moist places and higher by hot places from this experiment: take two cups of the same weight fired equally in the same furnace and which sound the same when struck; one of these should be immersed in water, then taken out, and then both should be struck. When this has been done, the sound each cup produces will be entirely different and they will not weigh the same. Similarly, even though men's bodies are created with the same type of physique and by the same relationship to the heavens, some emit high-pitched sounds on contact with the air because of the heat in their areas, while others emit very low-pitched sounds because of the abundance of moisture.

9. Further, the southern nations, with their minds made acute by heat, can arrive more easily and rapidly at the resolution of problems because of the lack of atmospheric density. But northern peoples, infused with a heavy atmosphere and chilled by moisture because of the density of the air, have sluggish

minds. We can see that this is the case by observing snakes; when heat removes the chill from the moisture in them they move very rapidly: but when chilled by the change in climate during autumnal and winter weather they remain motionless in a torpor. One should not be surprised, then, that warm air makes men's minds more acute, but that cold air makes them sluggish.

10. But though southern populations are endowed with very sharp minds and great resourcefulness in evolving plans, they fail as soon as they have to face up to acts of physical courage because the spirit of audacity has been sucked out of them by the sun. By contrast, those raised in cold regions are much more ready for the clash of arms and are endowed with boundless, fearless bravery; but because of their slowness of wit they are reckless and act against their own best interests by rushing in without thinking. Since, therefore, things in the world are arranged like this by nature and all nations vary because of their highly dissimilar temperaments, it is certainly true that the Roman people occupy the territory right at the centre of the earth in relation to the total extent of the lands of the world and of its regions.

11. In fact, people in Italy are the most balanced with respect to both north and south in terms of bodily form and the spiritual rigour required for decisive action. For exactly as the planet Jupiter is temperate, running in the middle between the sweltering planet Mars and the freezing planet Saturn, so, for the same reason, Italy has the unbeatable advantage of being balanced between the southern and northern regions, but with admixtures from both. And so she shatters the courage of barbarians by intelligent planning and foils the plots of southerners by force of arms.[3] Thus the divine mind allocated to the city of the Roman people a superb, temperate region in order that it could acquire governance of the whole world.

12. Now if it is true that the various regions have been created with different characteristics depending on the latitude, and that people's natures develop with different personalities, physiques and qualities, we cannot doubt that we should also plan various types of buildings to match the particular characteristics of nations and peoples appropriately, since we have at hand an accomplished and immediate demonstration of what to do from nature herself.

I have now explained as precisely as I could my understanding of the characteristics of the sites created by nature and have spoken about how various types of houses should be built bearing in mind the course of the sun, the latitude and the physical characteristics of people. Now I shall explain briefly the relative dimensions on both a large and small scale of the modular systems of the different types of houses.

CHAPTER II

Modularity and Optics

1. Nothing should preoccupy the architect more than that buildings should incorporate exactly the measurements implicit in the proportional system based on a predetermined unit. Therefore, when the system of modular relationships has been established and the relative dimensions worked out by calculation, it requires an acute mind to consider the nature of the site, its use and appearance, and to make adjustments by means of additions or subtractions[4] when something in the modular system must be augmented or reduced so that it can be seen to be designed correctly without damaging its appearance.

2. This is because the appearance of a building seen close up is clearly different from that of a building on a high site; it is not the same if it is in an enclosed site, and different again if it is in the open: in these cases it takes fine judgement to decide, ultimately, how to proceed. For our sense of vision does not seem to produce reliable results, and the mind is often misled by it to arrive at faulty conclusions; in painted scenery, for example, there appear to be columns that project, mutules that jut out and figures of statues that stand forward, when the picture is obviously absolutely flat; similarly, the oars of ships, which are in fact perfectly straight, look bent to the eye when immersed in water; up to the point when parts of the oars touch the surface of the water, they appear to be straight, as, of course, they are: but when submerged in water, these parts project from themselves images which float through the naturally transparent water up to the surface, where they fluctuate and seem to create for the eyes the appearance of broken oars.

3. But whether we see this because of emanations from the images or instead, as natural scientists maintain, because of the emission of rays from the eyes, either way it seems that what we see with our eyes can produce false impressions.

4. Therefore, since what is real may seem false and some things may turn out to be different from how they appear to the eyes, I do not think that there is any room to doubt that subtractions or additions should be made to cater for the characteristics and exigencies of sites, but in such a way that the buildings leave nothing to be desired. But such results cannot be achieved by training alone but also require fine judgement.

5. So, first and foremost, the rationale behind the modular system must be established and from it every adjustment can be derived without room for equivocation. Then, a site as long and wide as the spaces to be incorporated in the future building should be laid out, and once its full extent has been established the preparation of the proportional system should follow what is appropriate so that the appearance of harmony should be absolutely clear to observers. In this context I must now make clear the means by which this should be achieved, and first I will speak about how the courtyards of houses [*cava aedium*] should be designed.

CHAPTER III

Plans of Houses

1. The courtyards of houses are divided into five types, and these are the terms for the different designs: Tuscan, Corinthian, tetrastyle [four-columned], displuviate[5] and roofed [*testudinatum*].

Tuscan courtyards are those in which the joists taken across the whole breadth of the *atrium* should be provided with suspended cross-beams [*interpensiva*] and water-channels [*colliciae*] running down from the corners of the walls to the angles formed by the joists; in addition, rainwater is taken down common rafters into the aperture of the *compluvium*.[6] In Corinthian courtyards, the beams and *compluvia* are constructed on the same principles, but the joists are set on columns on all four sides at a certain distance from the walls. Tetrastyle courtyards are those in which four angle-columns set under the joists ensure their serviceability and stability because the joists are not required to support a great load and do not bear the weight of suspended cross-beams [Figs. 18, 19, 20].[7]

2. By contrast, displuviate courtyards are those in which upward sloping rafters support the frame of the *compluvium* and send the rainwater outwards. These are most advantageous for winter apartments because their raised *compluvia* do not obstruct the illumination of the dining rooms [*triclinia*]. But they are extremely difficult to repair because the pipes that take the rainwater as it flows from around the walls cannot accommodate the water pouring from the channels quickly enough, and consequently back up and overflow, and damage the internal woodwork and walls in these types of buildings [Fig. 21].

Roofed courtyards are in fact built where the spans are not great and where spacious rooms are provided in the upper storeys [Fig. 22].

3. The lengths and breadths of *atria* fall into three categories: the first category is laid out so that when the length has been divided into five units, three should be allocated to the breadth;[8] in the second, the length should be divided into three units and two assigned to the width;[9] and in the third, the breadth of the *atrium* should be incorporated in a square in which a diagonal should be drawn; the length of the diagonal should be allocated to the *atrium*.[10]

4. Their height, which should be taken up to the soffits of the joists, should equal the length minus a quarter, and the rest should be allocated to the panelled ceiling [*lacunar*] and the frame of the central aperture [of the *compluvium*] above the joists. The breadth of the side-rooms[11] to right and left should be set at a third of the length of the *atrium*, if the *atrium* measures from thirty to forty feet in length; if the length of the *atrium* ranges from forty to fifty feet, it should be divided into three and a half units, of which one should be given to the side-rooms. But when the length varies from fifty to sixty feet, a quarter of that length should be given to the side-rooms. Lengths of sixty to eighty feet should be divided into four and a half units, one of which should become the breadth of the side-rooms; *atria* from eighty to a hundred feet in length should be divided into five units, which will establish the appropriate breadth for the side-rooms. The beams of the lintels of the side-rooms should be set high enough for their height to match their breadth.

5. With regard to the *tablinum*; if the breadth of the *atrium* is to be twenty feet, a third should be subtracted from it and the rest given to the *tablinum*. If the breadth of the *atrium* is to be between thirty and forty feet, half should be allocated to the *tablinum*. When its breadth is between forty and sixty feet, it should be divided into five units, allocating two to the *tablinum*. Smaller *atria*, though, cannot incorporate modular systems transferred from larger *atria*, because if we were to use a system of larger modules in smaller *atria*, the *tablinum* and the side-rooms would be useless; but if we were to use the modular system of smaller *atria* for the larger ones, the same spaces would turn out vast and monstrous. So I thought it best to describe the dimensional systems of each type in great detail with regard to both function and appearance.

6. The height of the *tablinum* up to the beam [of the lintel] should be set at an eighth more than its breadth. Its panelled ceiling should be raised by the addition of a third of the breadth to the height. In smaller *atria*, the entrance corridor [*fauces*] should be the breadth of the *tablinum* minus a third, and in larger *atria* they should be half the breadth of the *tablinum*. Portraits with their insignia of office should be set up [in the *atrium*] in such a way that their height corresponds to the breadth of the side-rooms.

Plate 31. Section of Roman house. Palladio reconstructs the sequence of spaces in the Roman house. At right, a two-storeyed colonnaded *loggia* presumably intended to represent the *vestibulum*; next left, evidently a roofed variant of a Tuscan *atrium* with two large beams supporting the floor above; next left, the *tablinum*, erroneously in-

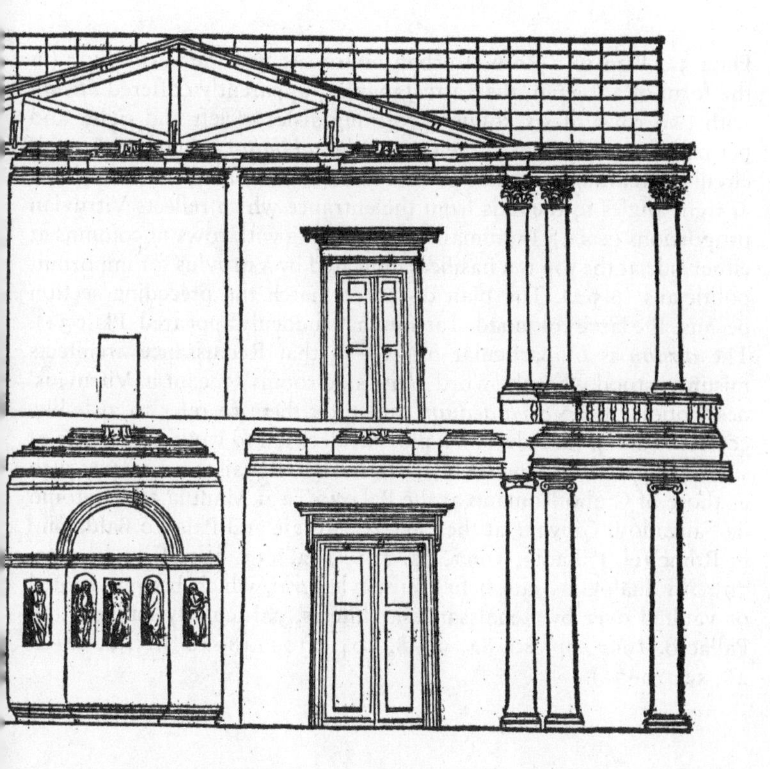

cluding statues of family members; then the colonnaded courtyard (*peristylium*) with Corinthian columns closer even than pycnostyle, diverging from the Vitruvian requirement of not less than three nor more than four diameters as intercolumniations (6.3.7).

Plate 32. Plan of a Roman urban house. At bottom: *vestibulum* (in the form of a *loggia*), then a rectangular, apparently coffered *atrium* with two rows of six columns forming aisles to left and right, and perhaps a *compluvium* in the centre; the *tablinum* (square with semi-circular *exedrae*); colonnaded courtyard (*peristylium*) with long side at right angles to the axis from the entrance which reflects Vitruvian proportions (2 × 3). Presumably the building with rows of columns at either side at the top is a basilica, indicated by Vitruvius for important politicians (6.5.2). The plan does not match the preceding section because the large colonnaded atrium has suddenly appeared (Plate 31). The *atrium* is of particular interest in that Renaissance architects misunderstood what the word *alae* ('side-rooms') meant in Vitruvius' description of the *cava aedium* and took them to refer to aisle-like corridors along the sides; this, plus the decision to vault the centres of such *atria*, produced spectacular entrances to Renaissance palaces such as those of Giulio Romano at the Palazzo Tè at Mantua and Antonio da Sangallo il Giovane at the Palazzo Farnese and Palazzo Baldassini in Rome (cf. Palladio, 1997, pp. 106, 152, 153, 157; II.30, 74, 75, 79). An analogous fate befell tetrastyle *atria*, which became roofed or vaulted over by Renaissance architects, particularly Palladio (cf. Palladio, 1997, pp. 80, 84, 97–8, 104, 130, 148–50: II.5, 8, 21–2, 28, 52, 70–72).

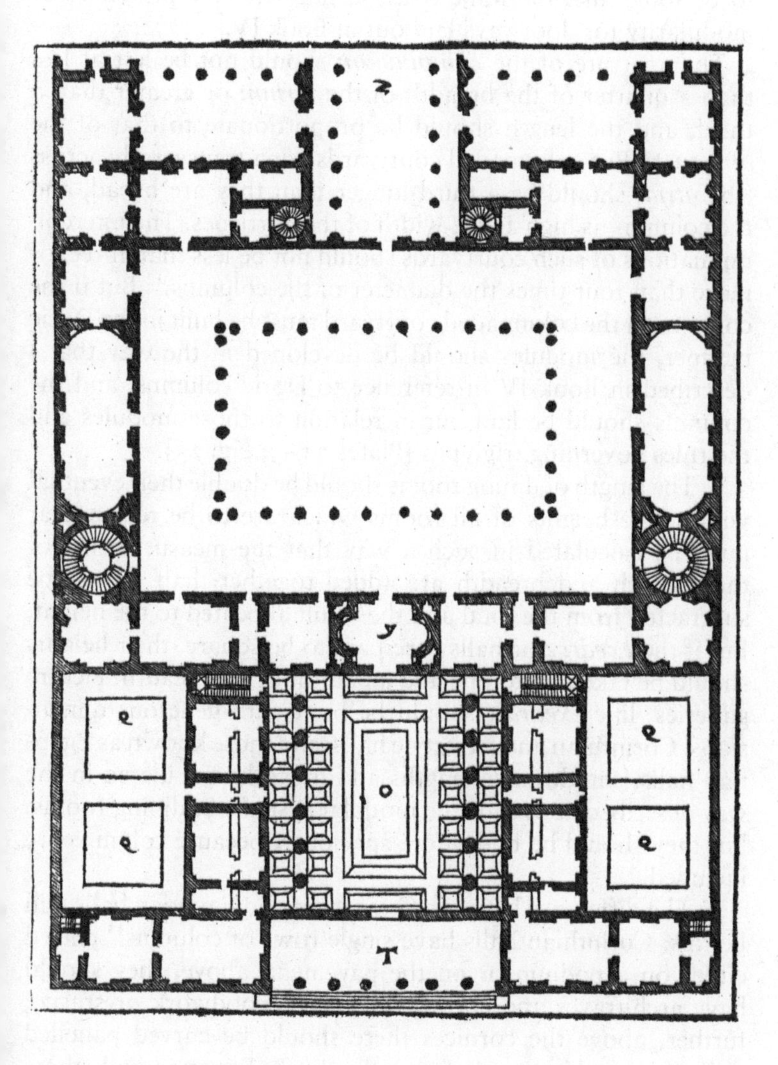
z
y
e
e
o
e
e
T

If the doors are to be Doric, their breadth in relation to their height should be worked out as for Doric doors, and if they are to be Ionic, like the Ionic ones, in line with the principles of modularity for doorways laid out in Book IV.

The aperture of the *compluvium* should not be left at less than a quarter of the breadth of the *atrium* or greater than a third, and the length should be proportionate to that of the *atrium*. **7.** But colonnaded courtyards lying transversely across [the *atria*] should be a third longer than they are broad, and the columns as high as the width of the porticoes. The intercolumniations of such courtyards should not be less than three, or more than four times the diameter of the columns.[12] But if the columns of the colonnaded courtyard must be built in the Doric manner, the modules should be developed in the way that I described in Book IV in reference to Doric columns, and the columns should be laid out in relation to those modules and the rules governing triglyphs [Plates 31–3; Fig. 23].

8. The length of dining rooms should be double their eventual width. The heights of all rooms which are to be rectangular must be calculated in such a way that the measurements of their length and breadth are added together: half should be subtracted from the total and the result allocated to the height. But if the *exedrae* or halls [*oeci*] are to be square, their heights should be taken up to one and a half times the breadth. Picture galleries, like *exedrae*, should be built with generous dimensions. Corinthian and tetrastyle halls, and those known as Egyptian halls, should have lengths and breadths calculated in the way described above for the modular systems of dining rooms, but they should be built more spaciously because columns are included.

9. The difference between Corinthian and Egyptian halls will be this: Corinthian halls have single rows of columns[13] placed either on a podium or on the pavement; above, they should have architraves and cornices of either woodwork or stucco; further, above the cornices there should be curved panelled ceilings describing part of a circle. But in Egyptian halls there are architraves over the columns and a wooden floor should be set on top of them and across to the surrounding walls; and

Plate 33. Façade of a Roman house. Vitruvius says nothing about the exteriors of Roman private houses. Giuliano da Sangallo, the first Renaissance architect to construct a façade for a country house which looked like a temple, made his version, at Poggio a Caiano, comprise an Ionic portico, tetrastyle *in antis* with spacing greater than areostyle, with a very flat pediment. Here Palladio produces a portico by transferring the components of an octastyle Corinthian temple plus pediment with spacing less than pycnostyle into another context, that of domestic housing, one of his most successful inventions, following an idea already expressed by Alberti.

above the floorboards, a pavement to create a walkway open to the sky around the hall. Then other columns, shorter by a quarter, should be placed on top of the architraves and on an axis with the lower columns. Above the architraves and their mouldings, they are decorated with panelled ceilings, and windows should be placed between the upper columns so that they resemble basilicas and not Corinthian dining rooms.

10. But in fact other types of hall are built which are not customary in Italy, and which the Greeks call Cyzicene. These are sited facing north and, particularly, overlooking gardens, and include folding doors in the middle. Also, these halls are long and wide enough for two dining-couches, facing each other and with room to walk around them, to be put in them; to right and left they have French windows with folding leaves providing views of the gardens through their openings for those on the couches. The height of these halls is one and a half times their width.

11. In these types of buildings all the principles of the modular systems which can be incorporated without interference from the sites should be put into practice, and the apertures will be located easily if they are not blocked by high walls; but if they should be obstructed by lack of space or other exigencies, then it will take acuteness and ingenuity to introduce additions and subtractions to the modular system so as to generate elegant solutions that do not diverge too much from ideal modularity.

CHAPTER IV

Orientations of Rooms

1. Now we will explain how the various types of rooms should be orientated correctly with respect to their function and the regions of the sky. Winter dining rooms and bathrooms should face the south-west because they need to receive the evening light; and also because the setting sun, shining splendidly in their direction and radiating heat, makes that orientation

warmer in the evening. Bedrooms and libraries should face east since their function requires morning light, and again, so that the books in libraries will not rot. For in the case of libraries facing south and west, books are damaged by bookworm and the damp because the arrival of moist winds generates and nourishes them and destroys the books with mould by diffusing damp air over them.

2. Spring and autumn dining rooms should face east. For the trajectory of the sun as it passes to the west opposite dining rooms exposed to the impact of its light makes their temperatures comfortable at the time when they are usually in use. Summer dining rooms should face north, since that orientation is not, like the others, made sweltering by the heat during the solstice but, being turned away from the course of the sun, is always cool and provides an environment that is healthy and pleasant to work in. It is no less important that picture galleries, embroiderers' workshops and painters' studios [should face north] so that the colours incorporated in their works will remain invariable in quality thanks to the consistency of the light.

CHAPTER V

Social Status and Houses

1. Once rooms have been orientated in this way, we must then pay attention to the principles by which spaces in private apartments reserved for heads of families and those shared with visitors should be built. Not everyone is allowed the privilege of entering reserved spaces, such as bedrooms, dining rooms, bathrooms and others used in the same way, unless they are invited. But communal spaces are those which any member of the public can enter by right even when not invited, for example, vestibules, courtyards with and without porticoes, and other spaces with similar functions. Therefore, for those who are only moderately well off, magnificent vestibules, *tabulina* and *atria*

are unnecessary because they fulfil their social obligations by canvassing others for patronage rather than by being canvassed.

2. However, those who do business in country produce must build stables and shops in their entrance courts, and, in their farm-buildings, cellars, granaries, storerooms and other such spaces more suited to the preservation of produce than to creating an elegant effect. Again, more comfortable and impressive rooms safe from robbery must be built for bankers and tax-collectors; but for lawyers and orators they must be more elegant and spacious for the reception of groups of people; and for important dignitaries who hold high office and magistracies and are obliged to serve the state, lofty and regal vestibules, grand *atria* and colonnaded courtyards should be built, as well as plantations of trees and broad avenues finished off so as to match their social standing; not to mention libraries, picture galleries and basilicas prepared with a splendour consonant with that of great public buildings, since public councils as well as private trials and arbitrations are often held in their houses.

3. If, therefore, houses are to be planned with these principles to suit different classes of persons in line with what I wrote on appropriateness in Book I, there will be nothing to criticize, for they will be planned conveniently and impeccably in every respect. But with regard to this subject, the same principles will apply not only to buildings in the city, but also to those in the country, except that in the city, *atria* are usually next to the front doors, while in the country, colonnaded courtyards in houses imitating those in the city come first, and then, consequently, *atria* surrounded by paved porticoes looking out onto gymnasia and walks.

I have now set out summarily as best I could the principles underlying urban buildings. Now I will speak of the plans of country buildings, so that they may be adequate to their function, and of the principles by which their sites should be organized.

CHAPTER VI

The Farmhouse

1. First of all the orientations of sites should be examined with respect to their healthiness, as I wrote in Book I on the siting of city-walls, and farmhouses [*villae*] should be located accordingly. Their dimensions should be appropriate to the size of the farm and the quantity of produce. The farmyards and their dimensions should be determined in relation to the number of cattle and yoke of oxen that will have to be accommodated there. The kitchen should be located in the warmest spot in the yard, and the cattle-sheds should be adjacent to it: their stalls should face the fire and the eastern region of the sky, because the coats of cattle facing light and fire do not become shaggy. Again, farmers, who are certainly not ignorant[14] of the regions of the sky, do not think that cattle should face any other part of it except the east.

2. But the sheds for oxen should not be less than ten or more than fifteen feet wide, and long enough to allow each yoke to occupy not less than seven feet. The baths too should be connected to the kitchen, for in this way washing facilities for the farmworkers will not be far away. The oil-press should also be next to the kitchen since in this way it will be well placed to deal with the olive harvest; so too the wine-cellar with windows receiving light from the north, for if it faces in some other direction so that the sun could heat it up, the wine in such a cellar will be weak because it has been spoilt by the heat.

3. But the oil-cellar is to be located so that it receives light from the south and warm regions because oil must not be chilled but kept fluid by warm conditions. Its dimensions should be worked out bearing in mind the amount of the harvest and the number of jars, each of which should, on average, occupy four feet since they each contain a *culleus*.[15] The [room for] the oil-press should not be less than forty feet long if the press is operated by pressure applied by levers and a beam, rather than by screws, so that it will provide enough room for the man operating the lever. Its width should not be less than sixteen feet, for this will allow the men working at full swing to go round freely and easily. But if the place has to accommodate two presses, twenty-four feet should be allowed for the width.

4. The folds for sheep and goats should be made large enough for each animal to have a space no less than four and a half feet and no more than six. The granaries should be raised[16] and face north or north-east, because this way the grain will not be able to heat up quickly, but, cooled by the air, will keep for a long time. In fact different orientations generate weevils and other small insects that habitually damage grain. The warmest places in the farmhouse should be allocated to the stables so long as they do not face the kitchen fire; for when draught-animals are stabled near a fire, their coats become rough.[17]

5. Again, stalls located outside the kitchen and in the open facing east are not without their uses; when oxen are led to them during the wintry periods of the year under a clear sky in the morning, they become more healthy looking as they take their fodder in the sun. It seems that barns for grain, stores for hay and spelt, and bakeries should be built away from the farmhouse, so that the farm-buildings will be less at risk from fire. If more refined spaces are to be created in farmhouses, they should be organized in conformity with the modular system established for urban buildings described above so that they are built without interfering with the functioning of the farm.

6. We must take care that all the buildings are well lit; clearly this is much easier to arrange in the case of buildings attached to farms because no neighbour's wall can get in the way; but in the city obstructions caused either by the heights of party walls or restrictions of site create darkness. With regard to this problem we must carry out the following test: on the side from which it is appropriate to take the light, a piece of string should be stretched from the top of the wall which seems to obstruct the light to the location where it would be convenient to admit it, and, if a wide area of the open sky can be seen when one looks up from the line [of the piece of string], the light there will not be obstructed.

7. But if beams, lintels[18] or wooden floors get in the way, the opening should be made higher up and the light admitted in this way. As a rule, we must organize things so that from whichever sides the sky can be seen, spaces for windows should be left there so that buildings will be well lit.

While the necessity for illumination is at a maximum in dining rooms and other rooms, it is also essential in passages, ramps and stairs because people moving in opposite directions carrying loads often run into each other in such places.

I have explained as best I could the layouts of our native buildings so that they will be comprehensible to builders. Now I will explain rapidly how buildings should be laid out following Greek traditions so that they will not be unfamiliar.

CHAPTER VII

The Greek House

1. The Greeks do not build *atria* since they have no use for them; but they provide narrow passages for those entering from the front door, with stables on one side and the doormen's quarters on the other, immediately followed by the inner doors; the space between the two doors is called the θυρωρών [*thyroron*; doormen's quarters] in Greek. Then one enters the colonnaded courtyard, which has porticoes on three sides, and in that facing south it has two *antae* some distance apart on which beams [i.e. an architrave] are carried; two-thirds of the distance between the piers is allocated to the depth of the interior of the portico. Some call this space the *prostas* [space in front], others the *pastas* [space at the side].

2. The large halls in which matriarchs hold their work sessions with the female wool-workers are built in the spaces of the interior. But to right and left of the *prostas* the bedrooms are located, one of which is called the master bedroom [*thalamos*], the other, the daughters' room [?*amphithalamos*]. In the porticoes around the courtyard are the dining rooms for

daily use, bedrooms and small rooms for the slaves. This part of the house is called the women's quarters [*gynaeconitis*].

3. Connected to these are larger apartments with more splendid courtyards and four porticoes of equal height, unless the one facing south is built with taller columns, in which case it is called Rhodian. Such quarters have impressive vestibules and their own imposing front doors; the porticoes of the colonnaded courtyards are decorated with plaster, stucco and coffered wooden ceilings; in the porticoes facing north are the Cyzicene dining rooms and picture galleries; at the east, the libraries; at the west, the *exedrae*; and facing south, by contrast, square halls so large that there would easily be room in each to set out four dining-couches and provide food and games.

4. Men's dinner parties take place in these large halls; for the practice of matriarchs reclining [to eat dinner] was never established as a Greek custom. These colonnaded courtyards are called the men's quarters [*andronitides*] of the house because they spend their time there without interruptions from the women. Moreover, small apartments are built to right and left which have their own doors to the street, as well as dining rooms and comfortable bedrooms, so that when guests arrive they may be put up in the guests' quarters rather than in the colonnaded courtyards. For when the Greeks were more refined and prosperous, they prepared dining rooms, bedrooms and stores with provisions for guests on their arrival; on the first day they would invite them to dinner, and on the next, they would send them chickens, eggs, vegetables, fruit and other country produce. This is why painters who represent the presents sent to guests in their pictures called them *xenia* [gifts for guests]. Consequently the heads of families in such quarters did not feel that they were in fact away from home since they enjoyed private generosity in these guest-rooms.

5. And between the two colonnaded courtyards and the guest-rooms are passages called *mesauloe* because they are placed midway between two courtyards; yet our people called them *andrones*. This, however, is really curious because here Greek and Latin disagree: the Greeks call the halls in which men's dinner parties are usually held ἀνδρῶνες [*andrones*], since women do not go into them. There are similar examples, such as *xystus*, *prothyrum*, *telamones* and so on. In fact, in Greek terminology, a ξυστός [*xystos*] is a very broad portico in which athletes exercise in the winter: but our people call uncovered walks *xysta*, while the Greeks call them παραδρομίδες [*paradromides*; running tracks]. Again, in Greek, the term πρόθυρα [*prothyra*] refers to vestibules immediately in front of the doors of the house; instead, the spaces we call *prothyra* are called διάθυρα [*diathyra*] [between doors] by the Greeks.

6. Again, if statues of male figures support mutules or cornices, our people call them *telamones*, and the whys and wherefores of that term cannot be found in history books; but the Greeks, in fact, call them ἄτλαντες [*atlantes*]. For Atlas is portrayed in history as holding up the firmament because he was the first to ensure, by virtue of his powerful intellect and ingenuity, that men learnt about the trajectory of the sun and the principles by which the moon and all the planets rotate. In recognition of this benefit to mankind, he is represented by painters and sculptors holding up the firmament, and his daughters, the Atlantides, whom we call the Vergiliae, but the Greeks call Πλειάδες [*Pleiades*], were consecrated as stars in the heavens.

7. But I have not drawn attention to these differences in usage in order to change customary nomenclature or common language, but because I decided they should be aired so that they would not remain unfamiliar to students of the language.

I have now explained the traditions by which buildings conform to Italian custom and the practices of the Greeks and have set down in detail the proportions of each type in relation to their modular systems. Therefore, since I have already written on their beauty and appropriateness, let us now discuss, with respect to their durability, how such constructions can be built to last for a very long time.

CHAPTER VIII
Foundations and Substructures

1. If the foundations of buildings set at ground level are built in the way I described in the previous books in relation to city-walls and theatres, they will certainly last for a very long time. But if underground spaces [*hypogea*] and vaulted chambers are to be constructed, their foundations should be built thicker than the masonry walls that will be constructed in the buildings above them; and the walls, piers and columns of the latter should be located vertically on the central axes of the substructures so that they rest on solid support; for if the loads of walls or columns were to rest on unsupported spans they cannot remain stable for long.

2. Again, if vertical supports are inserted between lintels and sills in the manner of piers or *antae*,[19] the walls will not suffer damage since lintels and beams sag in the middle and destroy the walls by undermining them when they are loaded with masonry. But when vertical supports are inserted and wedged in place under them, they prevent the beams from settling or being damaged.

3. We must also make sure that arches can take the load of the walls above by means of their sets of voussoirs with joints converging on a central point. For when arches with voussoirs are sprung from outside beam-ends or the ends of lintels, first, the wood will not sag when relieved of the load, and second, if some structural failure starts occurring with the passage of time, the wood can easily be replaced without resorting to the construction of shoring.

4. Similarly, in the case of buildings supported on piers of which the vaults are closed with sets of voussoirs and joints converging on a centre, their outer piers must be made wider so that they have the rigidity to resist when the voussoirs, compressed by the weight of the walls, discharge their load towards the central point through the joints, and push the imposts of the pilasters outwards. If, therefore, the angle-piers are more massive, they will guarantee the stability of buildings by holding the voussoirs together.

5. Once we appreciate that great pains must be taken with regard to this subject, we must be equally alert to the fact that all masonry walls must be built perpendicular and should not lean forward at any point. The greatest care must be taken with substructures because the mass of earth piled up in them usually causes great damage. For the earth cannot invariably weigh the same as it usually does during the summer: for in the winter it increases in weight and mass by absorbing large quantities of water from showers and bursts and pushes out sections of the container-walls.

6. So, to remedy this defect, one must build in such a way that first, the thickness of the masonry should be determined in relation to the mass of the earth infill; second, buttresses [*anterides*] or spur-walls [*erismae*] should be built on the outside faces of the substructure simultaneously with the walls; the distance between the buttresses should equal the height of the substructure when completed, and the buttresses themselves should be as thick as the walls of the substructure. Again, at the bottom, the buttress should project by a distance equal to the thickness established for the substructure, and then it should taper upwards gradually so that at the top its projection is equal to the thickness of the building above.

7. Furthermore, to contain the mass of earth, saw-shaped structures joined to the wall of the substructure should be built on the inside in such a way that the projection of the individual teeth from the wall equals the height of the future substructure, and the teeth themselves should be as thick as the masonry wall. Again, points should be marked out on [the walls on] either side of the corners of the substructure after measuring out distances away from the right angle equal to the height of the substructure, and a diagonal masonry wall should be built between these points: and from the centre of the diagonal wall, another masonry wall should be connected to the corner of the substructure. In this way, the teeth and the diagonal masonry would prevent the thrust of the mass of earth to be contained

from exerting maximum pressure on the wall, but would dissipate it [Fig. 24].

8. I have now explained how buildings should be constructed without defects and how to take precautions against such defects when they begin to appear. Our preoccupation with replacing tiles, beams or common rafters is not as great as it is with regard to foundations, because even if the former are defective, they can easily be replaced. I have accordingly explained the methods by which those elements which are not regarded as structurally indispensable can be made stable and how they should be built.

9. It is not within the competence of the architect to determine what kinds of materials should be used, because not all kinds of materials are found naturally in all places, as has been shown in Book I. Furthermore, it is up to the patron whether he wishes to build with fired brick, quarry-stone or squared stone. And so the approval awarded to all buildings falls under three headings, that is, the skilfulness of the workmanship, its magnificence and its projection. When a building is seen to have been completed magnificently, the expenditure derived from the resources of the patron will be applauded; when it has been carried out skilfully, the precision of the master-mason will be commended; but if the building has an elegant authority based on its proportions and modularity, then the glory will be the architect's.

10. These things turn out well if the architect allows himself to accept advice from both workmen and laymen. For all men, not just architects, are capable of appreciating quality; but there is a difference between laymen and architects in that the former cannot know what the building will be like unless he has seen it completed; while the architect knows perfectly well what it will be like with respect to beauty, function and appropriateness from the instant he conceives it in his mind, and before he begins it.

I have written at length and as clearly as I was able about the rules I thought useful for private houses and how they should be built. In the following book I will discuss pavements and the revetments of walls[20] in buildings so that they may remain elegant and without defects over the long term.

BOOK VII

Introduction

1. Our ancestors established the intelligent and useful practice of transmitting their thoughts to future generations in the form of bodies of notes so that they would not be lost but, growing generation by generation once they had been published as books, they would gradually arrive at the highest level of scientific development in the course of time. So for this we owe them no half-hearted thanks but infinite gratitude, because they did not jealously pass over these matters in silence but took great care to hand on to posterity their insights of all kinds in written form.

2. For if they had not done so, we would not have known about the exploits at Troy, nor what Thales, Democritus, Anaxagoras, Xenophanes and the other natural philosophers thought about nature, nor what rules for human social behaviour Socrates, Plato, Aristotle, Zeno, Epicurus and other philosophers had defined; nor would we know of the achievements and motivation of Croesus, Alexander, Darius and the other kings, had our ancestors not gathered all their teachings and transmitted them in their publications so that posterity would remember them.

3. So just as these men deserve our gratitude, others who, by contrast, steal their work and pass it off as their own must be criticized; and those who do not rely on their own ideas when they write but who, driven by jealousy, pillage other people's work and boast about their handiwork not only deserve censure but should actually be condemned for their immoral conduct.

In fact it is recorded that these malpractices did not go without very strict punishment by the ancients; and it is not irrelevant to record what the results of these cases were as they have been handed down to us.

4. After the Attalid kings, enthused with the great delights of high culture, had founded an excellent library at Pergamum for the pleasure of the public, then Ptolemy too, driven by limitless rivalry and boundless ambition, made every effort to establish a library at Alexandria along the same lines and with just as much effort.[1] But when he had finished it with the utmost care, he still felt that this was not enough unless he could ensure its expansion by inseminating it with shoots, as it were. And so he dedicated games to the Muses and Apollo and established prizes and honours for the winners of literary contests, just like those for athletes.

5. When the appropriate arrangements had been made and the contests were imminent, they had to select literary experts to adjudicate the contests. When the king had six judges chosen from the city at his disposal, but could not readily find a competent seventh judge, he consulted those who ran the library and asked them if they knew of anyone qualified for the task. Then they told him that there was a certain Aristophanes who was systematically reading through all the books day after day with the greatest commitment and assiduity.[2] And so when everybody had assembled for the contest and the seats reserved for the judges had been allocated, Aristophanes, summoned with the others, sat in the place set aside for him.

6. The poets were brought on to compete first, and while they recited their compositions, the whole of the audience made it clear to the judges what decision they should make. So when each of the judges in turn was asked his opinion, the six declared themselves agreed and gave first prize to the poet whom they noticed had pleased the crowd most, and second prize to the one who came next. But when he was asked for his vote, Aristophanes insisted that the poet who had pleased the crowd least should be declared the winner.

7. Since the king and everybody else were vehemently indignant, Aristophanes got up and was granted permission to speak

when he asked for it. So silence fell: he explained that only one of the contestants was a real poet, that the others had recited other people's work and that it was the duty of the judges to assess original compositions, not work lifted from others. The people were astonished and the king sceptical, so Aristophanes, relying only on his memory, brought a vast number of volumes from the appropriate bookcases, and by comparing them with what had been recited, forced the fraudsters to confess that they were thieves. So the king ordered proceedings against them for plagiarism and, once they were found guilty, sent them off in disgrace. But he showered Aristophanes with more than generous rewards and put him in charge of the library.

8. A few years later Zoilus, who adopted a surname in order that he would be known as Homeromastix [Scourge of Homer], came from Macedonia to Alexandria and read the writings he had composed attacking the *Iliad* and the *Odyssey* to the king.[3] But Ptolemy, when he realized that the father of poets and the founder of all literature was being assailed in his absence, and that his works, which were admired by all peoples, were being denigrated by Zoilus, was so indignant that he made absolutely no reply. Zoilus, however, after he had stayed in the kingdom for some time, was afflicted by poverty and supplicated the king to give him some financial aid.

9. But the king is said to have replied that Homer, who had died a thousand years before, had in all that time provided a living for many thousands of men; so equally, anyone who boasted a greater genius than his should be capable of maintaining not just one person but many others as well. In brief, Zoilus' death as a convicted parricide is recorded in various ways, since some have written that he was crucified by Philadelphus, others that he was stoned to death on Chios and yet others that he was thrown alive onto a funeral pyre at Smyrna; whichever form of death he suffered, it was the punishment he richly deserved: for those who put in the dock people who cannot answer back in person to explain the sense of what they wrote do not deserve any other punishment.

10. But I, Caesar, do not present this comprehensive treatise after changing the titles of other people's books and substituting

my own name, nor have I set about earning approval by running down anyone else's ideas; rather I express infinite gratitude to all those writers who have provided us with rich materials gathered together from ancient times with extraordinary intellectual skill in their different fields, so that we, by adapting them to our own particular needs as though drawing water from springs, have the resources with which to write more rapidly and easily, and, relying on such authors, have the confidence to prepare new and instructive manuals.

11. Accordingly, because I realized that the intellectual foundations they had established were suitable for my own project, I started with them and began to go further. First: when Aeschylus was producing a tragedy in Athens, Agatharcus created the scenery and left a commentary on it.[4] Learning from him, Democritus and Anaxagoras wrote on the same subject, showing how, once a particular spot has been fixed as the central point, the lines of the design itself should match, according to natural proportion, the gaze of the spectator and the extension of the rays, so that beginning with some indistinct object, precise images would reproduce the correct appearance of buildings painted on the scenery; and some of the objects represented on vertical, flat surfaces would seem to recede and others to project [Fig. 25].[5] **12.** Afterwards Silenus published a volume on the modularity of Doric buildings; Theodorus published one on the Doric Temple of Juno on Samos; Chersiphron and Metagenes on the Ionic Temple of Diana at Ephesus; Pytheos on the Ionic Temple of Minerva at Priene; again, Ictinos and Carpion, on the Doric Temple of Minerva on the Acropolis of Athens; Theodorus the Phocian on the round building at Delphi; Philo on the modular systems of temples and on the naval arsenal he built at the port of Piraeus; Hermogenes on the pseudodipteral Ionic Temple of Diana at Magnesia and on the monopteral Temple of Father Liber at Teos; Arcesius on Corinthian modularity and the Ionic Temple of Aesculapius at Tralles, which, in fact, he is said to have built with his own hands; on the Mausoleum [at Halicarnassus], Satyrus and Pytheos, on whom fortune bestowed the greatest and most extraordinary rewards.

13. For the artistic talents of these men, who dedicated exceptional work to their projects, are judged to be eternally fresh and worthy of the utmost praise for ever. Single artists in competition undertook the decoration and appraisal[6] of each of the façades – Leochares, Bryaxis, Scopas and Praxiteles and, as some think, Timotheus; and the outstanding quality of their art made the fame of the building so great that it was included in the Seven Wonders of the World.

14. Again, many less well-known authors have written treatises on the rules of modularity, such as Nexaris, Theocydes, Demophilos, Pollis, Leonidas, Silanion, Melampus, Sarnacus and Euphranor; others again, such as Diades, Archytas, Archimedes, Ctesibius, Nymphodorus, Philo of Byzantium, Diphilos, Democles, Charias, Polyidos, Pyrrhos and Agesistratos have written on the construction of machines.

I have gathered together in one systematic work what I regarded as useful in their treatises for the present theme, and I did this above all because I realized that many books on the subject had been published by the Greeks but very few indeed by Romans. In fact Fuficius, oddly enough, was the first [Roman] to decide to publish a book on this subject; also Terentius Varro has one book on architecture in his work on 'The Nine Disciplines', and Publius Septimius has two.[7]

15. But up till now nobody else seems to have devoted their energies to furthering this category of writing, even though there were fellow-citizens long ago who were great architects and could certainly have composed equally elegant treatises on it. At Athens, for example, the architects Antistates, Callaeschros, Antimachides and Pormos laid the foundations when Pisistratus was building the Temple of Olympian Jupiter, but after his death they abandoned what they had begun because of political upheavals. So it was that about four hundred years later, when King Antiochus promised to bear the cost of the project, Cossutius, a Roman citizen, splendidly designed the immense *cella*, the placement of the double colonnade around it and the distribution of the architraves and the other mouldings all correctly located with extraordinary technical and theoretical skill according to the modular system. This work is

appreciated for its magnificence not only by the public but also by experts.[8]

16. For there are four places where the structures of temples have been beautifully carried out with work in marble, and they are very famous precisely because of this. Their great qualities, and the technical brilliance with which they were designed, excite admiration for the cult of the gods. First, the Temple of Diana at Ephesus in the Ionic order was begun by Chersiphron of Cnossus and his son Metagenes, and is said to have been completed afterwards by Demetrius, a slave of Diana herself, and by Paeonius of Ephesus. At Miletus, the same Paeonius, with Daphnis of Miletus, undertook the construction of the Temple of Apollo, again using Ionic modularity. At Eleusis, Ictinos roofed over the immense *cella* of the Temple of Ceres and Proserpina in Doric style but left it without external columns to allow more space for the sacrificial rituals.

17. But later, when Demetrius of Phalerum held power in Athens, Philo added columns at the front of the temple and made it prostyle. So, increasing the size of the vestibule,[9] he created more space for the initiates and made the building much more imposing. At Athens, as I have already said, Cossutius is recorded as having undertaken the design of the Olympeion using a system of larger modules following Corinthian commensurability and proportions, but no written account by him has been found. But it would be desirable to have writings on these subjects not just from Cossutius; there is also Gaius Mucius, who, relying on his great technical skill, perfected the modularity of the *cella*, the columns and the architraves of Marius' Temple of Honour and Valour according to the accepted rules of the art. Had this building been made of marble, so that its technical refinement matched the splendour it derived from its magnificence and great expense, it would be regarded as one of the foremost of the great masterpieces.

18. Since, then, we find that in the past our architects were no less distinguished than the Greeks and that there were quite a number of them within living memory, but that few have published the principles of their art, I decided not to stay silent but to discuss each subject methodically in separate books.

Accordingly, since I have dealt in detail with the conventions that apply to private houses in Book VI, I will deal in this one, Book VII, with the methods of ensuring the beauty and durability of pavements and of the revetments of walls.

CHAPTER I
Pavements

1. I will begin with the concrete floor [*ruderatio*], which is the most important of the surfaces to be applied, to ensure that the technique for guaranteeing its solidity with the greatest care and skill may be understood. If the concrete floor is to be laid at ground level, one must find out whether the earth is consistently compact, and if it is, it should be levelled off and concrete with a rubble bedding [*rudus cum statumine*] applied. If, however, all or part of the site is made of infill, it must be compacted down with great care by ramming. But with regard to the wooden upper floors, we must be very careful that walls that do not reach up to the top[10] should not be built right up under pavements; otherwise, walls should include a space above which the floor is suspended. If in fact a complete wall is built right up under the floor, the unyielding solidity of its structure produces cracks in the pavements to left and right of it when the floors dry out and settle as they warp.

2. We must also ensure that planks of winter oak are not mixed up with those of common oak, because as soon as common oak planks absorb moisture they warp and make cracks in pavements. But if no winter oak is available, and we have no option, for want of anything else, it seems sensible to make do with common oak beams cut very thin, because the thinner they are, the easier they are to hold in place once they have been nailed down. Then planks should be fixed at the ends of each joist with a couple of nails so that the joists cannot force up their corners at either end by warping. With regard to turkey oak, beech or ash, none of them can last for very long.

Once a wooden floor has been laid, it should be covered with fern, if there is any, or if not, with straw, so that the wood is protected from being damaged by the lime.

3. Then spread on top a layer of hardcore comprising stones no smaller than can fill the hand. Once the layer of small stones has been laid, the aggregate [*rudus*], if new, should be mixed with three parts to one of lime; if it is reused rubble, five parts should match two of lime. Then the concrete should be spread and compacted as tightly as possible by being flattened down repeatedly with wooden rams by the gang of workmen that has been called in; once it has been flattened it should not be less than three-quarters of a foot deep. Above this, the screed [*nucleus*], comprising three parts of crushed terracotta mixed with one of lime, should be laid, forming a layer no less than six digits thick. On top of the screed, the pavement, whether of strips of stone or mosaic, should be laid out with set-square and level.

4. After the pavements have been laid and the inclines incorporated, they should be rubbed down in such a way that, if they consist of pieces of cut stone, no part of the lozenges, triangles, squares or hexagons should stick up, but the joints should be fitted together so that they all lie on the same plane. If the pavement is built with mosaic, the separate pieces should be rubbed down so that all the corners are level with each other; for if they are not all equally horizontal, the rubbing down will not be as good as it should be. Again, herringbone pavements made of fired brick, like that from Tivoli, must be laid down carefully, so that they do not have gaps or bumps, but are flat and rubbed down to a level. When the rubbing down has been finished by smoothing and polishing, marble-powder should be sprinkled on it and coats of lime and sand spread above that.

5. In the open air specially adapted pavements must be constructed because their floorboards swell with moisture or shrink with the dryness, or they warp and settle, damaging the pavements when they move; moreover, ice and frost do not leave them unscathed. And so, if the situation demands it, we should make them like this to minimize damage to them as much as possible; once the wooden floor has been finished, another one should be laid on top of it at right angles and fixed with nails so as to provide the floor with a double layer of protection. Then new aggregate should be mixed with crushed terracotta in the ratio of two-thirds to one, and two parts of lime added to five of the mixture in the mortar-trough.

6. After the rubble bedding [*statuminatio*] has been laid, the concrete should be spread over it to no less than a foot thick after it has been pounded down. Then, after spreading on the screed, as described above, a pavement of large tiles should be laid down with sides cut to about two digits, with an inclination of two digits in every ten feet; if it is properly adjusted and correctly sanded down, it will be safe from all damage. And so that the mortar between the joints will not suffer from the effects of frost it should be soaked every year before winter with olive-oil lees, and after being treated like this will resist the infiltration of the icy cold of the frost.

7. But if it seems necessary to take further precautions, two-foot tiles, fitted together and provided with little channels of one finger's depth cut in all the faces of the joints, should be laid on a bed of mortar over the concrete; once the channels have been joined together they should be filled with lime mixed with olive oil, and the joints, once pressed together, should be rubbed over. In this way the lime sticking in the channels will prevent water or anything else passing through the joints by hardening and solidifying into a mass. So when this layer [of tiles] is finished, the screed should be spread over it and worked flat by being beaten with wooden rams. Above this the pavements comprising large tiles or bricks in herringbone pattern should be laid, with the inclinations described above; if they are built like this they will not deteriorate quickly.

CHAPTER II

Lime

1. Now that we have left the subject of the treatment of pavements, it is time to discuss stucco. This will be made correctly if lumps of the highest quality lime are slaked a long time before they are needed, so that even if some lump has not been thoroughly baked in the kiln, it will still become uniformly cooked since it has been forced by the liquid to ferment during the long period of slaking. For when fresh lime which has not been thoroughly slaked is used, it will create blisters when applied, since it contains uncooked nodules hidden from view. Once such nodules become completely slaked when they have been applied to the wall, they break up and destroy the finish of the plasterwork.

2. But when the procedure for slaking has been followed correctly and the material prepared with great care, one should take a trowel and chop through the slaked lime in the mortar in the same way that wood is chopped. If the trowel encounters lumps, the lime is not mixed properly; when the iron is pulled out dry and clean, that will show that the lime lacks consistency and is dry. But when the lime is dense and correctly slaked, it will stick like glue to the tool, which will show that it has the right consistency. Then, once the scaffolding has been put up, the construction of the vaults of the rooms should be started, unless they are to be decorated with coffered ceilings.

CHAPTER III

Vaults and Revetments

1. So when one needs to know about the technique of building vaulted ceilings, one should proceed like this: common rafters should be arranged in parallel no more than two feet apart: preferably these should be of cypress wood, since fir deteriorates

rapidly with rot or age. When the rafters have been formed into a curve, they should be secured with wooden ties and numerous iron nails to the joists of the floors above, or, if that is the case, to the roof; these ties should be made from a type of wood that is not damaged by rot, age or moisture, such as box, juniper, olive, oak, cypress or others like them, but not from the common oak, since it warps and creates cracks in the structures in which it is used.

2. Once the rafters have been put in place, then Greek reeds, flattened into the required shape, are to be bound to them with string made of Spanish broom. Then mortar made of a mixture of lime and sand should be spread immediately above the top of the vault to block any drops that might fall from the wooden floors or roof. But if there is no supply of Greek reed, then thinner reeds should be collected from the marshes and made into bundles of the right length and uniform thickness bound with a rough thread, ensuring that no more than two feet separate the knots of the bindings: and, as described above, these bundles should be tied with string to the rafters and wooden pegs driven into them. All other operations should be carried out as described above.

3. Once the vaults have been put in place and woven with reeds, the soffit should be rough-plastered, then spread with sand-mortar, after which it should be finished off with chalk or powdered marble.

When the vaults have been given their final coat, cornices should be set in place under them: clearly these should be as slight and delicate as possible, for when they are large, their weight pulls them down and they cannot support themselves. With respect to cornices, gypsum is the very last thing that should be mixed in; rather, powdered marble-mortar should be spread evenly over them, lest the gypsum dry first and prevent the rest from drying out uniformly. Again, we must avoid the ancient methods of constructing vaults, because the projecting surfaces of these cornices are dangerous on account of their great weight.

4. The mouldings of some cornices are plain, others are decorated with reliefs. But in rooms where a fire or a large number of lamps are to be installed, the cornices should be made without decoration so that they can be cleaned more easily. In summer-rooms and *exedrae* where smoke or soot can hardly damage them at all, the cornices should be decorated with reliefs. For stucco always absorbs smoke, not only from the building in which it is located but also from others, because of its brilliant whiteness.

5. Once the cornices have been deployed, the walls should be rendered with the coarsest possible plaster, and then, while it dries, carefully aligned layers of sand-mortar should be applied and laid out precisely so that the lengths correspond to rule and line, the heights to the plumb-bob and the angles to the set-square, for in this way the revetment will look perfect when it comes to painting it. As it dries, a second and a third layer should be applied. So the thicker the layer of sand-mortar, the more resistant the hardness of the revetment will be.

6. When no fewer than three layers of sand-mortar have been laid on in addition to the coarse rendering, then layers of coarse powdered marble should be applied, and the mortar mixed so that when it is applied it does not stick to the trowel, but the tool comes out of the trough clean. After the coarse marble-mortar has been spread on and is drying out, another, thinner coat should be applied; when it has been smoothed down and well rubbed over, another, finer coat should be applied. In this way the walls, reinforced by three layers of sand-mortar and powdered marble-mortar, will not be susceptible to cracks or any other damage. 7. But once the durability of such revetments has been ensured by being worked over with plasterer's floats and polished with bright and stable marble-powder, they will be brilliantly luminous when the colours have been applied with the final surface.

With regard to the colours, they do not fade when they have been applied carefully to the damp plaster surface, but stay fixed permanently, because the lime becomes porous and loses its consistency once the moisture has been cooked out of it in the kiln, and is compelled by its aridity to absorb anything that encounters it; as it solidifies to a mass by mixing with the components of other substances or elements brought into contact with it, it dries, and, whatever the ingredients of which it is composed, is reconstituted in such a way that it seems to have unique and particular qualities of its own type.

8. So revetments which have been made correctly do not become rough over the course of time, nor do the colours fade when they are wiped over, unless the latter have been laid on carelessly and on a dry surface. And so when revetments have been applied to walls as described above, they will be strong, brilliant and capable of lasting a very long time. But when only one coat of sand-mortar and one of powdered marble-mortar have been spread on, the thin layer cracks readily since it is less robust and will not take on the appropriate shine when it is polished because of its lack of thickness. **9.** For, just as a silver mirror made of a thin lamina reflects images indistinctly and weakly, but one that is made more robustly and can take much vigorous polishing reflects images which are splendid to the eye and clearly distinguishable by observers, so too revetments laid on with a thin coat of mortar are not only prone to cracks but also fade rapidly; but dense and deep revetments prepared with a solid foundation of sand- and marble-mortar not only remain shiny after repeated polishing but also reflect back clear images from their surface to observers.

10. In fact, Greek plasterers not only employ these methods to make their work durable, but they also set up a mortar-trough, mix lime and sand in it, bring on a gang of workmen, and beat the mortar with wooden rams; then they use it after it has been vigorously worked. And in fact some workers cut sheets of plaster out of old walls and use them as wall-panels, and so the revetments, divided up by the panels and the mirror-like polished surfaces, have projecting borders all around them.

11. But if revetments are to be applied to half-timbered walls, cracks will certainly appear along the uprights and cross-bars because they necessarily absorb moisture when they are coated with clay-mortar and cause cracks in the revetments when they dry and shrink. This is the procedure that avoids the problem: after the whole wall has been coated with clay-mortar, then reeds should be fixed all along it with broad-headed nails; then after another coat of clay-mortar has been added, if the first layer of reeds has been fixed in place horizontally, the second layer should be fixed in place vertically, and the sand-mortar, marble-mortar and all the revetment should be spread on as described above. In this way the double layers of reeds fixed in the walls by the transverse reeds will prevent the walls from flaking or cracking.

CHAPTER IV

Revetments in Damp Locations and in Dining Rooms

1. I have discussed the methods by which revetments should be made in dry locations; now I will explain how plasterwork is to be made in damp locations so that it can last without deteriorating. First, in the case of rooms on the ground floor, a rough rendering [of mortar mixed with] powdered fired brick instead of sand should be applied for about three feet up from the pavement and smoothed off to prevent those areas of the plasterwork from being damaged by moisture. But if some wall suffers from damp all over, another thin wall should be built[11] a little way from it [on the inside], at a distance suited to the circumstances; and below the level of the room a channel with vents leading out into the open air should be led between the two walls. Similarly, when the wall is completed up to the top, air-holes should be left in place, for if the moisture has no means of escape through the vents at the bottom and top, it will certainly spread itself just as much through the new masonry wall. This done, the wall should be rendered with a rough layer of mortar made with powdered fired brick, smoothed off and then finished with plaster.

2. But if there is no room for a second masonry wall, channels with vents leading to the open air should be built. Then, on one side, two-foot tiles should be laid above the edge of the channel, and below, on the other side [of the channel], pillars made of eight-inch bricks should be built on which the corners of two tiles can rest in such a way that the pillars[12] stand away from the wall no more than a hand's breadth. Then above, tiles with corner hooks [*tegulae hamatae*] should be fastened vertically from the bottom of the wall to the top and their inner faces carefully covered with pitch so that they will repel moisture. Again, the walls should have air-holes at the bottom and above the vault [Fig. 26].

3. Then the tiles should be whitewashed with lime mixed with water so that they will not reject the rendering of powdered fired brick; because of the dryness produced by burning in the kiln, the tiles cannot grip or hold the rendering unless lime, which sticks the two materials together and forces them to cohere, has been applied. After the rendering has been applied, a mortar of powdered fired brick should be spread on instead of sand-mortar, and everything else carried out as has been described above in the discussion of the procedures for applying plaster revetments.

4. But the decorations to be applied as the final surface must observe their own principles of propriety so that they present an appearance which is both suitable to their locales and does not contravene the distinctions between the genres of painting. Neither paintings on elevated subjects [*megalographia*] nor delicately moulded cornices are practicable for the vaults in winter dining rooms on account of their structure, since they are ruined by smoke from fires and the soot constantly given off by lamps. In fact, above the dados in these rooms black varnished and polished panels should be applied, with borders [*cunei*] of yellow ochre or vermilion inserted between them. Once the vault has been left undecorated and polished, it will not be at all displeasing to consider the very inexpensive but practical method of preparing Greek pavements for use in winter rooms.

5. For one digs below the level of the dining room to a depth of about two feet, and, after the ground has been flattened down, a pavement of concrete or fired bricks is laid on it and inclined in such a way that it has outlets into a drain. Next, a layer of charcoal, piled up then trampled to compactness, is spread on it, and a mortar comprising coarse sand, lime and ashes is laid on top to a depth of half a foot. When the upper surface has been smoothed off to rule and level with a whetstone, it looks

like a black pavement. So during dinner parties whatever is spilt out of cups or spat out dries as soon as it falls onto it, and the servants coming and going there do not get cold from that kind of floor even if they go barefoot.

CHAPTER V

Decadent Painting

1. Particular conventions for particular subjects in painting were established by the ancients for other rooms, that is, those used in the spring, autumn and summer, as well as for *atria* and colonnaded courtyards. For a painting is a representation of something that exists or could exist, such as a man, a building, a ship or other things from whose finite and ascertainable shapes copies are derived by an imitation of their forms. Consequently the ancients who introduced the use of wall-decoration first imitated the varied patterns and shapes of stuccoes made with powdered marble and then various combinations of garlands, decorative mouldings and borders.

2. Afterwards they made so much progress that they were able to imitate the forms of buildings and the three-dimensional projections of columns and pediments; in open-sided spaces such as *exedrae* they painted the façades of stage-sets in the tragic, comic or satiric style because of the ample wall-space available: and they decorated covered walks, on account of their great length, with a variety of landscapes, of which they derived the representations from the characteristics of particular locations. For harbours, promontories, seashores, rivers, springs, canals, sanctuaries, sacred woods, mountains, flocks and shepherds are painted; again, in some places there are paintings with the grandeur of statues depicting images of the gods or sequences of mythological narratives, as well as the battles at Troy, or the wanderings of Ulysses from country to country, and other subjects generated by the natural world reproduced on similar principles.

3. But these images, which were modelled on reality, are now condemned in the light of current depraved tastes; now monstrosities rather than faithful representations of definable entities are painted in frescoes. For example, reeds are put up in place of columns, fluted stems with curly leaves and volutes instead of pediments, as well as candelabra supporting representations of shrines, above the pediments of which tender flowers with volutes rise up from roots and include figures senselessly seated on them, and even stalks with half-length figures, some with human heads, others with the heads of animals.

4. These things do not exist, cannot exist and never have existed. For how, in the real world, could a reed possibly support a roof, or a candelabrum the mouldings of a pediment, or such a thin and flexible stalk support a little figure sitting on it, or roots and stalks generate flowers or half-figures? But when people see these falsities they do not criticize them but find them delightful, ignoring the problem of whether any of them can exist or not. Therefore new tastes have forced on us a situation in which bad judges condemn artistic excellence as incompetence. But minds obscured by faulty taste are incapable of appreciating that which really can exist in accordance with convention and criteria of appropriateness. For pictures which are unlike reality should not be approved of; nor, even if they are technically accomplished, should they immediately be judged favourably if their subjects do not conform to ascertainable criteria developed without offending established conventions.

5. For example, at Tralles, when Apaturius of Alabanda had made, with great skill, the scenery for the little theatre (known locally as the ἐκκλησιαστήριον [*ekklesiasterion*]), he painted it with columns, statues, centaurs supporting architraves, the circular roofs of rotundas, the projecting angles of pediments and cornices decorated with lions' heads functioning as spouts for the water from roofs; and besides all this, he even made another storey on top of the stage-set [*episcaenium*] in which there were cupolas, porticoes of temples, half-pediments, and the decoration of all the roofs was provided by a variety of different colours. So when the appearance of his scenery delighted the gaze of all because of its perspectival effects, and everybody was ready to give it their approval, Licynos the mathematician presented himself at that point. He declared that the Alabandians are regarded as being shrewd enough when it came to politics, [6] but are seen as obtuse owing to their lack of a sense of the appropriate – not a great vice – because the statues in their gymnasium all look like lawyers presenting cases, while in the forum the statues hold discuses, run or throw the javelin. 'So' – Licynos went on – 'this inappropriate placement of statues with respect to what the sites require has landed us with a widespread reputation for poor judgement. So now let us be careful that Apaturius' stage-set does not make Alabandians or Abderites of us as well. For which of you can put houses or columns or elaborate pediments on top of your tiled roofs? Things like this are built on floors, not above roof-tiles. So, if we are going to approve of things represented in pictures that have no way of existing in reality, we too would join those communities which are regarded as insensitive because of similar defects.'

7. Apaturius did not dare reply to this, but removed the scenery and, once he had changed it to conform with reality, won approval for the corrected version. If only the immortal gods had arranged for Licynos to come back to life to correct the present madness and the aberrant practices introduced into our frescoes. It will not be irrelevant, however, to explain why false reasoning prevails over the truth. For what the ancients tried to win approval for with their artistic skills with the expenditure of great pains and energy is now achieved with colours and their beguiling appearance, and the prestige which the sensitivity of the artist used to confer on his work is now no longer wanted as a result of the lavish expenditure of patrons.

8. For example, which ancient painter seems to have used cinnabar other than very sparingly, as though it were a drug? But now walls everywhere are decorated with it, and most of them entirely covered with it; then too there are malachite green, purple and Armenian blue. When these colours are applied, even though without much skill, they present a dazzling

sight to the eyes, and, since they are expensive, they are subject to special stipulations in contracts to the effect that they should not be paid for by the contractor but by the patron.

I have now talked enough about the advice I am able to impart about how to avoid errors in plasterwork: now I will discuss the materials as they come to mind, and first I will describe marble, since I spoke about lime at the beginning.

CHAPTER VI
Marble-powder

1. The same kind of marble does not occur in all regions, but in some places blocks are found which have translucent granules in them, like those in salt, and when crushed and milled, these blocks are usable in revetments. But in places where supplies of such blocks are not found, marble-rubble, called chips, which the marble-workers discard when they work, is crushed, ground up and used in revetments after being sifted. Again, in other places, such as on the border between Magnesia and Ephesus, there are areas where marble-powder can be excavated ready for use without the need for grinding or sifting, and which is as fine as any that has been ground by hand and sieved.

CHAPTER VII
Natural Pigments

1. With regard to pigments: some occur naturally in particular areas where they can be excavated, but others are artificial compounds obtained from different substances treated and mixed together in appropriate proportions so that they are equally serviceable in revetments.

But first we shall discuss those which are excavated in their natural state, such as yellow ochre [*sil*], which is called ὤχρα [*ochra*] in Greek. This, in fact, is found in many places, including Italy, but the best, which was the Attic, is unobtainable now because when the silver mines in Athens were worked by slaves, underground shafts were dug in order to find silver; when a vein of ochre happened to be found there they would still deal with it as though it were silver, and so the ancients had at their disposal a copious supply of ochre for the decoration of walls.

2. Again, red earths are extracted in great quantity in many places, but the best in only a few, such as Sinope in Pontus, Egypt and the Balearic Islands of Spain as well as on Lemnos: the Senate and Roman People granted the Athenians the right to enjoy the revenue from this island.

3. In fact paraetonian white derives its name from the places where it is mined: for the same reason, white pigment is called *melinum* because the mine is said to be on the island of Melos in the Cyclades.

4. Again, green chalk is found in many places, but the best is at Smyrna. The Greeks call it θεοδότιον [*theodotion*] because Theodotus was the name of the man on whose land this kind of chalk was discovered first.

5. Orpiment [arsenic sulphate], which is called ἀρσενικόν [*arsenikon*] in Greek, is mined in Pontus: again sandarach [red arsenic] is found in many places, but the best mine for it is in Pontus next to the river Hypanis.

CHAPTER VIII

Cinnabar and Mercury

1. I shall now proceed to explain the characteristics of cinnabar [mercury sulphide]. This is reported to have been discovered first in the Cilbian fields belonging to the Ephesians, and both the pigment and its method of preparation cause great surprise: before they arrive at the cinnabar itself after various procedures, the so-called lump, an ore like iron, but redder in colour and enveloped in red dust, is dug out. While being excavated it releases copious drops of mercury resulting from the blows of the picks, which are immediately collected up by the miners.

2. When these lumps have been collected in the workshop they are put in a kiln to dry out since they are saturated with moisture, and one finds that, when the vapour forced out of them by the heat of the fire condenses on the floor of the kiln, it is in fact mercury. When the lumps are taken out of the kiln, the drops which will be deposited are too small to be collected and are swept into a water-container where they merge and coalesce into a mass; when a volume of four *sextarii*[13] of mercury is weighed it will come to one hundred pounds.

3. If the mercury has been poured into some water-container and a stone weighing a hundred pounds is put on top of it, the stone floats on the surface and cannot compress the liquid with its weight or expel or displace it. If we remove the hundred-pound weight, and replace it with a scruple of gold, it will not float but will sink to the bottom of its own accord. Therefore it cannot be denied that the weight of a substance does not depend on its volume but on its other properties.

4. Now mercury is useful for many purposes since, for example, neither silver nor copper can be gilded successfully without it. And when gold has been woven into a piece of clothing, and the clothing, worn with age, is too scruffy to wear, the pieces of cloth are put in terracotta vases over a fire and incinerated. The ash is put in water and mercury added; the mercury absorbs all the particles of gold and makes them coalesce with itself. Once the water has been drained off, what remains is deposited on a cloth and there squeezed by hand: then the mercury, since it is a liquid, slips through the loosely textured material, but pure gold, compacted by squeezing, is left on the cloth.

CHAPTER IX

Cinnabar (Vermilion)

1. I will now return to the preparation of cinnabar. When the lumps of ore are dry, they are pounded in iron mortars,[14] and brought to the state when the colours emerge after repeated washing and heating have removed the impurities. So when the cinnabar has lost its inherent natural characteristics because of the removal of the mercury, it becomes soft in texture and weak in consistency.

2. For this reason, when cinnabar has been applied to wall-surfaces in enclosed spaces it keeps its colour without any defects: but in open spaces such as colonnaded courtyards or *exedrae* or others buildings of that type where the brilliant rays of the sun and moon can penetrate, the area painted with cinnabar that is reached by them is damaged and goes black once its pigment has lost its strength. This is why, when Faberius the Secretary[15] wished, like many others, to have his house on the Aventine elegantly frescoed, he had cinnabar applied to all the walls of his colonnaded courtyard, which took on an ugly, mottled colour after thirty days. Therefore he immediately[16] made a contract to have other colours applied.

3. But if someone who is more discriminating wants the wall-surface painted with cinnabar to retain its original colour, he should apply Phoenician wax melted over a fire and mixed with a little oil with a brush once the wall has been painted and has dried; then, after packing charcoal into an iron brazier, he should make the wax melt by warming up the wall from close-quarters so that it can be smoothed off; finally he should polish it with a candle and pure linen cloth just as marble statues of nudes are treated.[17]

4. This process is called γάνωσις [*ganosis*] in Greek. In this way the protective coating of Phoenician wax prevents the light of the moon and the rays of the sun from playing on the paintings and extracting the colour from them. The workshops that were in the mines at Ephesus have now been transported

to Rome because this type of ore was later found in areas in Spain: the lumps are transported from these mines and treated by contractors in Rome. The workshops are located between the temples of Flora and of Quirinus.

5. Cinnabar is adulterated when it is mixed with lime. So if someone wants to test its purity, we must proceed like this: take an iron sheet, put cinnabar on it, and place it on the fire until the sheet becomes red-hot. When the colour has changed and goes black because of the heat, the sheet should be removed from the fire; if the original colour of the cinnabar is restored when it has cooled, this will prove that it is pure; but if it stays black, that means that it is adulterated.

6. I have now related all that I could think of in connection with cinnabar.

Malachite green is imported from Macedonia and is dug up in places near copper mines. Where Armenian blue and indigo are found is indicated by their names.

CHAPTER X

Black

1. I will now pass on to those materials which acquire their colouristic properties when transformed as a result of the alterations produced by their treatment with other substances. First I will deal with black, which is extraordinarily useful for wall-decorations, so as to make clear exactly how the workmen carry out the relevant procedures correctly.

2. An oven similar to a Spartan sweating-room should be built and reveted carefully with marble, and then polished. In front of it a small furnace should be built with vents leading into the oven; the mouth of the furnace should be closed very carefully to prevent the flames escaping. Resin should be put in the furynace. The intense fire burns the resin placed in the furnace and forces the soot through the vents into the oven where it sticks round the walls and the curve of the vault. The

soot is collected from there and some of it is mixed and worked up with gum for use as ink; painters mix the rest with size for use on walls.

3. But if these materials are in short supply, we should meet the difficulty like this to avoid delaying the work for long. Dry branches or small pieces of pitch pine should be burnt, then extinguished when they have turned to charcoal; then they should be ground up with size in a mortar. In this way the painters will be supplied with a fairly attractive black.

4. Again, if wine-lees, dried and cooked in an oven and then ground up with size, are applied to a revetment, they will produce a more pleasing colour than ordinary black: the better the wine from which the lees derive, the better will it imitate the colour not only of ordinary black but even of indigo.

CHAPTER XI

Blue and Burnt Ochre

1. Methods of making azure were first devised at Alexandria, and afterwards Vestorius started production at Pozzuoli. The method of extracting it from the substances in which it is found is really remarkable. Sand is ground up with natron flowers so finely that it becomes like flour, and Cypriot copper is grated with a coarse file over it, like sawdust, so that it sticks together. Then it is made into balls by being rolled in the hands and compressed so that it dries. The dry balls are put in an earthenware jar, and the jars in a furnace. As soon as the copper and sand grow hot and coalesce because of the heat of the fire, they give up their individual qualities by mutually exchanging liquids; once their essential elements have been consumed by the intensity of the fire, they take on a blue colour.

2. Burnt ochre, which is very useful for decorating wall-surfaces, is made as follows. A lump of good-quality yellow ochre is heated on a fire till it glows; it is then cooled off with vinegar and assumes a purple colour.

CHAPTER XII

White Lead, Verdigris and Sandarach

1. It is not irrelevant to describe how white lead and verdigris, which we call *aeruca* [copper rust], are prepared. In Rhodes they put vine-branches in large jars, pour vinegar over them and then place lumps of lead on top; then they close the jars with lids to prevent the contents from evaporating. After a certain period they open the jars to find that the lumps of lead have turned into white lead. If, using the same procedure, they put in strips of copper, they make verdigris, which is called *aeruca.*

2. When white lead is heated in a furnace, it changes colour on contact with the fire, and becomes sandarach [red arsenic] – men made this discovery by chance thanks to some fire that broke out – and it is much more manageable than the naturally occurring sandarach dug up in mines.

CHAPTER XIII

Purple

1. Now I would like to discuss purple, which has the most prized and most outstandingly beautiful appearance of all these colours. It is extracted from the marine shellfish from which purple dye is made, which is as amazing to the observer as anything else in nature, because it does not have a uniform colour in all the places where it is found, but is naturally altered by the trajectory of the sun.

2. The type gathered in Pontus and Gaul is black because those regions are nearest to the north. As one passes from north to west, it turns out to be rather blue; that which is collected at the eastern and western equinoxes is found to be a shade of violet. That which is extracted in southern countries is naturally red in quality, and is therefore found on the island of Rhodes and in other such countries nearest to the course of the sun.

3. Once these shellfish have been gathered, they are broken up all around the edges with iron implements, after which they are worked on by being ground in mortars, and the purple liquid, forced out from the wounds, flows out like tears:[18] and because it is extracted from the shells of marine molluscs it is therefore called *ostrum* [oyster-purple]. On account of its saltiness it dries rapidly unless it has honey poured over it.

CHAPTER XIV

Artificial Purple, Yellow Ochre, Malachite Green and Indigo

1. Purple colours are also made by dyeing chalk with madder root and hysginum: other colours are made from flowers. So when painters want to imitate Attic yellow ochre, they throw dried violets into a water-container and heat them on a fire; then, when the mixture is ready, they pour it onto a linen cloth, and, wringing it out by hand, collect the violet-coloured water in a mortar: then they pour chalk into the mortar and grind it in, so making Attic ochre pigment.

2. Using the same method they make an exquisite purple by preparing blueberries and mixing them with milk. Again, those who cannot use malachite green because of its cost mix blue with the plant called *luteum* [dyer's weed], and so create a very deep green, called dyer's green, ready for use. Again, in the absence of the colour indigo, they dye Selinusian chalk, or bead chalk, with woad, which the Greeks call ἴσατις [*isatis*], and make an imitation of indigo.

3. In this book I have described in detail, as they occurred to me, the methods and materials which should be used to ensure the durability and good taste of paintings as well as the characteristics that all colours should have. And so the account of all the techniques for the realization of buildings and of the principles that they should incorporate has been completed in seven books. In the next book I shall explain how water can be

found in locations where it is lacking [on the surface], the techniques by which it may be channelled to various places and how it can be tested to show that it is healthy and suitable for consumption.

BOOK VIII

Introduction

1. Thales of Miletus, one of the Seven Sages, announced that water is the fundamental element of all matter, Heraclitus chose fire, the Priests of the Magi, fire and water, Euripides, a disciple of Anaxagoras whom the Athenians called the Philosopher of the Stage, chose air and earth:[1] [his theory was that] the earth, impregnated by the fertilization caused by the rain from the sky, had given birth to the young of mankind and of all living creatures in the world, and that whatever was born of her returns to her when dissolved by the inevitable passage of time; and similarly whatever was born of the air returned to the regions of the sky without suffering annihilation, but, altered by the process of dissolution, reverted to the state in which it was originally. In fact Pythagoras, Empedocles, Epicharmus and other natural scientists and philosophers have argued that there are four elements – air, fire, earth and water – and that the mutual combination of one with another according to nature's schemes produces the characteristics of the different types of matter.

2. Indeed, we can see not only that everything born is created from these elements, but also that nothing can grow or be nourished or can keep itself alive without their natural sustenance. For bodies cannot live without a supply of air to breath, that is, unless an abundance of air flowing in produces constant inhalations and exhalations. In fact, if the right amount of heat is not present in the body, it will lack vital breath and an upright posture, and nourishing food cannot arrive at the right temperature to be digested. Again if the various parts of the body are not nourished by the produce of the earth they will waste away and so be deprived of the admixture of the element of earth.

3. In fact, if living organisms are deprived of the benefit of water, they will wither away when deprived of blood and when sucked dry of their liquid element. This is why the Divine Intelligence did not make the things that are absolutely essential for mankind difficult to find, or expensive, like pearls, gold and silver and the other things which neither the body nor nature needs, but has scattered everything without which the life of mortals cannot be assured in profusion over all the world within hand's reach. So, of these elements, it is air that supplies breath if a body happens to be short of it, since its function is to restore it. The heat of the sun is available to maintain bodily warmth and the discovery of fire makes life safer. Again, the fruits of the earth satisfy our needs with copious, even excessive, supplies of food, and continuously feed and nourish living beings. Finally, water, which not only provides us with what to drink but satisfies countless practical necessities, offers us advantages for which we are particularly grateful because it is free.

4. For this reason those who carry out their priestly duties according to Egyptian rites demonstrate that all things depend on the properties of water. So, when water is brought in a vase with the most reverent solemnity to the sanctuary and the temple, they prostrate themselves on the ground and, raising their hands to heaven, thank the divine beneficence for its creation.[2] Natural scientists, philosophers and priests maintain that the properties of water are the basis of all things, so I therefore decided that, since the principles of buildings have been explained in the previous seven books, I should write in this one about how to find water, on its qualities in relation to the characteristics of the places in which it occurs, on the methods of transporting it and, finally, on how to test its quality. For it is absolutely essential for life, pleasure and everyday use.

CHAPTER I
Finding Water

1. Finding water will be easier if there are sources of water flowing on the surface in the open. But if they do not flow in the open, their underground sources must be located and brought together. The investigation should be carried out like this: before sunrise one should lie face-down in the locations where one wants to make the search, and, propping one's chin firmly on the ground, survey the area. This way one's line of sight will not range higher than it should since the chin does not move, and the line of sight will mark out a consistent horizontal level within well-defined limits in the landscape. Then one should dig in those places where vapours will be seen curling up and rising into the air, for this type of indication cannot occur in a dry location.

2. Again, those searching for water should pay attention to the characteristics of the sites because water is found only in particular types of locale. In clay, the supply of water is scanty, meagre, not very deep and will not have the best taste. In coarse sand, too, the water is scarce, but can be found at a greater depth: it will be muddy and taste unpleasant. But in black earth some moisture and drops of water can be found which collect after winter storms and settle in compact, hard ground; this has the best taste. In gravel, the watercourses are mediocre and unreliable, but these too taste very good. Also in clayey sand, common sand and sandstone, the supply is more dependable, steady and tastes good. The supply in red tufa is both copious and of good quality if it does not trickle through the fissures and disappear. At the feet of mountains and in hard stone it is more copious and abundant, and is also colder and more beneficial. But spring water in flat country is salty, brackish, tepid and tastes bad except for that which runs underground from the mountains and springs up in the midst of flat terrain and there, protected by the shade of trees, offers the same pleasant taste as mountain springs.

3. By contrast, plants growing in the types of land described above should provide indications of the presence of water, such as slender rushes, wild willow, alder, agnus castus, reeds, ivy and other plants of the same sort that cannot grow on their own without moisture. In fact these plants usually grow in depressions that have formed in locations lower than the rest of the terrain, which receive water from showers and the surrounding fields during the winter and retain it longer, thanks to their capacity. These indications should not be entirely trusted, but we must search for water in those areas and locales, other than in depressions, where plants indicating the presence of water grow naturally of their own accord, and have not been planted.

4. If there are indications that water will be found in sites like this, one must test for it like this: a hole not less than three feet square nor less than five deep should be excavated and a bronze or lead vessel or basin placed in it around sunset. Whichever type of vessel is chosen should be coated with oil on the inside and placed upside down: the top of the excavation should be covered with reeds or leaves, and a layer of earth spread on top: it should be opened the next day, and if there are drops of water or moisture in the vessel, the place in question will contain water.

5. Again, if a vessel of unfired clay is put in the hole and covered in the same way, it will be damp when uncovered and will already have begun to dissolve because of the moisture if the place contains water. If a woollen fleece is placed in the hole and water can be wrung out of it the next day, this too will prove that the location has a supply of water. Furthermore, if a lamp is prepared, filled with oil and lighted, and is put in the hole and covered over, and the next day it has not burnt out but still contains a residue of oil and wick, and the lamp itself is found to be damp, this will indicate that the place contains water, since all heat sources attract moisture. Again, if a fire is lit in the hole and the earth releases a misty vapour when heated up and charred, the place will contain water.

6. After these experiments have been carefully carried out and the indications of the presence of water mentioned above

have been located, then a well must be sunk at the site in question; if a water source is found, a number of other wells must be dug around it and all of them joined up to one location by means of underground channels. But one should search for water particularly in the mountains and in areas exposed to the north because there the water tastes better and is more healthy and more plentiful. For such places are not exposed to the trajectory of the sun, and it is primarily in such areas that there are dense forest trees; and the mountains themselves stand in the way, creating their own shadows, and the rays of the sun do not arrive directly at the earth and cannot dry out the moisture.

7. Besides, valleys in the mountains receive the greatest amounts of rain, and because of the density of the forests, the snows stay there longer thanks to the shadows cast by the trees and mountains; then, when they melt, they percolate through the crevices in the earth and so arrive at the lowest slopes of the mountains, from which gushing springs pour out. By contrast, one cannot find plentiful supplies of water in flat terrain: and any water that is found there is not healthy because the fierce assault of the sun seizes on it in the absence of any protection from shade, and, extracting the moisture from the flat terrain, heats it up; the air takes up and disperses the lightest, purest and most delicately healthy elements of whatever water remains above ground in the vast expanse of the sky, while the heaviest, hardest and most unpleasant elements are left in the springs on the flat terrain.

CHAPTER II

Rain- and River-water

1. So water which is collected from showers has the most healthy qualities because it is drawn from the lightest and most delicately subtle of all sources, and then, filtered through the moving air during storms, it liquefies and lands on earth. Again,

rainfall is not frequent on the plains but it is in, or close to, the mountains, because when the vapours set in motion by the rising morning sun leave the earth and drive the air in front of them in whichever direction they are headed, they are then driven on by the vacuum in the area behind them and pull waves of rushing air in after them.

2. But the air, which rushes along driving the moisture everywhere, creates gales and expanding billows of wind with the force of its blast. Wherever the vapours rolled up from springs, rivers, marshes and the sea are taken by the winds, they are collected and sucked up by the heat of the sun and so carried aloft as clouds; then when the clouds, held up by the wave of air, arrive at, and collide with the mountains, they liquefy because they are full and heavy, and burst and pour over the land in the form of rainstorms.

3. That vapours, clouds and moisture are generated in the earth is evidently due to the fact that the earth contains burning heat, huge currents of air, cold zones and a vast supply of water. For this reason, when the rising sun strikes the surface of the earth, which has cooled off during the night, and gusts of wind begin to blow when it is still dark, clouds rise up into the sky from the damp areas. But then the air, once warmed by the sun, draws vapours from the earth, as is normal.

4. One may use the baths as an illustration of this. None of the vaults of hot baths can have a water source above them, but the air in the upper part, super-heated by the fire-vapour from the mouths of the furnace, draws the water up from the pavement and carries it up to the curves of the vaults and holds it there, since hot vapour always drives itself upwards. At first it[3] does not release the moisture because the amount is small, but as soon as it has collected a lot more, it cannot keep it suspended because of its weight, but sprinkles it on the heads of the bathers. Again, by the same principle, when the air in the sky absorbs the heat of the sun, it draws moisture from everywhere, hoists it up and collects it as clouds. So the earth, when touched by heat, releases moisture, just as the human body sweats when it is hot.

5. The winds provide demonstrations of this process: those that spring up and arrive from the coldest regions, Septentrio [N] and Aquilo [NE], blow out gusts that are rarefied by the dryness of the atmosphere: but the Auster [S] and the other winds which gain their impetus from the trajectory of the sun are moisture-laden and always bring showers because, as they arrive heated up by the warm regions, they soak up moisture from all the lands they touch, and consequently pour it out over the southern regions.

6. The sources of rivers can demonstrate that these things occur as we have described: the majority, and the most important ones represented on maps of the world and described by geographers, originate in the north. First, in India, the Ganges and the Indus originate in the Caucasus; in Syria, the Tigris and the Euphrates; again in Pontus in Asia, the Borysthenes [Dnieper], the Hypanis [Bug] and the Tanais [Don]; in Colchis, the Phasis; in Gaul, the Rhodanus [Rhône]; in Celtica, the Rhenus [Rhine]; on this side of the Alps, the Timavus and the Padus [Po]; in Italy, the Tiber; in Maurusia, which we call Mauretania, the Dyris originates at Mount Atlas; it starts in the northern region and proceeds west as far as Lake Eptagonus, where it changes its name to Ger; and then from Lake Eptabolus it runs under the desert mountains, and, flowing through the southern regions, debouches into the marsh called [. . .], it then winds around Meroë, which is the kingdom of the southern Ethiopians; from these marshlands, it winds along via the rivers Astasoba and Astoboa and many others, and passes through the mountains to the Cataract, from which it rushes on and drops down through the northern zones, arriving in Egypt between Elephantis and Syene and the Theban plain: and there it is called the Nile.

7. That the source of the Nile is in Mauretania can be deduced above all from the fact that on the other side of Mount Atlas there are other sources of rivers which flow to the Western Ocean, and there mongooses, crocodiles and other species of animals and fish originate, with the exception of hippopotamuses.

8. Since, to judge from descriptions of the world, all the great rivers evidently flow from the north and in the plains of Africa

exposed to the trajectory of the sun in the south the moisture is hidden deep down, springs are uncommon and rivers rare, one cannot but conclude that much better sources of water will be found in areas orientated towards the Septentrio [N] or Aquilo [NE], unless they encounter an area full of sulphur, alum or bitumen; for then they change completely and, whether the water is hot or cold, it pours out in springs that smell and taste bad.

9. For hot water has no distinctive quality,[4] but cold water, when it runs into a hot zone along its course, boils and comes up hot out of the earth through the fissures. Therefore the water cannot maintain this temperature very long, but soon goes cold; if it were naturally hot, its heat would not diminish. But it does not recover its taste, smell or colour since it has become contaminated and adulterated because of its lack of density.

CHAPTER III

Properties of Different Waters

1. However, there are some hot springs from which flows water that tastes very good and is so pleasant to drink that one would not miss even the Fountain of the Muses or the gushing Aqua Marcia.[5] These are produced by nature like this: when fire breaks out underground because of the presence of alum, bitumen or sulphur, it heats the earth above super-hot, and consequently releases hot vapour into the areas further above, so that if any sweet-water springs originate in the upper strata, they start boiling in the crevices of the earth when they are hit by this vapour and so flow out with their taste unimpaired.

2. There are also cold springs that smell and taste bad: they originate deep underground and pass through hot zones; and, traversing long tracts of earth from that point, they arrive at ground level much cooled and with a nasty taste, smell and colour, like the river Albula on the Tiburtine Way and the cold springs in the Ardea region which have the same odour and are

called 'sulphurated springs', and in a number of other such places. Now, although these springs are cold, they look as though they are boiling because when they run into a hot zone very deep down, the water and fire meet, and the water absorbs the powerful currents of air in the violent collision, and so, swollen and driven by the air-pressure, come out of the earth boiling vigorously. But some of these water sources which are not on the surface but obstructed by rocks are even pushed up through narrow fissures to the tops of hillocks by the force of the air.

3. Consequently those who think they can find the sources of springs at the height to which these hillocks arrive are disillusioned when they extend their trenches further. For example: when a bronze vessel, not filled up to the brim but to two-thirds of its capacity, has a lid placed on it and is subjected to the intense heat of fire, it forces the water to heat up; because of its natural lack of density, the water is subjected to a powerful expansion caused by the heat, and not only fills the vessel but lifts the lid up with its steam, and, increasing in volume, overflows; but once the lid is removed and the steam has escaped into the open air, the water returns to its original level. In the same way, then, when the sources of springs are restricted in narrow channels, the jets of air in the water rush to the surface as bubbles, but as soon as they are given wider outlets, they are deprived of air because of the lack of density peculiar to water and, subsiding, resume their normal level.

4. But all hot water has medicinal properties in that it acquires new and useful characteristics when boiled even in unhealthy conditions. For example, sulphurous springs make muscles work again by warming up and burning off unhealthy humours from the body with their heat. Alum water, when used in the treatment of the limbs of bodies weakened by paralysis or attacked by some other illness, cures them by warming them through their open pores and counteracting the chill with the opposite force of heat, and in this way the limbs gradually return to their original condition; and drinking bituminous water usually cures internal maladies thanks to its purgative qualities.

5. There is also a type of cold water containing alkalis

[natron], found at Pinna, the city of the Vestini, at Cutiliae and other similar places, which, if drunk, purges and even reduces scrofulous tumours when it passes through the bowels. Conversely, abundant springs are found where there are mines of gold, silver, iron, copper, lead and other similar metals, but they are extremely harmful. For like hot springs, they contain sulphur, alum, bitumen, asphalt, and [a deposit] which, on entering the body in the form of a drink, spreads through the veins and attacks muscles and joints, making them distend and harden. Therefore the muscles, swelling and expanding, contract in length and give men arthritis or rheumatism because the pores of their veins are impregnated with very hard, dense and cold substances.

6. There is also a sort of water which is not transparent enough and from which derives a bloom, similar in colour to purple glass. This can be observed particularly at Athens, since there water-ducts have been led to the city and to the port of Piraeus from places and springs like this which nobody drinks from for the reason just mentioned: rather they use them for washing and other purposes and drink from wells, so avoiding their ill effects. At Troezen it cannot be avoided, because absolutely no other kind of water can be found there except for that provided by the Cibdeli [Unhealthy Fountains], and so all or most people in this city have diseases of the feet. There is a river called the Cydnus in the city of Tarsus in Cilicia in which people with gout bathe their legs to relieve the pain.

7. There are, moreover, many other kinds of water with their own particular properties; for example, the river Himera in Sicily which divides into two branches after leaving its source; the one that flows in the direction of Etruria is infinitely sweet because it runs through sweet moisture in the earth; the other runs through country where salt is mined and therefore tastes saline. Again, at Paraetonium, on the road to the oracle at Ammon and at Casius in Egypt there are marshy lakes which are so saline that they have a crust of salt on the surface. And there are springs, rivers and lakes in many other places which inevitably become saline because they run through salt-deposits.

8. Again, other springs which flow through strata of viscous earth are mixed with oil when they burst out of the ground; at Soli, for example, a town in Cilicia, swimmers or bathers in the river called Liparis get covered in oil by the water itself. Similarly, there is a lake in Ethiopia which covers people who swim in it with oil, and one in India which emits a great quantity of oil when the sky is clear. Again, at Carthage there is a spring with oil floating on its surface – in fact, sheep are usually dipped in this oil – which smells like sawdust from cedar-wood. At Zacynthus, around Dyrrhachium and Apollonia there are springs which discharge a great quantity of pitch with the water. In Babylon, a lake of enormous extent, called Lake Bitumen, has liquid bitumen floating on its surface; Semiramis built a wall around Babylon made of this bitumen and fired bricks. Again at Jaffa in Syria and in nomadic Arabia there are lakes of immense size that yield enormous quantities of bitumen, which the local inhabitants make off with.

9. This need not cause surprise since there are many quarries of hard bitumen there. So when water forces its way through bituminous soil, it drags the bitumen up with it and, having sprung up to the surface, separates itself and so ejects it. Again, there is a large lake on the road between Mazaca and Tyana in Cappadocia: if part of a reed or of some other object is dipped in it and taken out the next day, the part taken out of the water will be found to have turned to stone, while the part which stayed above water remains in its original condition.

10. In the same way, at Hierapolis in Phrygia there are numerous boiling springs from which water is conducted through channels dug out there and is sent around gardens and vineyards: after a year the water turns into a stony crust. Therefore each year they build dykes of earth to right and left, let the water in and so make banks in the fields out of this crust. This seems to occur naturally since there is a fluid in the ground similar in composition to rennet in the areas and terrain where it occurs: then, when this concentrated mixture emerges through the springs onto the surface, it is forced by the heat of the sun and of the air to solidify, which is something we also see in salt-pans.

11. There are also springs which emerge from the earth

impregnated with very sour liquids, such as the river Hypanis in Pontus. This river is very sweet as it flows for about forty miles from its source; then it arrives at a place a hundred and sixty miles from its mouth, where it is joined by a very small stream. When this stream flows into it, it makes even a river of this volume bitter, because the stream becomes soured by flowing through the types of earth and strata in which they mine sandarach [red arsenic].

12. Waters acquire their different flavours from the characteristics of the earth, as can also be seen in the case of fruit. For if the roots of trees, vines and the other plants did not produce their fruit by absorbing liquid from soils with different properties, then all their tastes would be uniform across all districts and regions. But we observe that on the island of Lesbos there is a wine called *protropum* [wine from the first pressing]; in Maeonia, the *catacecaumenite* [wine from burnt earth]; in Lydia, the Tmolite; in Sicily, the Mamertine; in Campania, the Falernian and between Terracina and Fondi, the Caecuban: and in numerous other places a vast number of different varieties and types of wine with various characteristics are produced, which could not happen unless the liquid in the soil with its different types of flavour did not infiltrate the stem through the roots and, rising up through it to the top of the plant, diffuse the particular taste of its own locale and type.

13. If soils were not very varied with regard to the types of liquid they contain, Syria and Arabia would not be the only countries in which the reeds, rushes and all the plants are aromatic, where the trees produce incense or bear pepperberries and little lumps of myrrh, and where silphium-gum grows on the stalks of plants at Cyrene, but the produce of the earth of all the different regions would be of the same type. It is the latitude and the power of the sun in the various localities and regions, depending on whether its trajectory is closer or farther away, that determines these differences in the quality of the earth and the liquid it contains. But if the properties of the individual types of soil were not modified by the power of the sun these phenomena would not occur like this either in these cases or in those of sheep and cattle.

14. For example, in Boeotia there are the rivers Cephisos and Melas, in Lucania, the Crathis, at Troy, the Xanthus, and certain springs and streams in the country of the Clazomenians, the Erythraeans and the Laodiceans. When the sheep are ready for breeding at the appropriate time of year, they are driven daily to drink in this period: the result is that, however white they may be, they produce ash-grey lambs in some places, brown lambs in other places and lambs as black as ravens in yet others. So on entering their bodies, the singular property of the water reproduces itself once the sheep have been impregnated with its particular characteristics. Since the cattle are born red and the sheep grey near the river on the plains around Troy, the inhabitants of Ilium are said to have named the river 'Xanthus' [red-blond].

15. Furthermore, lethal varieties of water can also be found running through noxious liquid in the earth and absorbing its toxicity; for instance, they say that there was a spring at Terracina, called the Spring of Neptune, and that those who unwisely drank from it lost their lives: the ancients are said to have blocked it up because of this. At Chrobs in Thrace there is a lake which causes the death not only of those who drink from it but also of those who bathe in it. In Thessaly too there is a gushing spring which sheep never even taste and no other type of animal ever goes near; there is a tree which blooms with purple flowers near this spring.

16. Similarly, in Macedonia, at the spot where Euripides is buried, two streams, arriving to the left and right of his tomb, join each other and travellers often lie down and eat their picnics[6] next to one of them because its water is good; but nobody goes near the stream on the other side of the tomb because the water is said to be fatal. Again, in Arcadia there is a region called Nonacris, where very cold water trickles from the rocks in the mountains. This water is called Styx Water, and not even vessels of silver, bronze or iron can contain it without bursting apart and falling to bits. Nothing can hold and conserve it except a mule's hoof, and this was how, according to tradition, some of this water was carried for Antipater by his son Iollas to the province where Alexander was staying, which is how the king was murdered by Antipater using this water.[7]

17. Again, in the kingdom of Cottius in the Alps there is a type of water which makes those who taste it drop dead on the spot. On the Via Campana on the Cornetus plain in Faliscan country there is a grove in which a spring emerges and the bones of birds, lizards and other reptiles can be seen lying around it. Also, some sources of springs are acidic, as, for example, at Lyncestus, in Velian country in Italy, at Teanum in Campania and in many other places, and have the capacity, when drunk, to break up the stones which form in men's bodies.

18. This seems to happen naturally because a pungent and acidic liquid is present in the soil which impregnates the streams of water with bitterness as they emerge from it; when the liquids enter the body, they dissolve whatever has been deposited by water and coagulates in our bodies when they encounter it. We can see why these things are dispersed by acidic liquid from these experiments: if an egg is left for some time in vinegar, its shell will soften and dissolve; again, if a piece of lead, which is very resistant and extremely heavy, is put in a vase and vinegar is poured over it and then the vase is covered and sealed, the result will be that the lead will dissolve and turn into white lead.

19. For the same reasons, if copper, which is naturally even more dense, is treated in the same way, it will dissolve and turn into verdigris; the same applies to pearls and even hard limestone, which neither fire nor iron can dissolve by themselves, but which, on being heated up by fire, fall to bits and disintegrate when sprinkled with vinegar. And so when we see these phenomena occurring in front of our very eyes, we must conclude that for the same reason, even those suffering from stones can be cured by acidic water in a similar way by natural means thanks to the sharpness of the liquid.

20. There are also springs that seem to be mixed with wine, such as the one in Paphlagonia; the water from this spring makes those drinking from it tipsy, even when they have had no wine. Again, in the land of the Aequiculi in Italy and in Medullian territory in the Alps there is a type of water which causes the development of goitres in those who drink it.

21. In Arcadia, in fact, is the very well-known town of Clitor, in whose territory there is a cave running with water which turns people who drink it into teetotallers. At this spring there is an epigram in Greek verse inscribed on a stone which says that the water is unsuitable for bathing and also damaging to vines because it was at this spring that Melampus purged the madness of Proetus' daughters by making sacrifices, and restored these virgins' minds to their former sanity. This is the text of the epigram:

> Shepherd, if at midday thirst oppresses you as you arrive with your flock at the borders of Clitor, take water to drink from the fountain and rest all your flock near the water nymphs. But do not pour water over your body to wash yourself, lest even a drop harms you while you're sunk in sweet drunkenness. But keep away from my fountain, life's enemy, where Melampus, having liberated the daughters of Proetus from their terrible frenzy, sunk and hid all the sacrificial objects when he had arrived at the mountains of rocky Arcadia from Argos.

22. Again, there is a spring on the island of Cea and those who unwisely drink from it lose their minds; an epigram is inscribed there which says that a drink from the spring is delightful but that whoever drinks from it will end up with the brain of a stone. These are the verses:

> Sweet is the flow of the fresh drinking water which the spring pours out, but the mind of whoever drinks it will turn to stone.

23. And at Susa, capital city of the Persian kingdoms, there is a little spring which causes those who drink from it to lose their teeth. Again, there is an epigram there of which the gist is that the water is excellent for bathing, but if drunk, knocks the teeth out of their roots. These are the Greek verses of this epigram:

> Stranger, you see these spring waters with which men can wash their hands without ill effects; but if you take the clear water

from inside the cave, and just let it touch the edge of your extended lips, the same day your teeth, which chop up your food, will fall on the floor leaving empty sockets in your jaw.

24. There are also springs in a number of places which have the property of giving those born there good singing voices, such as Tarsus in Magnesia and in other such areas. Then again there is Zama, a city in Africa, whose walls King Juba enclosed with a double circuit, and where he built the royal residence for himself. Twenty miles away is the town of Ismuc, where the areas of cultivation are marked off by an incredible boundary. For although Africa is the mother and nurse of wild animals, especially snakes, none is born in the lands belonging to that town, and if one is ever brought in and left there, it immediately dies; not only this, but if earth from these areas is taken elsewhere, the same thing happens. It is said that this kind of soil can also be found in the Balearic Islands. But this earth has another astounding characteristic, about which I heard this story. **25.** Gaius Julius, Masinissa's son, who owned all the lands around the town, served in the army with Caesar, your father. He stayed with me as my guest, so in our daily encounters conversation inevitably turned to erudite matters.[8] In the course of a conversation about the efficacy and qualities of water, he explained that there were springs in that country which were such that people born there have excellent singing voices, and so they always bought handsome men and nubile girls from abroad[9] and married them so that their children would grow up not only with fine voices but good looks as well.

26. This immense variety has been distributed by nature among different things so that in the human body, which is in part earth, there are many types of liquid, such as blood, milk, sweat, urine and tears; therefore, if such a diversity of liquids is to be found in a tiny particle of earth, it is not surprising that in the vast expanse of the world innumerable varieties of liquids can be found, and that flowing water, impregnated by running through currents of these liquids, arrives at the outlets of springs: and this is the reason why springs become different and varied with their own characteristics because of diversities of locale, the qualities of the land and the dissimilar properties of the terrain.

27. I myself have observed a number of these phenomena; others I have found discussed in Greek texts, of which the authors are Theophrastus, Timaeus, Posidonius, Hegesias, Herodotus, Aristides and Metrodorus; writers who, with great diligence and infinite application, expounded on the peculiarities of various locales in their writings, on the properties of the waters and the distribution of various regional characteristics depending on the latitude. Following on from the progress they have made, I have recorded in detail in this book what I felt was sufficient information on the different kinds of water to make it easier, by means of this body of advice, for people to choose sources from which they can conduct running water for use in cities and municipalities.

28. For it seems that nothing in all the world has more essential, practical uses than water, since the constitutions of all living creatures, if deprived of the product of grain, could at least keep themselves alive using other types of food from trees, meat, fish or any other food: but without water no animal's body nor any nourishing food can be produced, cultivated or provided. Consequently we must take the greatest care and pains when searching for and selecting springs bearing in mind human well-being.

CHAPTER IV

Testing Water

1. Springs should be tried and tested like this: if they flow freely in the open, one should take a good look at the physiques of the people who live in their vicinity before one starts to take water from them: if the locals have robust bodies, fresh complexions, sound legs and eyes without inflammation, then the springs will have passed the test with flying colours. Again, if a new water source has just been dug out and water from it is poured into a Corinthian vase, or any other sort made of good bronze, without leaving a stain, then the water is excellent. Equally, if the water is boiled for a long time in a bronze cauldron then left to stand, and is then poured out without sand or mud being found at the bottom of the vessel, then that water too will have proved its excellence.

2. And again, if vegetables put in a pot with this water and placed over the fire are rapidly cooked right through, this will prove that the water is good and healthy. Moreover, if the same water in the spring is bright and transparent and, in the location at which it arrives and from which it flows out, there is no growth of moss or reeds, and if the area is not tainted by any pollution but looks pristine, these are all indications that the water is light and extremely wholesome.

CHAPTER V

Levelling Instruments, the chorobates

1. Now I will explain how water should be supplied to houses and towns. The first step is the taking of levels. Levels are taken either with *dioptrae*, water levels or the *chorobates*, but it is done more accurately with the latter, because *dioptrae* and water levels are liable to mistakes. The *chorobates* is a wooden beam about twenty feet long: at the ends it has arms, made identically and connected at right angles to the extremities of the beam; between the beam and the arms cross-pieces are fixed with hinges which have exactly perpendicular lines drawn on them and plumb-lines hang vertically from the wooden beam, one on each side.[10] When the beam has been put in position and the plumb-lines coincide precisely and equally with the drawn lines, they show that the *chorobates* is level.

2. But if the wind interferes with the operation and the lines cannot supply a definite reading because of the oscillations, then a groove five feet long, one digit wide and one and a half digits deep should be cut in the upper face [of the beam] and water poured into it; if the water touches the rims of the channel uniformly, then one knows that it is horizontal. Again when the level has been taken in this way with the *chorobates*, one will know the amount of the incline.

3. Perhaps someone who has read Archimedes' books would say that true levelling cannot be achieved using water, since Archimedes maintains that water is not horizontal but part of a sphere of which the centre coincides with that of the earth. But whether water is flat or spherical, it necessarily follows that when the beam is level, it will maintain water at the same level at its extremities to right and left; but if it slopes down at either end, the water will not reach the rims of the channel at the higher end: for wherever water is poured, it will necessarily assume a convex curvature rising to the centre, yet the ends to right and left will be level with each other. An illustration of the *chorobates* will be drawn at the end of the book [Fig. 27]. If there is a considerable incline the flow of water will be easier: but if the watercourse is interrupted by declivities, we must make use of substructures.

CHAPTER VI

Aqueducts, Wells and Cisterns

1. There are three ways of conducting water: in watercourses running through masonry channels, in lead pipes or in terracotta tubes. These are the methods of constructing them; with regard to channels, the masonry should be as solid as possible; the bed of the watercourse should have a smooth gradient of not less than a *sicilicus*[11] every hundred feet and the masonry structures should be vaulted over so that the sun hits the water as little as possible. When the water has arrived at the town, a reservoir should be built with a tank with three compartments connected to the reservoir to receive the water; three equally spaced pipes[12] should be fixed in the reservoir and connected

to the compartments so that when the water overflows from those at the sides, it runs into the one in the centre.

2. Again, the pipes leading to all the cisterns and fountains will be fixed in the middle compartment; from the second compartment, those leading to the baths, so that they may supply the state with annual returns; and from the third, those to private houses so that water for public use will not run short; for private citizens will not be able to divert [the water from public use] if they have their own supplies from the sources. These are the reasons why I established these divisions: also so that those leading water into their own houses would safeguard the maintenance of the aqueducts by the contractors through the water rates.

3. But if there are mountains between the city-walls and the source of the water, we must proceed like this: underground galleries should be excavated and their gradients made smooth, as mentioned above; if the terrain consists of tufa or stone, then the channel should be cut into the rock itself; but if the bed comprises earth or sand, masonry walls with vaults should be built in the tunnel and the water should be conducted in that way; air-shafts should be built every two hundred and forty feet. 4. But if the water is to be conducted in lead pipes, a reservoir should be built first at the source; then the bore of the pipes should be set in relation to the volume of water, and these pipes should be laid from the reservoir to another one which will be located inside the town-walls. The pipes should be cast in lengths of not less than ten feet. If they are to be hundred-inch pipes, the weight of each section should be 1,200 pounds; for eighty-inch pipes, 960 pounds; for fifty, 600 pounds; for forty, 480 pounds; for thirty, 360 pounds; for twenty, 240 pounds; for fifteen, 180 pounds; for ten, 120 pounds; for eight, 100 pounds; and for five-inch pipes, 60 pounds. The pipes acquire their names from the width of the sheets of lead expressed in digits before they are rolled into cylinders. So when a pipe is made from a sheet fifty digits wide, it will be called a fifty-digit pipe, and so on for the others.

5. If water is to be conducted through lead pipes then this technique should be used: if there is a regular incline from the

source to the city-walls, and any hills lying between them are not high enough to constitute an obstacle, but there are low-lying areas, it is essential to build substructures to bring it up to the required level, as in the case of conduits and channels. If the distance around the low-lying areas is not great, the water should be taken around them; but if the declivities are very extensive, the watercourses should be led straight down the slope. When it arrives at the lowest point in the valley, the substructure should not be built high, so that the level remains the same for as great a length as possible. This will be the 'stomach', which the Greeks call κοιλία [*koilia*]. Then, on reaching the upwards slope on the other side, the water gradually expands in the long 'stomach' so that it is forced up to the top of the hill.

6. But if no 'stomach' and no level substructure have been built in the declivities, but only an 'elbow joint', the water will burst it and break up the joints of the pipes. In the 'stomach' valves[13] must be made by which the air-pressure can be reduced. Again, those who want to conduct water through lead pipes can do so following these rules, with excellent results, because with this method they can manage the descents, the deviations of course, the 'stomachs' and the ascents, once the gradient from the source to the city has been levelled smoothly.

7. Again, it is certainly not a waste of time to construct cisterns every two hundred *actus*,[14] so that if some defect occurs anywhere, it will not ruin the structure in its entirety and the spot where it has happened can easily be found; but such cisterns should not be built on a descent, nor on the same level as a 'stomach' nor on an upward incline, and absolutely not in valleys, but only where there is continuous level ground.

8. But if we wish to spend less money, we must proceed as follows. Terracotta tubes should be made with walls no less than two digits thick; they should be tapered at one end so that one can fit into another and stay connected. The joints should be coated with quicklime mixed with oil, and where there are changes in the level of the 'stomach', a block of red tufa is to be placed at the 'elbow joint': this block is to be perforated so that the last length of the descending pipe will be jointed into

the stone as well as the first length of the level area of the 'stomach'; in the same way, towards the upward incline, the last length of the level area of the 'stomach' should be securely fixed into a hole in red tufa, and the first length of the tube on the upward incline should be jointed into it in the same way.

9. Once the pipes have been levelled to a plane in this way, they will not be forced out of position by the force of the ascents and descents of the water. For very strong air-pressure tends to build up when water is in motion, pressure so strong that it can even break up rocks unless the water is introduced gently and sparingly from the source and is contained at the 'elbow joints' or bends with collars or by heavy sand-ballast. Everything else should be installed in the same way as for lead pipes. Again, before the water is released from the source for the first time, ashes should be poured in so that any joints that have not been sealed enough will be grouted by the ash.

10. Conducting water in terracotta tubes has these advantages: first of all, if any damage occurs in the structure, anybody can repair it. Besides, water from terracotta pipes is much more healthy than that taken through lead pipes, which seems to be particularly damaging seeing that white lead, said to be harmful to the human organism, derives from lead. Hence, if its by-product is damaging, there can no doubt that lead itself is deleterious as well.

11. We can take the lead-workers as proof of this, since a deep pallor obscures their colouring. For when lead melts during casting, the fumes released by it penetrate the various parts of their bodies, and day after day burn out and drain away the life-saving properties of blood from their limbs. So it is obvious that water should never be conducted in lead pipes if we want it to be beneficial to our health. Our everyday food proves that water from terracotta pipes tastes better, for when all our tables are heaped with silver vessels, people still use the earthenware ones because of the purity of taste.

12. But if there are no springs from which to take water, we must dig wells. When digging wells, though, sound method is not to be ignored and we must consider the principles of nature with acumen and great sagacity, since the earth itself comprises many different substances. For like everything else, the earth itself is composed of the four elements: in the first place, it is itself earth and includes springs of water deriving from the liquid element, and again, it contains heat, from which sulphur, alum and bitumen originate; and finally, it contains great currents of air which make their way, impregnated, through the porous fissures of the earth to the places where wells are being dug out, and which attack the excavators there, blocking the life-giving breath in their nostrils with the force of their vapours. So those who do not get away from such places very quickly perish there.

13. One can prevent this from happening like this: a lighted lamp should be lowered into the well, and if it stays alight, one can go down into it safely. But if the flame is extinguished by the draught caused by the vapour, then air-shafts should be dug to left and right next to the well, so the fumes will be discharged through the vents like breath through the nose. When these operations have been completed like this and water has been reached, a dry-stone shaft should be built around it to ensure that the runs of water in the earth are not blocked.

14. But if the terrain is hard or if the runs of water are too deep, then supplies of water must be collected from roofs or higher ground in cisterns built with *opus signinum*. Structures of *opus signinum* are built as follows; first of all, collect the purest and roughest sand; then hard limestone should be ground up into rubble in such a way that no stone weighs more than a pound; five parts of sand should be mixed in a trough with two of the strongest possible lime. The excavation for the cistern should be packed down hard with wooden rams shod with iron to the level of the required depth.

15. After the walls have been flattened all round, the earth in the middle should be cleared out level with the bottom of the walls. When the earth has been levelled off, the floor should be flattened to the required depth. If these cisterns have two or three compartments so that their contents can be changed by successive filtering, they will make the water much more wholesome and pleasant when consumed. For if the sediment has a place to settle, the water becomes clearer and will retain its

taste without any smell; if not, then it will be essential to add salt and so purify it.

In this book I have set out what I could on the subject of the virtues and variety of water, the uses it has and the methods by which it should be conducted and tested; in the next volume I will examine carefully the subject of sundials and the principles behind clocks.

BOOK IX

Introduction

1. The ancient Greeks established such great rewards for the celebrated athletes who had won at the Olympic, Pythian, Isthmian and Nemean games that they not only receive acclamation as they stand amongst the crowd with palm and crown, but also, when they return to their own towns they are carried in victorious triumph inside the city-walls in four-horse chariots to their family homes and enjoy a pension for life deriving from revenues set aside for them by the state. When I consider this, I am amazed that the same or even greater honours are not granted to the authors who provide services of infinite value for all people in perpetuity. It would have been much more appropriate to establish a ceremony like this for them since athletes merely make their bodies stronger by training, while writers invigorate not only their own intellects but those of all other men by collecting in their books the teachings by which others learn and sharpen their minds.

2. For how does the fact that Milo of Croton[1] and other victors of that type were invincible benefit mankind, apart from the fact that while they were alive they enjoyed celebrity amongst their fellow-citizens? But the doctrines of Pythagoras, Democritus, Plato and Aristotle and of other philosophers, refined every day with ceaseless assiduity, produce fresh and ever-ripe fruit not only for their fellow-citizens but also for all mankind. And, thanks to them, those who from an early age fill their minds with this mass of knowledge and have acquired the most sophisticated understanding of wisdom, establish in their cities civilized customs, impartial justice and laws without which no community can be safe.

3. Therefore, since such extraordinary benefits for people, individually and collectively, derive from the wisdom of such writers, I maintain not only that palms and crowns should be granted to them, but that triumphs too should be decreed for them and that they should be judged worthy of consecration in the home of the gods. As examples I will present some of the many ideas of a select few of these philosophers useful for the development of human life; and those who recognize their worth will necessarily admit that these men must be honoured.

4. First of all, I will explain one of Plato's many exceptionally useful theorems as he formulated it.[2] If there is a site or square field, that is, one with equal sides, which we have to double, the solution can be found by drawing lines accurately, since we will need a type of number that cannot be arrived at by multiplication. The proof of this is as follows: a square site ten feet long and ten wide produces an area of a hundred square feet. If, then, we need to double it and produce a square of two hundred feet, we must find out how long the side of the square would be to obtain from it the two hundred feet corresponding to the doubling of the area. Nobody can discover this by calculation: for if we take the number fourteen, multiplication will give a hundred and ninety-six square feet; if we take fifteen, it will give two hundred and twenty-five square feet.

5. Therefore, since we cannot solve this problem arithmetically, a diagonal line should be drawn in the ten foot square from angle to angle so that it is divided into two triangles of equal size, each fifty feet in area; a square with equal sides should be drawn along the length of this diagonal. In this way four triangles will be produced in the larger square of the same size and number of feet as the two triangles of fifty square feet created by the diagonal in the smaller square. The problem of doubling an area was solved by Plato with this procedure using geometrical methods, as is shown in the diagram at the foot of the page.

6. Again, Pythagoras demonstrated how to devise a set-square without the intervention of workmen;[3] the results which

workmen arrive at when they make set-squares, with considerable effort but without great accuracy, can be arrived at with precision using the principles and methods derived from his teachings. For if we take three rulers, three, four and five feet long, and assemble them with their ends touching in the form of a triangle, they will form a perfect set-square. If squares with equal sides are drawn along the lengths of each ruler, the three-foot side will produce an area of nine square feet, the four-foot side an area of sixteen square feet and the five-foot side an area of twenty-five square feet.

7. In this way, the area in square feet generated by the two squares based on the sides three and four feet long is numerically equal to that produced by the square based on the side five feet long. When Pythagoras discovered this, he is said to have sacrificed victims to the Muses to express his boundless gratitude, never doubting that they had inspired him to make his discovery.

This theorem is not only handy for working out the dimensions of a number of objects, but it is perfectly adapted for finding the correct inclination of the steps when staircases are being built in houses. 8. For if the height of a storey from the ceiling above to the level of the floor below is divided into three units, five such units will provide the correct length for the incline of the string of the staircase; so whatever the height of the three units between the ceiling and the floor-level, four such units should be marked off from the [base of the] perpendicular [of three units] and there the lowest treads of the strings should be placed. In this way the configuration both of the steps and of the staircase itself will be established with precision. Again, a diagram of this technique will be given below [Plate 34].

9. Archimedes in fact made many extraordinary discoveries in different fields; of all of them, though, the one which I shall explain now seems to have been developed with truly infinite ingenuity.[4] It is hardly surprising that Hiero, on acquiring regal powers in Syracuse, decided to place a crown of gold which he had vowed to the immortal gods in a certain temple to celebrate his success; he let the contract for its production out at an agreed price and weighed the gold for the contractor in a balance. The

contractor presented the exquisitely made object to the king for his approval at the time established, and seemed to have presented a crown of exactly the right weight.

10. But afterwards there was a whisper to the effect that some gold had been removed and that an equal weight of silver had been mixed in during the making of the crown. Hiero, incensed that he had been made a fool of and not knowing how to prove fraud, asked Archimedes to consider the matter on his behalf. Then Archimedes, pondering on the problem, happened to go to the baths, and while immersing himself in a bath-tub, noticed that the volume of his body sunk in it equalled the volume of water spilling out of it. Since this showed him a way of resolving the problem, he ecstatically jumped out of the bath without a moment's delay and rushed off home, stark naked, announcing at the top of his voice that he had found what he was looking for, since as he ran along, he shouted repeatedly in Greek: 'Heureka, heureka' [I've found it, I've found it].

11. Then starting from this initial discovery, he is said to have made two ingots, one of gold, the other of silver, each equal in weight to the crown. When he had done this, he filled a large vessel with water up to the brim and put in the silver ingot: the volume of water that overflowed equalled that of the ingot in the vase. Then, taking the ingot out, he poured back the water that had overflowed, measuring it out with a *sextarius* jug,[5] so that the water reached the brim in the same way as before. In this way he discovered the weight of silver corresponding to a given measure of water.

12. After this experiment, he put the ingot of gold into the vessel full of water in the same way and, having taken it out and discovering its weight using the same method, found out that the amount of water displaced was not the same, but in fact less, by as much as the volume of an ingot of gold is less than that of an ingot of silver of the same weight. Finally, having filled the vessel again, he submerged the crown in the same quantity of water and found that more water overflowed with the crown than with the ingot of gold of the same weight, and so, continuing his line of reasoning from the fact that the crown caused more water-loss than the ingot of gold, he

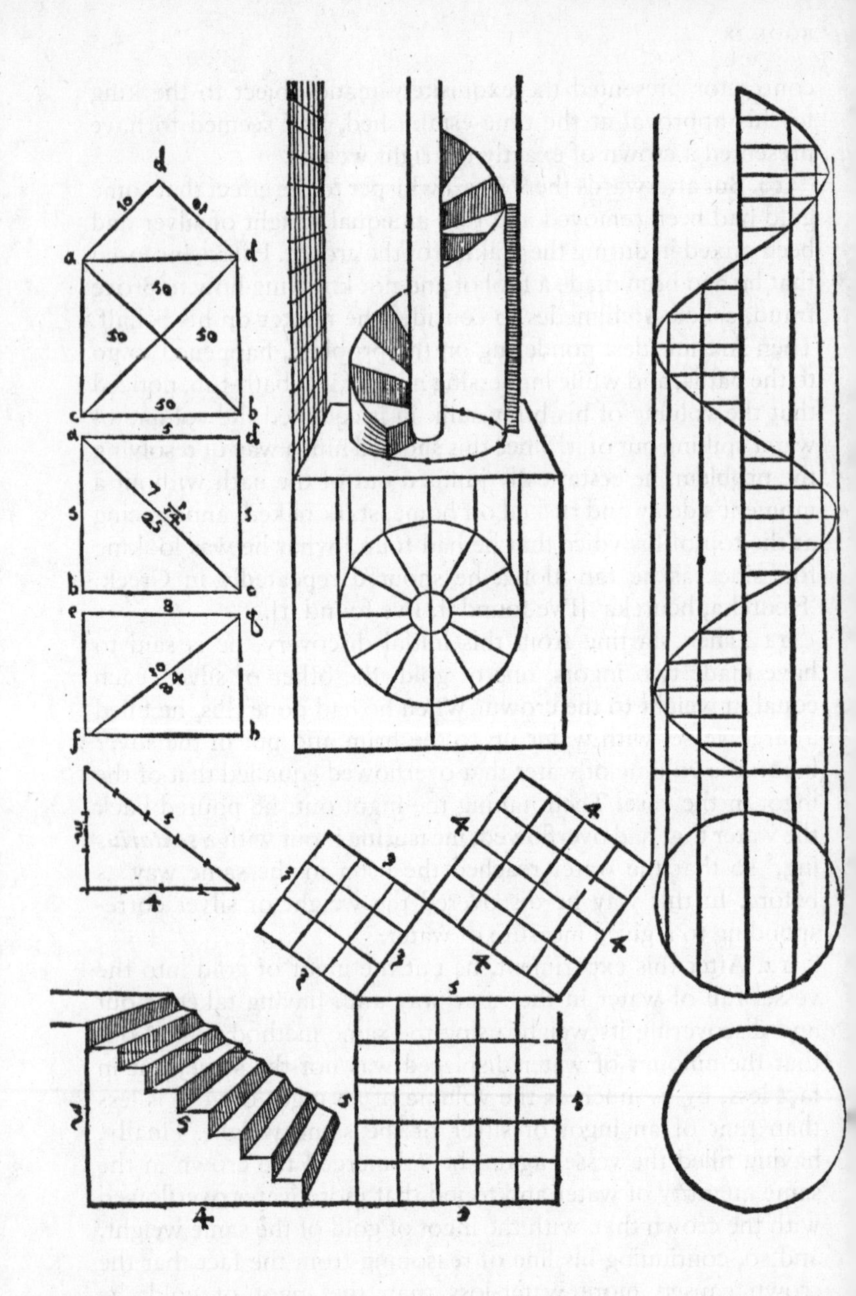

d
10
10
a
d
50
50
50
50
c
b
a
5
d
5
5
b
c
e
8
g
10
84
f
h
5
4
4
4
3
5
4
3

detected the presence of silver mixed with the gold, and so revealed the contractor's theft.

13. Now let us turn our attention to the ideas of Archytas of Tarentum and Eratosthenes of Cyrene, since they made many discoveries using mathematical techniques which are extremely valuable for mankind, and so, while they deserve recognition for their other inventions, they merit the greatest admiration for their theorizing[6] about the following problem: for each of them, using different methods, explained what Apollo had ordered [the islanders of] Delos to do in an oracular response: that the number of cubic feet his altars occupied should be doubled, and that in this way, the inhabitants of the island would be released from a divine curse. **14.** Accordingly Archytas solved it with his drawings of half-cylinders and Eratosthenes did the same using his device, the *mesolabe.*

Once we have appreciated these discoveries with the immense pleasure that derives from scientific erudition, we are naturally obliged to be deeply impressed by these inventions in disparate fields when we reflect on their results; contemplating many different subjects, I admire particularly the volumes of

Plate 34. Composite illustration. i. upper left margin: Plato's method of doubling the surface area of a square using equilateral triangles. ii. bottom left and centre: Pythagoras' theorem and its application to establishing the inclination of stairs; the same inclination is suggested for the water-screw described by Vitruvius (10.6). iii. upper centre and right: Construction of spiral staircase, based on Dürer; the method is the same as that described in 10.6 for the construction of the water-screw. The circular wall housing the stair forms a cylinder of which the ends are divided into twelve equal segments (eight in the case of the water-screw); running from these twelve segments along the length of the cylinder are equidistant longitudinal lines, as well as circumferential lines, which mark out a series of squares or rectangles; by tracing lines obliquely across the points of intersection of the circumferential and longitudinal lines, the curve of the spiral is arrived at. Altering the distances between the circumferential lines, so creating broader or smaller rectangles with longer or shorter diagonals, enables the inclination or steepness of the spiral to be adjusted.

Democritus on nature and his commentary entitled *Cheirometa* [Made by Hand], in which, in fact, he used a ring to mark in soft wax the principles which he himself had put to the test.[7]

15. Therefore the ideas of these men remain eternally valid not only with respect to the improvement of our morals but also because of their usefulness for all mankind. The fame of athletes, however, declines rapidly along with their bodily powers and so neither when they are at their most powerful nor in later ages [i.e. after their deaths] can they benefit the lives of men in the same way as the ideas of great intellectuals.

16. But although honours are not provided for the outstanding morals and teachings of writers,[8] their minds, looking up of their own accord to higher things, and raised up to heaven on the staircase of human recollection, ensure that not only their thoughts but also their likenesses are known to succeeding generations for an eternity. And so men whose minds are entranced with the delights of literature cannot fail to preserve the image of the poet Ennius dedicated in their hearts, as they do those of the gods. Those who are devotedly attached to the poetry of Accius seem to have with them not only the excellence of his language but even his portrait.

17. So, too, many born after our generation will feel as if they are disputing the natural world with Lucretius as though he were in front of them or the art of rhetoric with Cicero; and many of our successors will hold conversations with Varro about the Latin language. In the same way, there will be many men of culture who, considering numerous problems with the Greek philosophers, will feel that they are holding private discussions with them: and, in short, the thoughts of profound authors, despite their physical absence, flourish with the passage of time, and when they enter into debates and discussions enjoy greater authority than all the living disputants.[9]

18. And so, Caesar, I have written these books relying on these authorities and following their advice and opinions: I discussed buildings in the first seven books and water in the eighth: in this one I shall explain how the techniques for constructing sundials were discovered using the shadows cast by

the gnomon originating in the rays of the sun in the sky and by what principles such shadows lengthen and shorten.

CHAPTER I

The Zodiac and the Planets

1. These phenomena have been organized by the Divine Intelligence and those who contemplate them are absolutely astonished that the shadow of a gnomon at the equinox is of one length at Athens, another at Alexandria, yet another at Rome and different again at Piacenza, and so on in other places in the world. Accordingly the designs of sundials differ very widely when one passes from one region to another. For the configurations of *analemmata* are established on the basis of the lengths of the shadows at the equinox, and from these *analemmata* the demarcations of the hours are established according to the locales and the shadows of the gnomons. The ἀνάλημμα [*analemma*] is a system derived from the trajectory of the sun and from an observation of the shadow as it lengthens until the winter solstice; from this it has been possible to discover the sun's effect on the universe using architectural procedures and lines drawn with compasses.

2. Now the universe comprises the entire system of all natural phenomena, as well as the sky which is formed by the constellations and the courses of the stars. The sky revolves ceaselessly round the earth and sea on pivots at the extremities of its axis. For at these points the power of nature has located and structured the pivots as though they were the centres of circles: one at the highest point of the universe in relation to the earth and sea extending past the northern stars themselves; the other exactly opposite it under the earth in the southern regions; there, around the pivots, nature has fixed circles, called πόλοι [*poloi*] in Greek, as on a lathe, around which the sky flies repeatedly for ever. So the earth, as well as the sea, is naturally placed at the central point.

3. These things have been organized by nature in such a way that in the northern zone the central point is higher above the earth while in the southern region it has been set below in the lower regions under the earth, and is hidden by it. Then again there is a broad circular belt comprising the twelve signs [i.e. the constellations] located obliquely with respect to the centre and inclined towards the south; the appearance of each one of these, with the stars arranged in twelve equal sections, forms figures designed by nature. And so the constellations, glittering in the sky with the universe and the other stars and revolving quickly in splendid order around land and sea, complete their circuits following the sphere of the universe.

4. All of them are organized so that they are visible or not depending on the exigencies of the seasons. Six of these constellations revolve with the sky above the earth, and the others, which pass under the earth, are obscured by its shadow. But six of them are always flying above the earth; for however much of the most recently visible constellation is hidden under the earth, forced down by the descent caused by its rotation, the same amount of the opposite constellation, carried up by the inevitable circular movement and having completed its rotation, passes into the zones visible to us, and from the dark into the light. For one and the same irresistible force causes the simultaneous rising and setting of the constellations.

5. These constellations, of which there are twelve, with each one occupying a twelfth of the sky, revolve ceaselessly from east to west: but the moon, the planets Mercury and Venus, the sun itself, as well as Mars, Jupiter and Saturn, move from west to east incessantly in orbits of different size, as though running up stairs step by step, and pass through the constellations in exactly the opposite direction. The moon completes its circuit of the sky in twenty-eight days and about an hour, and, arriving at the constellation from which it set off, completes the lunar month.

6. The sun, however, crosses the expanse of a constellation, which is a twelfth of the sky, in the course of a month: so it travels the distance across the twelve constellations in twelve months and, returning to the one from which it began, com-

pletes the period of a full year. Consequently the circuit made by the moon thirteen times in twelve months is measured out by the sun only once in the same number of months. But the planets of Mercury and Venus, forming a crown around the rays of the sun at the centre with their trajectories, move backwards and slow down, and, because of their orbits, make stops and wait at certain points in the expanses of the constellations.

7. The existence of this phenomenon can be confirmed above all by the planet Venus, because when it follows the sun it appears in the sky after sunset shining brilliantly, and is called Vesperugo [Evening Star]: at other times, though, it runs in front of the sun and, rising before dawn, is called Lucifer [Morning Star]. Consequently, Mercury and Venus sometimes wait in one constellation for a number of days, and at other times move rapidly into the next sign. So, because they do not spend the same number of days in each constellation, the longer they have delayed at first, the more they rush at a much faster pace to the next, so completing the course as they should. The result is that although they are delayed in some constellations, they still complete their allotted orbit rapidly when they manage to pull themselves free from the enforced delay.

8. The planet Mercury flies through the universe in such a way that, running through the expanses of the constellations in three hundred and sixty days, it reaches the constellation from which it had begun its course during the preceding orbit, and in this way its journey on average equals, in terms of the number of days, about thirty in each constellation.

9. With regard to the planet Venus: when freed from the interference of the sun's rays, it runs across the space of a constellation in thirty days. Although it stays less than forty days in each constellation, it makes up that amount by remaining stationary in one constellation. Accordingly, when it has measured out the whole of its circuit in the sky on the four hundred and eighty-fifth day, it enters the constellation from which it began its journey once again.

10. And in fact the planet Mars, after moving across the expanses of the constellations for about six hundred and eighty-three days, arrives at the point from which it had previously

begun its course, and after it stops, makes up the required number of days in the constellations which it crosses more rapidly. The planet Jupiter, climbing at a more gentle pace against the rotation of the sky, measures out each constellation in about three hundred and sixty days, and, stopping after eleven years and three hundred and thirteen days, returns to the constellation in which it had been twelve years before. The planet Saturn, moving across the space of a constellation in twenty-nine months plus a few days, returns in the twenty-ninth year plus about a hundred and sixty days to the constellation where it had been thirty years before; because of this, the less Saturn's distance from the edge of the universe, the greater the orbit which it runs and the slower it appears to be.

11. These planets, making their circuits above the course of the sun, do not move forward, especially when they are in the triangle which the sun has entered, but pull back and wait until the sun has passed from the triangle into another constellation. Some maintain that this occurs because, they claim, when the sun is very far away, it hinders the stars moving across this expanse in badly lit trajectories with delays caused by the lack of light. But this does not seem right to me. The brilliance of the sun is visible and clear without being obscured in any way throughout the whole universe so that even we can see clearly when these planets move backwards and delay.

12. Therefore, if human eyesight can observe this phenomenon at such great distances, why should we think that anything could obscure the divine splendours of the stars? Instead, this is the explanation that makes sense to me: that heat attracts and draws everything to itself. We see, for example, crops rising out of the earth because of heat, not to mention water-vapour from springs driven up to the clouds along rainbows; on the same principle, the powerful energy of the sun, with its rays extended in the form of a triangle, draws to itself the planets that follow it, and, with regard to those running in front of it, prevents them from moving forward, as though reining them in and restraining them, and compels them to move back towards itself until it moves into the constellation of another triangle.

13. Perhaps the question will be raised of why the sun keeps the planets in the fifth constellation away from itself by enveloping them in heat in this way, rather than in the second or third, which are closer. I shall therefore explain how this seems to happen: the sun's rays extend in straight lines through the universe in the form of an equilateral triangle, that is, to the fifth constellation away from it and no more or less. Therefore if the rays diffused through the whole universe moved in circular orbits rather than being extended in straight lines forming a triangle, they would burn everything nearer to them. This is something that the Greek poet Euripides seems to have been aware of; for he says that zones at a greater distance from the sun burn violently while it maintains those that are closer at a more moderate temperature. So he wrote this in his tragedy *Phaethon*:[10] 'He burns things that are far away, and keeps things that are close temperate.'

14. If then, the facts themselves, our analysis and the evidence of an ancient poet demonstrate this point, I do not think that one should hold an opinion different from what I have written above on this subject.

The planet Jupiter traverses an orbit between the planets Mars and Saturn greater than that of Mars but smaller than that of Saturn. Again, with regard to the other planets, the further their distance from the edge of the universe and the closer their orbits to the earth, the more rapidly they seem to complete their circuits, because some of them, running through a smaller orbit, often go under the planets they find above them and overtake them.

15. In the same way, if one were to put seven ants on a potter's wheel, making the same number of concentric grooves on the wheel around its centre expanding progressively from that point towards the circumference in which the ants are obliged to make circuits; and if the wheel were to be revolved in the opposite direction, the ants will necessarily have to run against its direction of rotation; and the ant nearest the centre will proceed more rapidly than that in the circuit furthest from the centre of the wheel; and even though the ants run at the same speed, the latter will complete the circuit much more

slowly because of the extent of the circle. In the same way the planets, struggling on in the opposite direction to that of the universe, complete their circuits along their own trajectories, but because of the rotation of the universe are held back and delayed by its daily revolution.

16. The fact that some stars are temperate, others very hot and yet others cold seems to be due to the fact that all fires produce flames that rise to higher regions. Therefore the sun burns the atmosphere above it with its rays, making it incandescent in the very regions which the planet Mars passes through, and consequently the planet is burnt by the sun's heat. But the planet Saturn is intensely cold because it is nearest to the limit of the universe and touches the frozen regions of the sky. As a consequence, the planet Jupiter seems to enjoy a very balanced temperature corresponding to a median between the cold and heat of these planets since its course runs between the orbits of both of them.

I have now explained what I learnt from my teachers with regard to the belt of the twelve constellations and the contrary movements of the Seven Planets, about the principles and numerical relations according to which they pass from constellation to constellation, and about their orbits. Now I shall speak of the waxing and waning of the light of the moon according to what our ancestors have told us.

CHAPTER II

The Phases of the Moon

1. Berosus, who left the city, or rather nation, of the Chaldeans and even spread Chaldean learning in Asia, propounded the following:[11] that the moon is a ball of which one half is luminous white, the other half dark blue. When making its way through its orbit it passes under the disc of the sun and is then seized upon by its rays and the force of its heat, and turns its shining side towards the light of the sun because of its luminous properties. When it has been attracted to the disc of the sun, and its upper parts face upwards, then the lower half, which is not luminous, seems to be obscured because of its similarity to the surrounding air. When the moon is perpendicular to the rays of the sun, all the light is held on its upper surface; it is then called the new moon.

2. When the moon moves further on and arrives in the eastern zone of the sky, it frees itself from the force of the sun, and the periphery of its radiant half transmits its luminosity to the earth in the form of an extremely thin strip: for this reason it is called the second moon. Each day the moon frees itself as it revolves; the third moon, the fourth, and so on, are counted out each day. When the sun is in the west on the seventh day, the moon occupies the middle of the sky between east and west, and, since it is half the distance across the sky from the sun, presents half its luminous side to the earth. But when the entire distance across the sky lies between the sun and the moon, and at the rising of the latter the sun is at the extreme opposite in the west, the moon, liberated more and more from the attraction of its rays as the distance increases, transmits luminosity from the whole of its surface on the fourteenth day when its disc is complete. During the remaining days, it wanes daily until it completes the lunar month, and, undergoing once again the attraction of the sun in the course of its orbit, passes under its disc and rays and fulfils its monthly quota of days.

3. I shall now discuss the different explanation of these phenomena which Aristarchus of Samos,[12] a mathematician of great intellectual powers, left for us in his teachings: his theory is that it is obvious that the moon is not itself a light source but is like a mirror, of which the luminosity derives from the force of the sun. For of all the Seven Planets, the moon completes the shortest orbit, and its course is the one nearest the earth. So for the one day each month that the moon is under the sun's disc and rays before overtaking it, it is hidden and invisible; when it is in conjunction with the sun, it is called the new moon. On the next day, counted as the second, the moon overtakes the sun and allows a glimpse of the periphery of its sphere. When it has retreated from the sun for three days, the moon waxes and receives more light. Moving further away every day until it arrives at the seventh day, when its distance from the setting sun is about half the distance across the sky, half the moon is luminous, that is, the half facing the sun is illuminated by it.

4. On the fourteenth day, when the moon is diametrically opposite the sun across the whole expanse of the sky, it waxes and rises when the sun is in the west because it stops opposite it, separated from it by the distance across the universe, and receives the brilliance thrown by the force of the sun over the whole of its disc. When the sun rises on the seventeenth day, the moon is inclined towards the west. When the sun has risen on the twenty-first day, the moon occupies the regions roughly in the middle of the sky, and the side exposed to the sun is luminous and the rest dark. And so, continuing its orbit each day, the moon passes under the rays of the sun on about the twenty-eighth day, and so completes its lunar functions.

CHAPTER III

The Sun and the Constellations

1. Now I will describe how the sun, passing through a different constellation each month, makes the duration of days and hours increase and diminish. For when the sun enters the sign of Aries and passes through the eighth degree, it completes the vernal equinox. When it arrives at the tail of Taurus and the constellation of the Pleiades, from which the front half of Taurus projects, it runs along for a distance greater than half the universe as it moves north. When it progresses from Taurus into Gemini when the Pleiades are rising, it moves higher above the earth and increases the length of days. Then, from Gemini it enters Cancer, which occupies the least space in the sky, and, traversing the eighth degree, concludes the summer solstice and, moving on, arrives at the head and chest of Leo, because these parts of the sign are assigned to Cancer.

2. The sun, leaving the chest of Leo and the boundaries of

Cancer, and crossing the rest of Leo, reduces the length of the days and the extent of its own orbit, then goes back to a course identical to that which it had run in Gemini. Then crossing from Leo into Virgo and progressing as far as the fold of her dress, it shortens its circuit and follows the same type of course as it did in Taurus. Moving on from Virgo through the fold of her dress, which comprises the first degree of Libra, it completes the autumnal equinox in the eighth degree of Libra. This course has a circuit equal to what it was in the sign of Aries.

3. When the sun has entered Scorpio as the Pleiades are setting, it makes the days shorter as it moves towards the southern regions. Running on from Scorpio, it enters Sagittarius at the thighs and flies through an even shorter daily course. When it runs on from Sagittarius' thighs, which are allocated to Capricorn, it arrives at the eighth degree of this sign and flies through the shortest course in the sky. For this reason, the winter solstice [*bruma*] and the winter days [*dies brumales*] derive their names from their brevity.[13] But moving on from Capricorn into Aquarius, the sun increases the length of the day, making it the same as it was when it was in Sagittarius. When it has entered Pisces from Aquarius, and the Favonius [W] is blowing, it makes its course equal to what it was in Scorpio. In this way the sun orbits through the constellations, extending or shortening the length of the days and hours at certain times.

Now I would like to discuss the other constellations at the left and right of the zodiac in the southern and northern regions of the universe and which are organized in groups of stars formed into images.

CHAPTER IV

The Northern Constellations

1. The northern constellation [Great Bear], called Ἄρκτος [*Arktos*; Bear] or Ἑλίκη [*Helike*; Spiral] in Greek, has the Guardian [*Custos*] behind her. Not far from the Guardian is the

figure of Virgo, on whose right shoulder rests an extremely bright star which we call the Forerunner of the Vintage and the Greeks, Προτρυγητής [*Protrygetes*]. But in that constellation, Spica [Ear of Grain] is much brighter again. Opposite and between the knees of the Guardian of the Bear is yet another coloured star which has been enshrined there under the name Arcturus.

2. On an oblique line from the location of the head of the northern constellation down to the feet of Gemini, the Charioteer [Auriga] stands on the tip of the horn of Taurus; the Charioteer, in fact, holds the sole of his foot on the tip of Taurus' left horn. The stars supported by the hands of the Charioteer are called the Kids [Haedi], and the one at his left shoulder is the She-Goat [Capra]. Above Taurus and Aries is Perseus, who has the Pleiades running under his feet at his right, and at his left, the head of the Charioteer; his right hand rests on the figure of Cassiopeia, and with his left he holds the head of the Gorgon over Aries, throwing it at the feet of Andromeda.

3. Again, Andromeda is above the Pisces, as well as the stomach of the Horse [Equus] and its wings, located above its backbone; an extremely bright star separates its stomach from Andromeda's head. Andromeda's right hand is set above the figure of Cassiopeia, and her left is next to the northern Pisces. Again, the Horse's head is above the figure of Aquarius. The Horse's hoofs touch Aquarius' knees. Cassiopeia is consecrated in the middle. High above the image of Capricorn are the Eagle [Aquila] and the Dolphin [Delphinus], and nearby them is the Arrow [Sagitta]. Next to the Arrow is the Bird [Volucris], whose right wing touches the hand and sceptre of Cepheus, its left wing resting on Cassiopeia. The hoofs of the Horse are placed under the tail of the Bird.

4. Then the Serpent [Serpens], touching the Crown [Corona] with the tip of its snout, stands above Sagittarius, Scorpio and Libra. But the Snake-Handler [Ophiuchus] holds the Serpent by the middle with his hands, and treads on the face of Scorpio with his left foot. Not far from the Snake-Handler's head is that of the so-called Kneeler [Ingeniculatus]. The crowns of their heads are very easily observed because they comprise stars which are not at all dim.

5. One foot of the Kneeler rests on the temple of the head of the Dragon which is entwined with the She-Bears called the Northerners. In front of the beak of the Bird is placed the Lyre. The badly lit Dolphin curves between them. The Crown is located between the shoulders of the Guardian and those of the Kneeler. In fact, in the northern circle, the two She-Bears are placed back to back with their chests pointing in opposite directions; in Greek the smaller of the two is called Κυνόσουρα [*Kynosoura*; the Lesser Bear] and the larger, Ἑλίκη [Helike]: their heads are so arranged that they look in opposite directions. Their tails are represented in different directions with each opposite the head of the other; for both tails rise up and stand in the upper part of the sky.

6. Again, the Dragon is said to be stretched out between their tails, and in it, near the head of the Greater Northerner, twinkles a star called the Pole Star. The Dragon [Draco] winds itself around the head of the star nearest to it [The Greater Bear] but at the same time throws itself in a loop round the head of the Lesser Bear and stretches down very close to its paws. But there the Dragon, twisting back and coiling on itself, rises up and curves his snout and right temple back from the head of the Lesser Bear to the head of the Greater Bear. Again, Cepheus' feet are above the tail of the Lesser Bear, and there, at the very top, are stars forming an equilateral triangle. A fair number of stars are mixed haphazardly in with the Lesser Bear and the figure of Cassiopeia.

I have talked about the constellations located in the sky to the right of the east and between the belt of constellations and the north. Now I shall describe those that have been distributed by nature to the left of the east and in the southern regions.

CHAPTER V

The Southern Constellations

1. First, the southern Pisces lie under Capricorn facing the tail of the Whale [Cetus]: there is empty space between it and Sagittarius. The Altar [Turibulum] is under Scorpio's sting. The front limbs of the Centaur are very close to Libra and Scorpio, and he holds in his hands a figure which astronomers have named the Beast [Bestia]. The Hydra, extending itself in a long line of stars, twists itself round and binds together Virgo, Leo and Cancer, raising its snout up near the latter: it supports the Cup [Crater] with the central part of its body near Leo and slides its tail, on which the Crow [Corvus] stands, under the hand of Virgo. The stars located above the Snake's back are equally splendid.

2. The Centaur is located below the lower part of the Hydra's belly under its tail. Near the Cup and Leo is the ship called the Argo, of which the prow is invisible, but the mast and the higher parts around the rudder are easily recognized, while the poop itself of the craft is joined to the tip of the Dog's tail. The Little Dog follows the Twins opposite the Hydra's head. Again, the Greater Dog follows the Lesser. Orion lies obliquely below, pressed under Taurus' hoof, holding a sword in his left hand and brandishing a club in the direction of the Twins.[14]

3. At his feet is the Dog which follows the Hare at a close distance. The Whale lies beneath Aries and the Pisces, and from his dorsal crest a thin strip of stars, called in Greek Ἁρπεδόναι [*Harpedonai*; ropes, cords], is laid out regularly in the direction of the two Pisces. Moving inwards across a great expanse of space, the tight knot of winding stars touches the upper part of the dorsal crest of the Whale. The river Eridanus flows out in the form of stars from its source at the left foot of Orion. But Water [Aqua] which is recorded as being poured out by Aquarius runs between the head of the southern Fish and the tail of the Whale.

4. I have described the shapes and forms of the represen-

tations of the constellations in the universe designed by nature and by the Divine Intelligence, following the theories of the natural philosopher Democritus, but only those whose risings and settings we can observe and see with the naked eye. For just as the northerners [the Greater and Lesser Bears], gyrating around the pole of the axis, do not set or sink below the earth, so too there are stars that revolve around the south pole, which is always below the earth because of the inclination of the universe, and remain hidden there with no possibility of rising above the earth. So their configurations are unknown because the earth stands in the way. The example of the star Canopus proves this, for it is unknown in this part of the world, but merchants who visited the remotest areas of Egypt and the borders at the extreme limits of the earth report on it.

CHAPTER VI

Astrology and Astronomy

1. I have explained the perpetual motion of the universe around the earth, and the disposition of the Twelve Signs and of the constellations in the northern and southern hemispheres so that they may be distinguished easily. For it is from this gyration of the universe, from the contrary trajectory of the sun through the signs and from the shadows of gnomons during the equinox that we develop the design of *analemmata*.

2. With respect to other aspects of astronomy concerning the influence which the twelve zodiacal signs, the Five Planets and the sun and the moon have on the course of human life, we should leave them to the calculations of the Chaldeans, because the method of casting horoscopes by which they are capable of explaining the past and future using computations based on the stars is entirely theirs. For men of wisdom and great acuity descending from the nation of the Chaldeans itself have left us their discoveries; first, Berosus, who settled in the city of Cos on the homonymous island, where he disseminated the discipline;

afterwards Antipater studied the subject, as well as Athenodorus, who also left us techniques for devising horoscopes which were not derived from the moment of birth but from that of conception.

3. But with regard to the natural sciences, Thales of Miletus, Anaxagoras of Clazomenae, Pythagoras of Samos, Xenophanes of Colophon and Democritus of Abdera have handed on to us the theories they had developed about how natural phenomena are governed, and the ways in which they produce their effects. Following on from these discoveries, Eudoxus, Euctemon, Callippus, Meton, Philippus, Hipparchus and Aratus and others discovered the laws concerning the risings and settings of the constellations and of the weather on the basis of astronomy using solar calendars, and left the principles they had established for later generations. Their expertise deserves the admiration of mankind for it was so precise that they are able, apparently endowed with divine intelligence, to make predictions about what the future seasons will be in advance. For this reason, this subject must be left to their scrupulous researches.

CHAPTER VII
The analemma

1. But these subjects apart, it is our task to present clear explanations of the principles governing the shortening and lengthening of the days each month. For when the sun passes through Aries or Libra during the equinox, it makes the gnomon cast a shadow eight-ninths of its own length at the latitude of Rome. Similarly in Athens, the shadow is three-quarters the length of the gnomon, at Rhodes, five-sevenths, at Tarentum, nine-elevenths and at Alexandria, three-fifths; and so in other places one finds that all the shadows of the gnomons at the equinox differ from each other as a result of natural phenomena.

2. In which case, the length of the shadow during the equinox must be taken wherever sundials are to be laid out. If, as in Rome, the shadows equal eight-ninths of the gnomon, a line should be drawn on a flat surface, and at its central point a perpendicular, called a gnomon, should be raised with the help of a set-square: then, starting from the line drawn on the flat surface, nine divisions should be measured off on the line of the gnomon with compasses, and, at the point that will mark the ninth division, the centre should be fixed where the letter A will be: then, with the compasses opened, the circumference of a circle, which we shall call the meridian, should be drawn from this central point to the line on the flat surface where the letter B will be (Fig. 28).

3. Then eight of the nine divisions between the flat surface and the centre of the gnomon should be taken and marked off on the line drawn on the plane to where the letter C will be. This will be the shadow of the gnomon at the equinox. From the point indicated by the letter C, a line should be drawn through the centre indicated by the letter A, which will represent a ray of the sun at the equinox. Then, with the compasses opened up, equidistant points farthest from the gnomon should be marked off on the circumference drawn from the centre to the line on the horizontal surface, indicated by the letter E at the left and I on the right, and a line should be drawn through the centre such that it divides the circle into two equal semicircles; this line is called the horizon by mathematicians.

4. Then a fifteenth of the entire circumference should be divided off and the centre of the compasses placed on it where the equinoctial ray intersects the line where the letter F will be, and the letters G and H should be marked off to right and left. Then lines must be traced from these points [through the centre] to the line on the flat surface where the letters T and R will be. These will represent the rays of the sun, one for the winter, the other for the summer. Opposite E will be the letter I, at the point where the line that runs through the centre where the letter A is cuts the circumference; and opposite G and H will be the letters L and K, and opposite C, F and A will be the letter N.

5. Then, diameters should be drawn from G to L and from

H to K. The upper part will indicate the summer zone, the lower, the winter zone. These diameters are to be divided into two equal parts at the points in the middle which will be indicated by the letters M and O: a line will be drawn which passes through these two points and the centre A to cut the circumference at letters P and Q. This line will be at right angles to the equinoctial ray and is called 'the axis' in mathematical parlance. And starting from these same centres [M, O], the compasses should be opened up as far as the ends of the diameters, and semicircles should be drawn there, one for summer, the other for winter.

6. Next: at the points where the parallel lines intersect the line called the horizon, the letter S will be at the right and the letter V on the left; from the extremity of the semicircle where the letter G is, a line parallel to the axis should be drawn to the semicircle at the left where the letter H is. This parallel line is called the *logotomus*. Then one must centre the compasses at the point where the equinoctial ray intersects the line where the letter D will be, then open them up to the point where the summer ray intersects the circumference where the letter H is. From the equinoctial centre, at a distance equivalent to the summer ray, the circumference of the circle of the months called the *menaeus* should be drawn. This is how we shall establish the configuration of the *analemma*.

7. Once it has been drawn and put in order like this, the system of the hours must be worked out following the *analemma* and projected onto a horizontal plane using either winter, summer or equinoctial lines or those of the months as well. Many different types of sundials derive from the *analemma* and are set out with these technical procedures: but the result of all these designs and their realizations is the same: the division of the day at the equinox and at the winter and summer solstices into twelve equal parts. This is why I have left this subject out, not because I was put off by laziness,[15] but so as not to bore readers by writing too much about the inventors of the various types and layouts of sundials. For in fact I am not in a position to invent new types now nor does it seem right to pass off others' inventions as my own. I shall therefore talk about the types of sundials that have been handed down to us and of their respective inventors.

CHAPTER VIII
Sundials and Water Clocks

1. The half-cylindrical type hollowed out of a square block of stone and sliced according to the inclination of the pole is said to have been invented by Berosus the Chaldean; the concave quadrant or hemisphere, as well as the flat disc, by Aristarchus of Samos; the inventor of the *arachne* [spider] was the astronomer Eudoxus or, as others maintain, Apollonius; the *plinthium* [plinth] or *lacunar* [coffer], like the one, for example, in the Circus Flaminius, was invented by Scopinas of Syracuse; the quadrant called πρὸς τα ἱστορούμενα [*pros ta historoumena*; for places studied] by Parmenion; that called πρὸς πᾶν κλῖμα [*pros pan klima*; for every latitude] by Theodosius and Andrias; the *pelecinum* [double axe-head] by Patrocles; the *cone*, by Dionysodorus; the *quiver*, by Apollonius; and other types again have been invented and handed down both by the men just mentioned and by many others, for example, the so-called *conical spider*, the *conical plinth*, and the one facing north. Again, many astronomers have also left us written instructions for making other types based on these, such as portable ones that can be hung up. Whoever wishes to find out about their various horizontal projections from the writings of these men will be able to do so provided he understands the layout of the *analemma*.

2. Again, these same writers have researched techniques for making clocks running on water, above all Ctesibius the Alexandrian, who also discovered natural air-pressure and the principles of pneumatics. But it is worthwhile for students of the subject to appreciate how these discoveries were made. Ctesibius, then, the son of a barber, was born in Alexandria:[16] he stood out from his contemporaries for his natural talents and great industry and was well known for his passionate interest in mechanical devices. For when he wanted to hang up a mirror in his father's shop in such a way that, when it was lowered and raised, a hidden weight on a piece of string would take it back up, he set up a device of this type.

3. He fixed a wooden channel under a beam and housed a set of pulleys in it. He led a cord along the channel to the corner of the room where he placed some narrow tubes. He arranged that a ball of lead attached to the string could be let down in the tubes. In this way the weight falling through the narrow pipes compressed the air, and, because of its violent descent through the narrow apertures, forced the air, now denser because of compression, into the open, producing a high-pitched sound because of the impact [with the air outside].

4. So Ctesibius, realizing that jets of compressed air in contact with the atmosphere produce noisy blasts of air, was the first to construct hydraulic machines using such principles. He also devised jets of water, automata and playthings of various kinds, which included mechanisms for water clocks. First he made an aperture in a piece of gold or in a gem by drilling through it, because these materials do not suffer from erosion from the flow of water or attract deposits that could cause blockages. **5.** For water flowing regularly through such an orifice can keep afloat an inverted bowl, which clockmakers call a 'cork' or 'drum': a vertical bar which makes contact with a revolving disc is fixed to the 'cork' and the bar and disc have matching teeth. These teeth engage with each other and produce precisely calibrated rotations and movements. Again, other bars and discs provided with matching teeth produce various kinds of motion when they turn and are subjected to the same thrust: figurines that move, cones that revolve, pebbles or eggs which are thrown, trumpets which sound and other amusing little things.

6. In these clocks the hours are marked on a column or pilaster, and a figurine emerging from the bottom indicates them with a pointer throughout the day. Wedges are inserted or extracted each day of every month to reproduce the decrease or increase in the duration of the days. The stopcocks for controlling the flow of water are constructed like this: two cones are made, one solid and the other hollow, and finished on a lathe so that one will go into and fit perfectly with the other; they open or close because of the action of the same bar causing the water to run quickly or slowly into the vessels. So, employing these principles and mechanisms, one can assemble the workings of winter clocks running on water.

7. But if one finds that the reproduction of the shortening or lengthening of the days by the insertion or extraction of wedges is unsatisfactory since they are often defective, one must proceed like this. The hours taken from the *analemma* should be drawn transversely on the little column and the lines of the months should also be marked on it. The column should be made to revolve so that, as it turns continuously in the direction of the figurine emerging with its pointer and indicating the hours, it would reproduce the shortening or lengthening of the hours for each month by revolving continuously [Plate 35].

8. Other types of winter clocks, called 'anaphoric' [indicating the rising of the stars], are constructed like this: the hours are indicated by bronze rods placed at the front and radiating from the centre following the scheme of the *analemma*. On this surface circles are marked indicating the extent of each month. Behind these rods is a disc on which the universe and the zodiac are laid out and represented; starting from the centre, images of the Twelve Signs should also be shown, some represented on a large scale, others on a smaller scale. An axle is inserted in the centre of the back of the disc and a flexible bronze chain wound around it: at one end of the chain hangs the 'cork' raised by the water, and at the other a counter-weight of sand equal in weight to the 'cork'.

9. Hence, the cork is raised by the water by the same amount that the counter-weight of sand falls, and so turns the axle, which then turns the disc. The rotation of this disc sometimes causes a larger, and sometimes a smaller section of the zodiac to indicate the length of the hours corresponding to their seasons in the course of its rotations. For holes have been made in each sign which correspond to the number of days of the month, and a boss, evidently used to represent the sun in clocks,

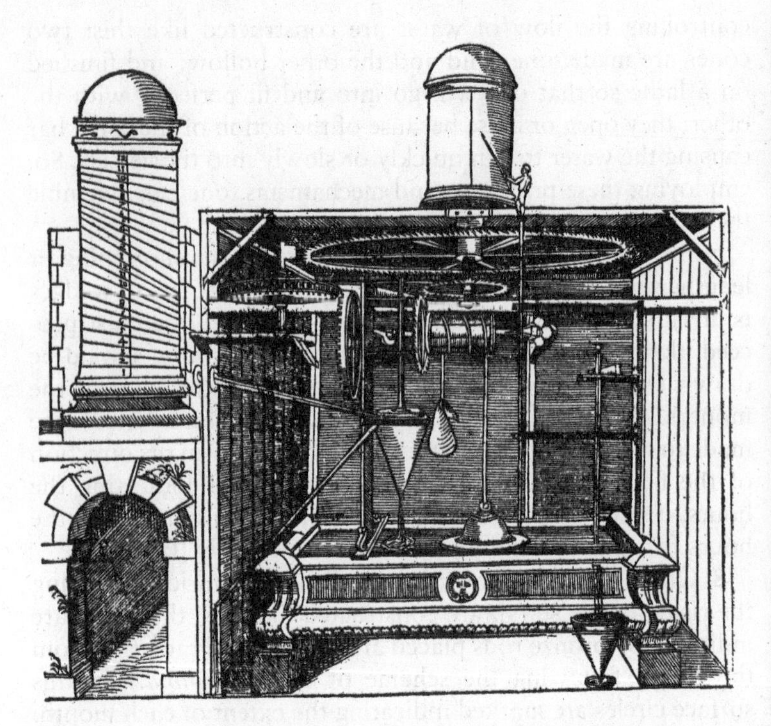

Plate 35. Water clock. Vitruvius explains that an object like an inverted bowl floats in a water-tank and its rises and falls move a vertical bar with teeth which mesh with those on a horizontal disc causing objects attached to the disc to revolve. The hours are marked on a cylinder and indicated by a figurine with a baton at the bottom of the column. The water supply is controlled by raising or lowering a bar of which the height is adjusted by placing wedges of varying sizes under it (not shown here); stopcocks which control the flow of water are suspended from the bars; the stopcocks consist of two cones fitted together, and raising or lowering the upper one lets more or less water through and hence governs the rise or fall of a vertical bar with teeth attached to the upper cone. Barbaro's version is an imaginative concoction of these elements; he shows the two cones, as well as a vertical bar, but does not connect the vertical bar with the floating inverted bowl and adds a weight on an axle as a counterbalance taken from the 'anaphoric' clock.

indicates the duration of the hours. This boss, passing from hole to hole, completes the circuit of the month as it turns.

10. So it is that, just as the sun, travelling through the expanses of the constellations, makes the days and hours longer or shorter, so the boss on the clock, advancing from hole to hole in a circuit contrary to that of the centre of the disc, moves itself each day over sometimes wider and sometimes narrower distances at different periods, and so provides a visualization of the hours and days within the divisions of each month. With regard to the supply of water and the way in which it should be regulated, we must proceed as follows.

11. On the inside and behind the face of the clock a tank should be installed into which water can run down through a pipe: in the bottom it should have a hole, to which a bronze disc is attached with an aperture through which water from the tank can flow through the disc. A smaller disc should be fixed inside the larger disc with hinges and sockets worked on a lathe and fitted together in such a way that the smaller disc, revolving inside the larger one like a valve, rotates smoothly.

12. The rim of the larger disc should be marked off with three hundred and sixty-five equidistant points, while the smaller disc should have a little pointer fixed on the edge of its circumference, the tip of which should be directed to the area with the points; in this little disc there should also be an aperture of the appropriate size. Since the heavenly constellations will be represented on the rim of the larger disc, but the disc itself should remain motionless, the sign for Cancer should be represented at the top, with that for Capricorn vertically below it at the bottom; Libra at the spectator's right; the sign for Aries at the left, and all the other signs should be represented in the spaces between them in the order in which they appear in the sky.

13. As a result of this, when the sun is in Capricorn, the pointer of the small disc touches daily each of the points of Capricorn in the section reserved for it on the larger disc, and is perpendicular to the vertical pressure of the running water; this pushes the water rapidly through the hole of the small disc into the container, which takes in the water and soon fills up,

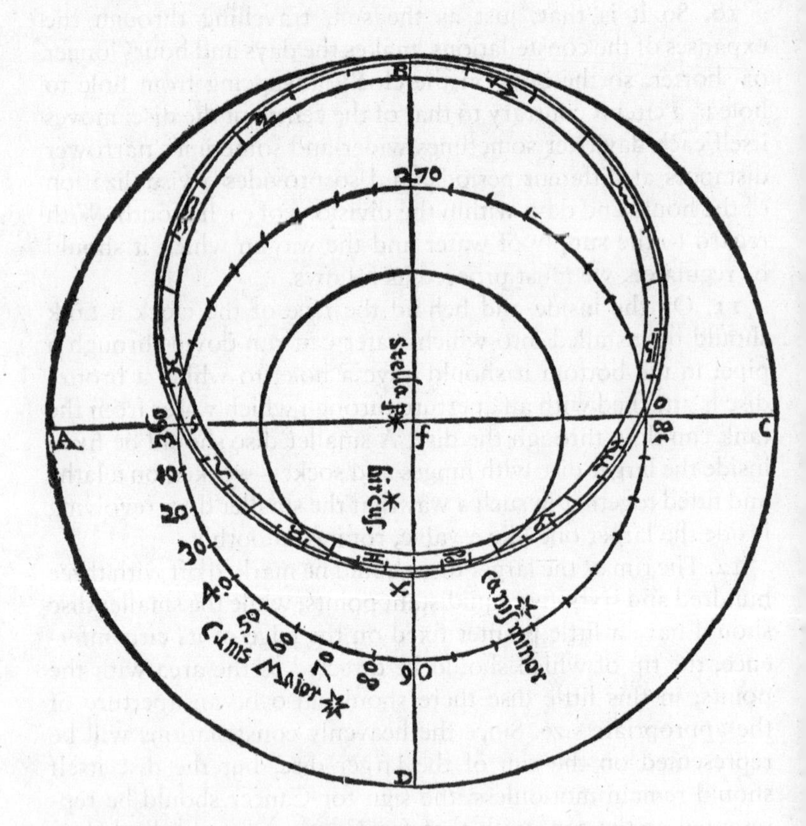

Plate 36. Face of water clock. The grille at the right fits over the representation of the constellations at the left.

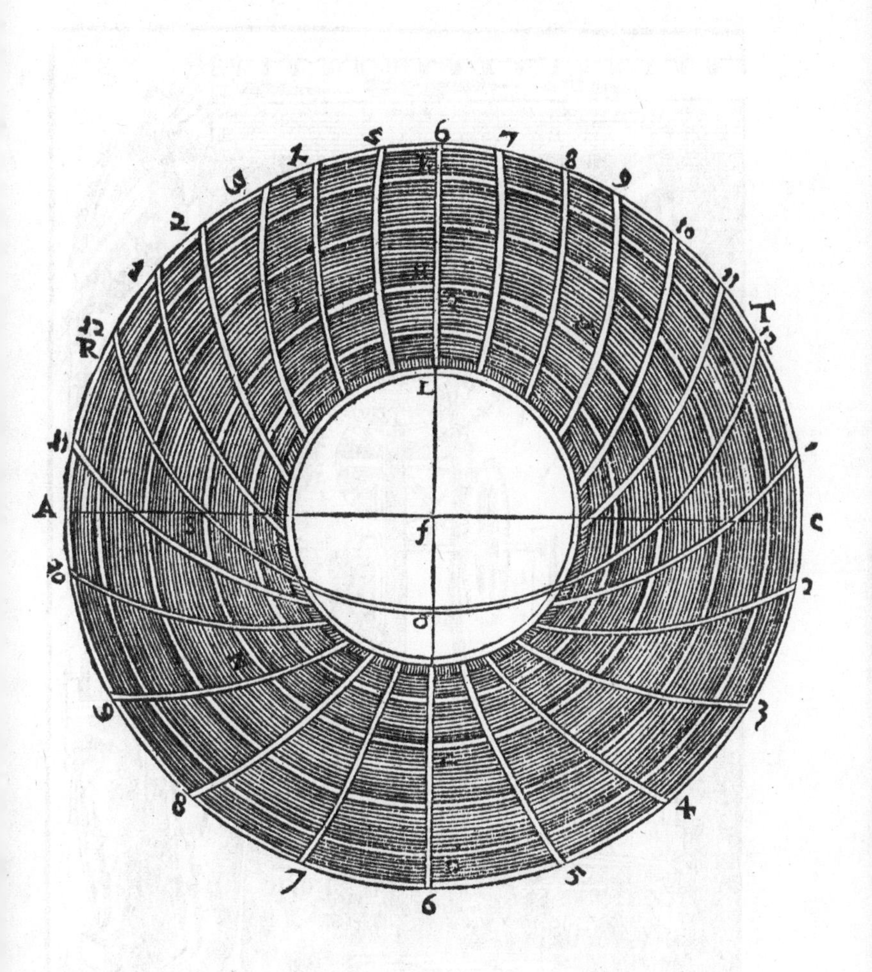

6
7
8
9
10
11
T
12
1
C
2
3
4
5
6
7
8
9
10
11
A
R
12
1
2
3
4
5
L
f
D

Plate 37. Mechanism of water clocks. Barbaro shows the axle inserted into the back of the clock face and the chain wound around it, with the inverted bowl at one end and a counter-weight at the other. Inside the building he shows the winter clock, regulated by a variable stream of water from a tank behind the clock face running through an aperture in the inner moving facing (illustrated next). *Telamones*, mentioned by Vitruvius at 6.7.6, appear supporting the structure at the right.

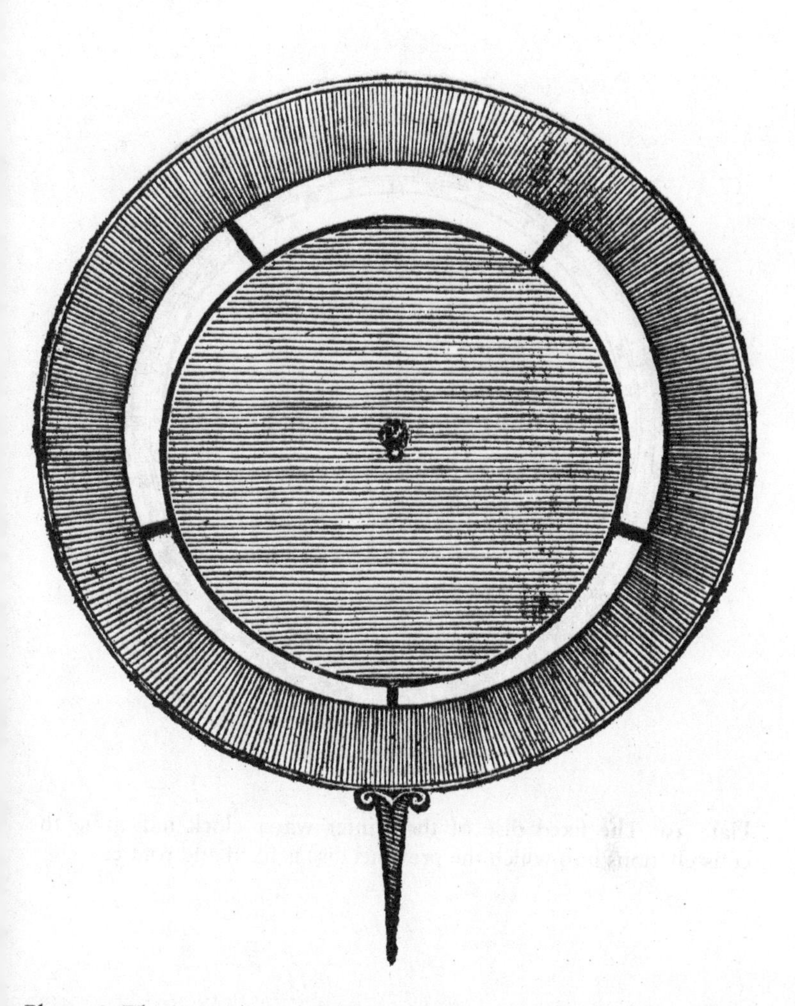

Plate 38. The inner disc of the face of the winter water clock with its pointer.

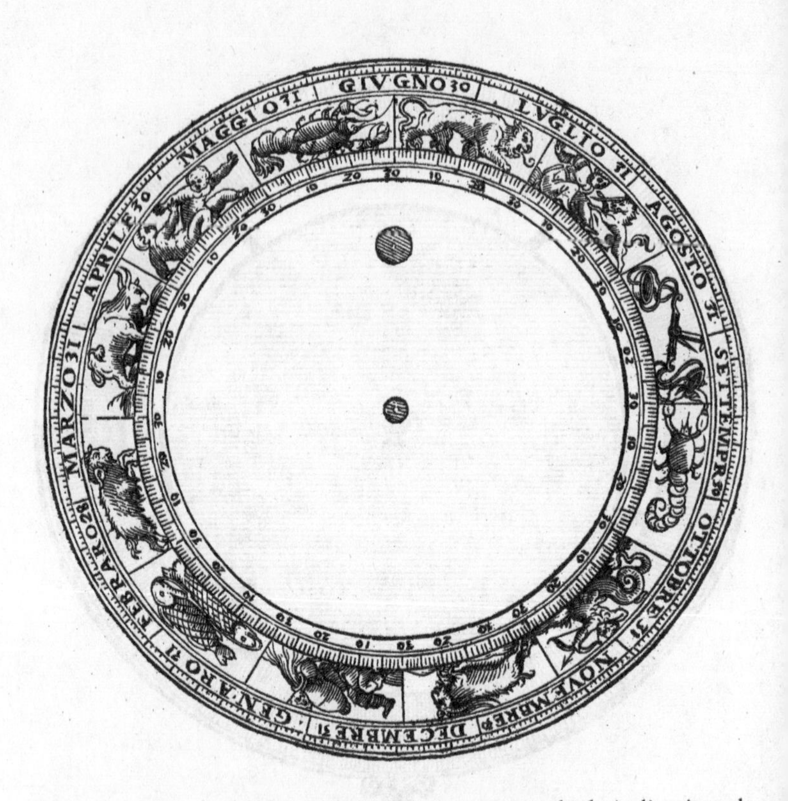

Plate 39. The fixed disc of the winter water clock indicating the constellations into which the previous disc is fixed and rotates.

so reducing and contracting the duration of the days and hours. But when, during the daily revolution of the smaller disc, the pointer enters the group of points of Aquarius, the aperture will incline away from the vertical and the water will be compelled to flow out more slowly instead of in a violent rush. So the lower the velocity of the water received by the vessel, the more the length of the hours increases.

14. The hole in the small disc ascends through the points of Aquarius and the Pisces, as though going upstairs, and, touching the eighth degree of Aries, indicates the equinoctial hours thanks to a moderate flow of water. When, as a result of the rotations of the disc, the hole moves from Aries through the areas of Taurus and Gemini up to the highest points of Cancer corresponding to the eighth degree, and in this way goes back up, the force of the water grows weaker; and the slower flow of water causes delays and, by lengthening the days, produces the hours of the summer solstice in the constellation of Cancer. When it descends from Cancer and passes through Leo and Virgo, it turns back as far as the points indicating the eighth degree of Libra, and gradually shortens and reduces the lengths of the hours, so arriving at the points in Libra, where it creates the equinoctial hours once again. **15.** The aperture descends more steeply through the sectors of Scorpio and Sagittarius, and on its return, having completed its revolution to the eighth degree of Capricorn, the impetus of the flow of water restores it to the short hours of the winter [Plates 36–9].

I have described in detail as best I could the principles and components involved in designing clocks in order to make them easier to use. Now it remains for me to discuss mechanical devices and their principles. In order to make my treatise on architecture comprehensive and free from oversights, I shall start writing on these subjects in the following book.

BOOK X

Introduction

1. It is said that in the famous and magnificent Greek city of Ephesus a law with stringent but not unjust stipulations was established long ago by their ancestors: when an architect accepts the commission for some public building, he declares in advance what the cost of the operation will be. Once his estimate has been handed over to a magistrate, his property is pledged as collateral until the work has been completed. Once it is finished, if the cost corresponds to the estimate, he is rewarded with official decrees and honours. Again, if no more than a quarter has to be added to his estimate, it is made up from public funds and no penalty is inflicted on the architect. But if more than a quarter extra is spent on the work, the money needed to make up the difference is taken from the architect's own resources.

2. If only the immortal gods had arranged that a similar law had been established for the Roman people not only for public but also for private buildings. For the incompetent would no longer run riot unpunished, but only those who are trained most rigorously in the subtleties of the discipline would practise the profession of architecture without diffidence:[1] nor would the heads of families be misled into wasting such enormous sums of money that they find themselves evicted from their own properties; and architects themselves would be forced by their fear of the penalty to calculate the amount of the expenditure more carefully so that heads of households could complete their buildings for the amount they had fixed in advance, or with only a small addition. For men who can afford to spend four hundred thousand *sestertii* on a building are delighted by the prospect of the completion of the work, even if they have to put in another hundred thousand; but if they are weighed down with an increase of half the cost again or more, they give up hope and, having wasted the money spent so far, are forced to abandon the project with their finances and morale devastated.

3. This is a problem which affects not only buildings but also shows laid on by magistrates, whether they are gladiatorial contests in the forum or stage-plays: in these cases no delay or postponement is allowed and absolute necessity demands that everything should be ready for a specific time: the seats for the audience, the devices for pulling the awnings across and all the equipment that is prepared for the delight of the public using machinery following the traditions of the theatre. In this field one needs a thorough training coupled with a highly qualified mind because none of these things can be carried out without the use of machines and competent and versatile application to one's studies.

4. Therefore, since these are our traditions and practices, it does not seem inopportune that, before work is begun, the requisite procedures should be set out in advance carefully and with the greatest diligence. Consequently, since we have no law or established custom to enforce this practice and every year the praetors and aediles must provide machinery for the games,[2] I thought that it was not inappropriate, Supreme Ruler, to explain in this volume, which constitutes the conclusion of the treatise, the principles behind machines in order of subject, since I have discussed buildings in the previous books.

CHAPTER I

Machines and Instruments

1. A machine is a self-contained combination of wooden elements fastened together with the greatest capacity to move weights. It is driven by rotations according to the principle of circular movement that the Greeks call κυκλικὴ κίνησις [*kyklike kinesis*]. There is a type for climbing, which the Greeks call ἀκροβατικόν [*akrobatikon*], another which operates using air, which they call πνευματικόν [*pneumatikon*], and a third for hauling objects which they call βαρουλκόν [*baroulkon*]. But climbing machines are those that have been constructed with beams arranged vertically and bound together by cross-bars[3] so that one can safely ascend them to a considerable height in order to oversee operations. By contrast, pneumatic machines are those in which the air is forced out under pressure or by a blow, and sounds are emitted by means of an instrument [*organikos*].

2. A hauling system is one by which loads are pulled by machines in such a way that they can be raised up high and put in place. The system used for climbing machines cannot be credited with any scientific principle except personal bravery: it is held together by dowels, cross-beams, entwined ropes and supporting props. A machine that is set in motion by pneumatic pressure produces attractive results thanks to technical refinements. But hauling machines have a greater and more spectacular potential for work and are immensely useful if operated with skill.

3. Some of these machines operate mechanically [*mechanikos*], others instrumentally [*organikos*]. Evidently the difference between machines and instruments is that machines require more operatives and greater power to make them produce a result, such as *ballistae* and the beams of presses: but instruments [*organa*][4] fulfil their required function at the skilled touch of a single operative, such as rotating a *scorpio* or systems of gears of different sizes. Both instruments and mechanical systems are therefore essential for practical purposes and without them nothing can be done without difficulty.

4. All mechanisms owe their origins to nature and are made following the guidance and instruction of the rotation of the universe. First of all let us consider and examine the system comprising the sun, the moon and the Five Planets; if they had not revolved in accordance with the laws of mechanics we would not have had regular periods of light or the ripening of fruit. When, therefore, our ancestors had understood the nature of these phenomena, they selected examples from nature, and by copying them, were inspired by such divine exemplars to perfect versions useful for their way of life. To make them more readily useful, they made some systems in the form of machines with rotating mechanisms and others in the form of instruments: and so with study, technical skills and gradual improvements in scientific knowledge, they gradually perfected the things which they had realized were practically useful.

5. First, let us consider an invention initially brought about by necessity, such as clothing, for example: how weaving together the warp and the woof of cloth, thanks to the work of instruments, not only ensures that our bodies are protected simply by being covered but also provides them with attractive clothing. We would certainly not have had rich supplies of food if yokes and ploughs for oxen and all the draught-animals had not been invented. If windlasses, beams and levers had not been invented for presses, we would not have been able to enjoy the gleam of oil nor the fruit of the vine; and these things could not have been transported on land or water without the invention of the mechanisms for carts or wagons on land and ships at sea.

6. The discovery of methods for weighing with scales and balances safeguards our lives from fraud by introducing honest practices. Again, there are innumerable varieties of mechanical devices about which it seems unnecessary to speak, since they are to hand every day, such as mills, blacksmiths' bellows, four-wheeled wagons and two-wheeled carts, lathes and other devices that fulfil ordinary practical functions in everyday life. So let us begin with those which we encounter rarely so that they may become better known.

CHAPTER II

Hoisting Machines: The trispastos, pentaspastos *and* polyspastos

1. First of all, let us deal with those machines which must be prepared for sacred temples and the construction of public buildings, which should be assembled like this: two wooden beams [*tigna*] proportional to the size of the load are procured. They are erected and fastened together at the top with a bolt [*fibula*] and spread apart at the bottom; they are held upright by ropes tied to them at the top and fixed in the ground at intervals around them. A pulley-block [*troclea*], which some also call a *rechamus*, is fastened at the top. Two pulleys [*orbiculi*] revolving on axles [*axiculi*] are inserted into the pulley-block: a traction-rope [*ductarius funis*] is passed around the upper pulley, then is let down and taken round the pulley in the lower pulley-block; then it is brought up again to the lowest pulley in the pulley-block at the top, and once again goes back down to the lower pulley-block, where it is tied off in a hole [*foramen*]. The other end of the rope is brought back down between the feet of the machine.

2. The socket-pieces [*chelonia*] in which the ends of the windlasses [*suculae*] are fixed so that the axles may turn freely are secured to the flat, rear surfaces of the beams at the point where they spread apart.[5] Near the ends of these windlasses are two holes arranged so that hand-spikes [*vectes*] can be fitted into them. Iron pincers [*ferrei forfices*], of which the points fit into the holes drilled in stones, are fastened at the bottom of the lower *rechamus*. When one end of the rope is fastened to the windlass and the latter is turned using the hand-spikes, the rope winds itself round the windlass and becomes taut, and so raises the loads to the height and location of the work being undertaken [Fig. 29].

3. This type of machinery, which depends on the rotations of three pulleys, is called a *trispastos*. When in fact there are two pulleys revolving in the lower pulley-block and three in the

upper, the machine is called a *pentaspastos*. But if machines have to be prepared for heavier loads, then longer and thicker beams must be used, and one must proceed, using the same methods, with bolts at the top and windlasses that turn at the bottom. When these are ready, support-ropes [*antarii funes*], initially left slack, should be attached, and control ropes [*retinacula*] should be let out over the shoulders of the machine for a considerable distance; if there is nowhere where they can be tied off, inclined stakes [*pali*] to which the ropes can be tied should be fixed into the ground and made secure by ramming down the earth around them

4. A pulley-block should then be attached by a hawser [*rudens*] to the top of the machine, and a rope should be led from it to a stake with a pulley-block tied to it. The rope should be passed round the pulley and taken back to the pulley-block that will be tied to the top of the machine. Then the rope, having been passed around the pulley, should go down from the top and return to the windlass at the foot of the machine so that it can be tied off there. The windlass will now be turned by means of hand-spikes and will raise the machine by itself without any danger. In this way, with ropes and hawsers arranged around it and attached to stakes, a machine of greater dimensions can be set up. The pulley-blocks and traction-ropes are arranged as described above.

5. However, if the loads to be used for the work happen to be absolutely colossal in size and weight, we should not rely on the windlass, but, in the same way that a windlass is held in place by its sockets, one should install an axle with a large drum in the middle, which some call a wheel [*rota*], and the Greeks call an ἀμφίεσις [*amphiesis*; drum], others, a περιθήκιον [*perithekion*; container].

6. Not all the pulley-blocks in these machines are arranged in the same way, and there are other methods; for they have double rows of pulleys at the bottom and at the top. The traction-rope is passed through a hole in the lower block so that the two ends of the rope are of equal length when it is taut; and both parts of the rope are held in place at the lower pulley-block by a cord [*resticula*] passed around them and tied

so that they cannot come adrift to right or left. Then the ends of the rope are taken up into the upper pulley-block from the outside and passed down around its lower pulleys; then they return to the bottom, where they are passed from the inside to the pulleys in the lowest pulley-block, then return to the top and around the highest set of pulleys.

7. After they have been passed from the outside, they are taken to right and left of the drum on the axle where they are tied so that they stay fast. Then another rope which has been wound round the drum is taken to a capstan [*ergata*], and when the capstan[6] is turned, it makes the drum and axle revolve: the ropes tighten equally as they wind round and so gradually raise the loads without danger. But if a larger drum is placed in the middle or far to one side, the men treading round it will be able to carry out the work more quickly without a capstan.

8. There is another kind of machine, ingenious enough and devised for rapid use, but which only skilled men can operate. It comprises, in fact, a single beam erected and held in place with ropes at four points in the ground. Two socket-pieces are fixed under the retaining ropes and a pulley-block is bound with ropes above the socket-pieces; a straight piece of wood [*regula*] about two feet long, six digits wide and four thick is set under the pulley-block. Pulley-blocks with three rows of pulleys placed in line are put in position. In this way three traction-ropes are attached to the machine and are then led down to the pulley-block at the bottom and passed round its upper pulleys from the inside. Then they are taken back to the upper pulley-block and passed from the outside to the inside through the lowest pulleys.

9. When they have gone down to the bottom, they are pulled across the second row of pulleys from the inside to the outside, and taken back up to the second row of pulleys at the top: having been passed through them they go back down to the bottom, and from the bottom are pulled back up to the top; once passed through the pulleys at the top they drop down to the base of the machine. A third pulley-block, which the Greeks call an ἐπάγων [*epagon*] and we call an *artemon*, is installed at

the foot of the machine. This pulley-block, which should be bound to the base of the machine, comprises three pulleys through which the ropes are passed for the operators to pull on. In this way three crews of men hauling are able to raise a load high up quickly without a capstan [Fig. 30].

10. This kind of machine is called a *polyspastos* because it owes its exceptional ease of handling and speed to its multiple revolving pulleys. Erecting just one beam has the advantage that, by inclining it, one can let the load down as far forward and sideways to left and right as one wants.

All the mechanisms described above operate on principles appropriate not only for the functions just mentioned but also for loading and unloading ships, since some are set up vertically and others placed horizontally on revolving platforms [*carchesia*].[7] Similarly, ships can be pulled ashore using the same principles by adjusting ropes and pulley-blocks used on level ground without the erection of beams.

11. It is not irrelevant either to explain the ingenious system devised by Chersiphron.[8] When he wanted to transport the shafts of columns from the quarries to the Temple of Diana at Ephesus, he had no confidence in carts because of the magnitude of the loads and the softness of the country roads, so he tried out the following system to prevent the wheels from sinking in: he fitted and jointed together [a rectangular frame of] four wooden beams measuring a third of a foot, that is, two as long as the column-shafts with two cross-bars placed between them; he leaded in metal pivots like dowels at the ends of the column-shafts and fixed rings into the wood to house the pivots: then he tied the ends of the frames to wooden steering-beams. The pivots housed in the rings could revolve so freely that when the oxen yoked together pulled the frame, the column-shafts, revolving on the pivots and rings, turned continuously.

12. When they had transported all the column-shafts like this and it became urgent to bring up the architraves, Metagenes, Chersiphron's son, adapted the system of transporting column-shafts for moving the architraves as well. He made wheels about twelve feet in diameter and inserted the ends of the architraves at the centres of the wheels; using the same technique, he fixed pivots and rings at their ends. So when the frames of a third of a foot were pulled by the oxen, the pivots housed in the rings made the wheels go round,[9] and the architraves, housed like axles in the wheels in the same way as the columns-shafts, arrived at the building without a hitch. A good example of this method is provided by the way in which rollers are used for flattening the walks in gymnastic complexes. But this could not have been done were it not for the fact that, above all, the distance was short, there were no more than eight miles between the quarries and the temple, and there was no incline but uninterrupted flat country.

13. But within my own memory, when the pedestal of the colossal Apollo in his temple had developed bad cracks with age and they were afraid that the statue would fall and be smashed to pieces, they made a contract for a pedestal to be cut in the same quarries. A certain Paconius took the contract. This pedestal was twelve feet long, eight feet wide and six feet tall. Paconius, confident that it would make his name,[10] did not use Metagenes' method to transport it, but decided to make a different type of machine, although using the same principle.

14. For he made wheels about fifteen feet in diameter and enclosed the ends of the stone block in the wheels; then he fastened battens of wood of a sixth of a foot around the stone in a circle from wheel to wheel in such a way that there was a gap of no more than a foot between each batten. Then he wound a rope round the battens, and having yoked up the oxen, began to pull on it: consequently the rope made the wheels revolve as it unwound, but it could not pull them in a straight line and kept slipping to the side; so the machine had to be brought back on course. Consequently Paconius squandered such so money pulling the machine back and forth like this that he went bankrupt.

15. I will make a very brief digression to explain how these stone-quarries were discovered. Pixodarus was a shepherd who lived in this area. When the citizens of Ephesus were planning to build the temple dedicated to Diana of marble and were trying to decide whether to procure it from Paros, Proconnesus, Heraclea or Thasos, Pixodarus drove his sheep along and fed

his flock in that very area: there two rams charged at each other, but each ran past the other; one of them, carried on by the momentum, hit a rock with his horns, knocking off a brilliant white flake of marble. So it is said that Pixodarus left his sheep in the mountains and immediately carried the flake to Ephesus because that was the burning question of the day.[11] So they immediately voted him honours and changed his name from Pixodarus to Evangelus. And to this day the chief magistrate goes each month to that very spot to make a sacrifice to him, and faces a fine if he fails to do so.

CHAPTER III

Types of Motion

1. I have explained briefly what I felt to be the essential points regarding hauling machines. With respect to their motion and their efficacy there are two different and distinct factors which act together as basic principles and produce the desired results: the first is that of the straight line, which the Greeks call ἐυθεῖα [*eutheia*], and the other is that of the circle, which the Greeks call κυκλωτή [*kyklote*]: but in fact neither motion in a straight line without circular movement, nor circular movement without rectilinear motion can succeed in raising weights. I will explain this concept so that it is comprehensible.

2. Little axles, functioning as axes, are inserted in pulleys and fixed into pulley-blocks; a rope passed round these pulleys is pulled with rectilinear force and fixed to a windlass and, as the hand-spikes are turned, makes the loads lift off. The pivots of the windlass, placed in a straight line like axes in their sockets, and the hand-spikes inserted in the holes [of the windlass], rotate at their ends with a circular motion like a lathe, causing the loads to be raised. And again, when an iron bar is brought to bear on a weight that many hands cannot budge, and a rectilinear support, which the Greeks call a ὑπομόχλιον [*hypomochlion*; fulcrum], is pushed under it as a fulcrum, and the

short end of the lever is eased under the load, the force of just one man pressing down on the long end raises the load.

3. This is because the shorter end of the lever at the front slips under the load beyond from the support which constitutes the fulcrum, and when the end which is further from that point is brought to bear on it, it makes a circular movement which allows a very heavy load to be counterbalanced by a few hands exerting downward pressure. Again, if the short end of an iron lever is slid under a weight and the long end is not pressed down but, on the contrary, is pushed up, the short end, pushing against the surface of the ground, will treat the ground as the weight and the corner of the actual load as the fulcrum. So the weight of the load will still be raised by motion in the opposite direction, though not as easily as by downward pressure. Therefore, if the short end of a lever placed on a *hypomochlion* slips too far forward under the weight, and its long end is pressed down too close to the fulcrum, it will not be able to lift the weight unless, as described above, equilibrium has been achieved along the length of the lever by putting downward pressure on its end.

4. This phenomenon may be observed in the type of balances called steelyards. For when the handle has been placed close to the end from which the scale hangs and functions in that position as the fulcrum, and the counterbalance at the other end of the beam is positioned further away, moving along point by point, or is even taken to the end, it makes a very large weight equal to a much smaller weight because of the equilibrium brought about by the balancing of the beam. A small and relatively light counter-weight moving further away from the fulcrum in this way moves a heavier weight without difficulty and causes it to rise gently upwards.

5. Similarly the pilot of a very large cargo ship, holding the handle of the tiller, which the Greeks call οἴαξ [*oiax*], and manoeuvring it skilfully with one hand by pushing it around its axis, alters the ship's course even though it is laden with a huge, or even enormous weight of merchandise and provisions. And when the sails are set at half mast, a ship cannot sail quickly; but when the yards have been hoisted up to the top,

then the ship makes much better speed because the sails cannot catch the wind when they are close to the heel of the mast, which is the location of the fulcrum, but can when they have been moved further away from it to the top.

6. Just as a lever pushed under a weight offers greater resistance and does not budge if pressure is exerted at its centre, but when it is pushed down at the end raises the weight easily, in the same way sails adjusted to half mast are much less effective; but when they are hoisted to the very top of the mast and much farther away from their fulcrum, they drive the ship along more rapidly because of the force at the top, even when the wind is no stronger than it was before. Again: when the oars bound to the thole-pins by cords are pushed forward and pulled back by the oarsmen, and the tips of the blades move forward at some distance from the axis, the oars, foamy from the waves of the sea, push the ship forward in a straight line with great force, while the prow slices easily through the water.[12]

7. When very heavy loads are carried on poles by four or six porters, their points of equilibrium pass through an axis at the centres of the poles so that the total weight of the load is shared according to a definite principle of distribution by which each porter should carry an equal weight on his shoulder. For the centres of the poles, to which the straps used by the porters are attached, are fixed with nails to prevent the straps from slipping sideways; when they move forward beyond the limit of the axis, they weigh down on the point which they have moved closer to, just as a weight on a balance does when it moves to the points at the end of the yard during balancing.

8. By the same principle, beasts of burden pull equal loads when the yokes are adjusted at the middle with the appropriate straps; but when their strength is unequal and the stronger one shoves the other, one side of the yoke should be made longer by repositioning the strap, which would help the weaker animal. So, when the straps on poles and yokes are not tied in the middle but to one side, they make one side shorter and the other longer depending on how far they move from the centre. For this reason, if both ends are made to revolve around the axis at the point to which the strap has been brought, the longer

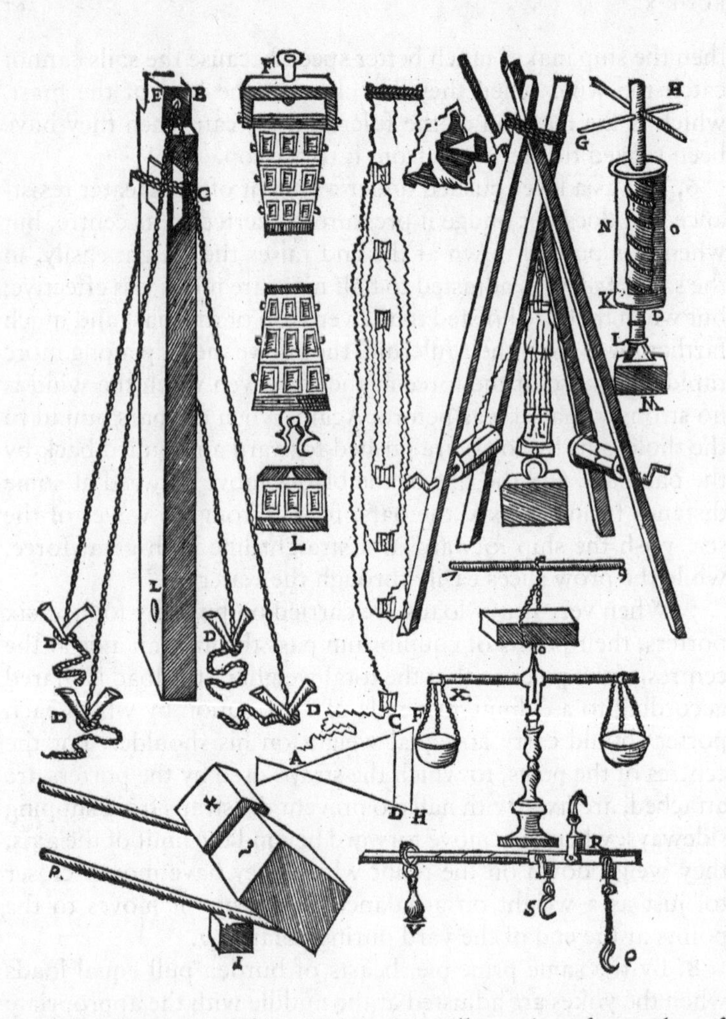

Plate 40. Composite and in part imaginary illustration of a number of Vitruvius' machines. Upper centre: Upper and lower pulley-blocks (10.2.1). Left: Single beam crane illustrating 10.2.8–9. Bottom left: The principles of the fulcrum, illustrating 10.3.1–3. Bottom right: A weight carried on a pole, a balance and a steelyard (10.3.2–4). Upper right: An imaginary crane with four main beams and two windlasses, not the one- or two-beam types with single windlasses described by Vitruvius (10.2.1–4), and a screw for lifting weights (not in Vitruvius either).

section would describe more of a circle than the shorter section.

9. In the same way that smaller wheels move less readily and with greater difficulty, so poles and yokes exert heavy pressure on the neck when the distance between the axis and their ends is less; but when the distance from the same axis is greater, they relieve both draught-animals and porters of some of the weight. Since these devices have been set in motion by rectilinear and circular motion around an axis, so too wagons, four-wheel carriages, discs, millstones, screws, *scorpiones*, *ballistae*, the beams of presses and other machines produce the required results on the same principles by revolving around a rectilinear axis and by circular rotation [Plate 40].

CHAPTER IV

Devices for Raising Water

1. Now I will explain how the different types of devices invented for drawing up water are constructed; first I will talk about the drum. This does not lift water up very high but can extract large quantities of it very efficiently. An axle [*axis*] made on a lathe or with compasses, with its ends shod with iron hoops, should be enclosed at its centre by a drum made of wooden boards jointed together and mounted on posts sheathed with iron under the ends of the axle. Eight cross-beams extending from the axis to the circumference are inserted in the cavity of the drum, dividing its interior into equal compartments.

2. Boards are nailed round the circumference of the drum leaving apertures six inches wide so that water can be pulled inside it: and again, small holes should be cut in each compartment near the axle on one side. When the drum has been covered with pitch following the method used for ships, it can be made to revolve by men pushing it round with their feet; drawing water in through the apertures around the circumference of the drum, it sends it back through the holes near the axle into a wooden trough placed below which has a water-channel

connected to it. In this way large quantities of water are supplied for irrigating gardens or for diluting saline water in the salt-works.

3. But when water has to be raised higher, the same mechanism will be adapted like this: a wheel with a diameter great enough to arrive at the required height should be built around the axle. Square buckets, sealed with pitch and wax, should be attached round the circumference of the wheel. So when the wheel is turned by the men pushing it with their feet, the buckets full of water lifted to the top then turn down again towards the bottom and should automatically pour the water that they have extracted into a tank.

4. But if the water is to be supplied to much higher locations, a double iron chain from which hang bronze buckets with a capacity of a *congius*[13] will be taken around the axis of the same wheel and will be set in place so as to reach as low as possible.[14] So the turning of the wheel winding the chain around the axle will carry the buckets to the top, and as they are carried over the axle they will necessarily invert themselves and pour the water they have extracted into a tank.

CHAPTER V

Waterwheels and Watermills

1. Wheels in rivers are also built with the same techniques just described. Paddles are fixed around their perimeters which force the wheels to rotate when they move forward, impelled by the river's current; the wheels, turned in this way without men pushing them round with their feet but by the river's impetus, extract the water with their buckets, carry it aloft and supply what is required.

2. Watermills are turned on the same principle: all their components are the same except that a disc with teeth is fitted onto one end of the axle. This disc is set vertically on its edge and revolves simultaneously with the wheel. Adjoining it there is a larger disc, set horizontally and also equipped with teeth, which engages with the first disc. So the teeth of the vertical disc fitted onto the axle force the teeth of the horizontal disc to move and cause the millstones to revolve. A hopper suspended above the machine supplies the millstones with the corn which is ground to flour by their rotation.

CHAPTER VI

The Water-screw

1. There is also a system using a screw which raises a great quantity of water but cannot carry it as high as the wheel does. This device is constructed as follows: a wooden beam is selected and its length in feet should be measured out so as to equal its diameter in inches. It should then be made circular in section using compasses. The circumferences at the ends [of the beam] should be divided with compasses into eight segments comprising quadrants and octants; the lines should be located so that when the beam is laid down horizontally, those at either end line up exactly with each other; and the spacings marked along the length of the beam should equal an eighth of the circumference at the ends. Again, when the beam has been laid flat, perfectly straight lines should be drawn from one end of it to the other. In this way equal spaces between the lines will be generated both around the circumference and along the length of the beam. So where the lines have been drawn along the length of the beam they will create intersections [with the circumferential lines], and precise points at the intersections.

2. When these lines have been drawn correctly, one takes a thin strip of willow or one cut from agnus castus which, after being smeared with liquid pitch, is fixed at the first point of intersection. Then it is taken diagonally across to the next points of intersection of the longitudinal and circumferential lines; then, progressing in due order, it goes through each point and, winding around the beam, should be attached to each intersection; in this way, moving back from the first to the eighth point of intersection, it arrives at and is secured to the same line to which its end was fastened at the beginning. In this way the distance it covers by proceeding obliquely through the eight points is the same as the distance it covers longitudinally as far as the eighth point. Using the same procedure, wooden strips fixed obliquely at each intersection across all the spaces of the length and circumference make channels winding round the eight divisions of the diameter and an accurate and realistic imitation of a snail-shell.

3. More and more strips of wood coated with liquid pitch are fixed one over the other along the same grid, and should be built up until the total thickness is an eighth of the length. Wooden boards are placed above and around these strips and secured so as to protect the spiral. Then these boards are soaked with pitch and bound together with iron strips so that the impact of the water will not force them apart. The ends of the beam are shod with iron. To right and left of the screw, vertical supports are put in position at either end,[15] with cross-beams fixed on both sides. In these cross-beams iron sockets are

Plate 41. Waterwheels and watermills. Top left: A drum with apertures in the periphery scoops up water and sends it out via smaller holes (*columbaria*, on the side not visible to the viewer) around the axle (*axis*). It is not clear how men using their feet could drive the machine as Vitruvius says: perhaps they trod on the wheel itself at the top from a platform, but the holes in the periphery would cause them difficulties. Instead, Barbaro's illustration shows two men turning a handle (10.4.1–2). Top right: Wheel with square buckets (*modioli quadrati*) fitted around the periphery: again, it is not obvious how men could tread round the wheel, and Barbaro does not illustrate this aspect of it (10.4.3). Bottom left: Watermill driving millstones with grain fed from above; Barbaro changes the second toothed wheel, which is horizontal, into a vertical cylinder with bars meshing with the teeth of the vertical wheel (10.5.2). Bottom right: Water-screw fitted here with a drum at the centre, copying the illustration in Fra Giocondo (1511), presumably to be turned by men using their hands: in fact, there must have been treads of some kind on the exterior of the housing of the screw near the top since Vitruvius mentions men pushing it round with their feet (10.6.1–4).

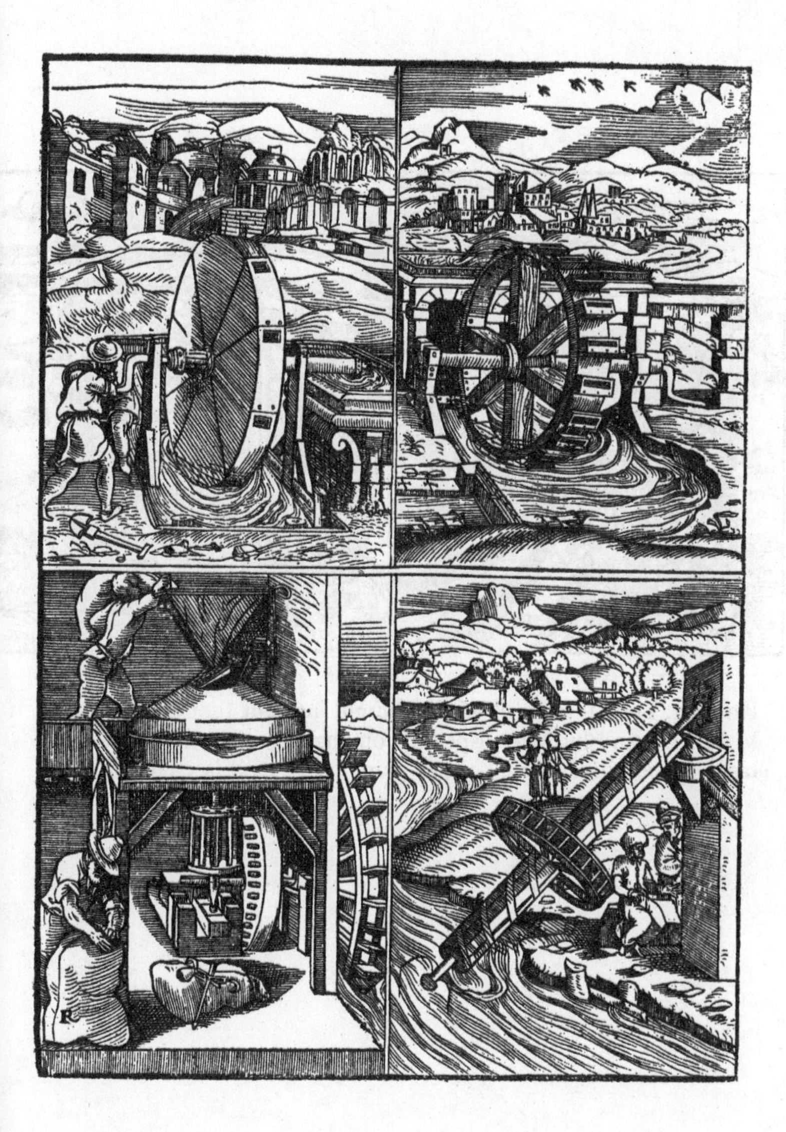

Plate 42. Left: Watermill with treadmill inside it (not sanctioned by Vitruvius), and buckets suspended on chains (10.4.4). Right: A fantastic version of Ctesibius' water-pump (10.7.1–3) (see Fig. 31).

inserted in which the pivots are housed; and so the screws turn, powered by the men pushing with their feet.

4. The screw will be set up with an inclination that should correspond to the way in which a Pythagorean right-angled triangle is drawn; that is, its length should be divided into five units, of which three indicate the height of the upper end of the screw, so that the distance from the perpendicular to the apertures at the bottom will be four units. A diagram is drawn at the appropriate place at the end of the book showing the procedure which should be followed.[16]

I have now described in as much detail and as clearly as I possibly could – in order to make them more widely known – the methods by which devices made of wood for extracting water should be made and the means by which they are set in motion so that by rotating they perform an infinity of useful functions.

CHAPTER VII

Ctesibius' Pump

1. We now proceed with an explanation of Ctesibius' device for raising water. It should be made of bronze: at the bottom twin cylinders [*modioli*] set a small distance apart should be constructed with tubes [*fistulae*] attached to them which converge rather like a fork into a tank [*catinus*] in the middle. In this tank valves [*asses*] should be made and jointed precisely into the upper outlets of the tubes and these, by blocking the holes of the outlets, prevent anything [i.e. water] that has been forced into the tank by the air-pressure from going back down.

2. A hood [*paenula*] like an inverted funnel [*infundibulum*] is fitted on top of the tank and securely attached to it by a pin driven through a clasp [*fibula*] to prevent the water-pressure from forcing it up. At the top of the hood, a pipe, called a trumpet [*tuba*], should be attached and fixed in place so that it is vertical. In the twin cylinders, valves are inserted under the lower outlets of the tubes [leading into the tank] and above their apertures at the bottom.

3. Pistons [*emboli masculi*], turned on a lathe and rubbed with oil, are fitted into the cylinders from above; by means of piston-rods [*regulae*] and levers [*vectes*] they set in motion whatever air is present there with the water, and, with the valves blocking the outlets [in the bottom of the cylinders], the pistons force the water, by compression then decompression, through the outlets of the pipes into the tank [*castellum*]; the hood takes in the water from the tank and it is driven upwards by air-pressure through the pipe; and this is how a flow of running water can be supplied from a water-tank placed at a lower level [Plates 41, 42; Fig. 31].

4. However, this is not the only invention that Ctesibius is said to have devised; many other different types can be observed which produce results borrowed from nature thanks to the pressure exerted by the compressed air generated by this liquid, such as the songs of blackbirds created by the movement of water, *angobatae*,[17] figurines that drink and move, and other devices of which the function is to please the senses by delighting our eyes and ears.

5. I have chosen from these devices those which are, in my view, the most useful and necessary: in the previous book I decided to discuss clocks and, in this one, devices for raising water. As for other inventions which do not have practical functions but satisfy a desire for pleasure and amusement, those who are particularly fascinated by such ingenuity can find out about them from the writings of Ctesibius himself.

CHAPTER VIII

The Water-organ

1. With regard to water-organs, I will not neglect to mention the mechanisms by which they operate, and to explain them in written form as briefly and accurately as possible. After a

wooden base [*basis*] has been put together, an altar-shaped container [*ara*] made of bronze is placed on it. Above the base, struts of wood [*regulae*] put together like a ladder are set up at the sides to right and left and house bronze cylinders: the cylinders contain movable pistons [*ambulatiles funduli*] carefully turned on a lathe which have iron piston-rods [*ancones*] fastened to their centres which are connected by hinges [*verticulae*] to levers [*vectes*]; they are also wrapped in sheepskins. There are also apertures in the flat tops of the cylinders about three digits in diameter. Just above these apertures are bronze dolphins [*delphini*] mounted on pivots which have cymbal-shaped valves [*cymbala*] hanging from their mouths on chains let down below the apertures of the cylinders.

2. In the altar-shaped container containing the water, there is a regulator [*pnigeus*] shaped like an inverted funnel, under which have been fitted little blocks [*taxilli*] about three digits high which maintain an even space at the bottom between the rim of the regulator and the floor of the container. Above the neck [*cervicula*] of the regulator a small chamber [*arcula*] is jointed on which supports the main component of the device [*caput machinae*], which is called the κανὼν μουσικός [*kanon mousikos*] in Greek. There are channels [*canales*] running along its length, four, if the instrument is to be tetrachord, six, if it is to be hexachord and eight, if it is to be octochord.

3. A stopcock [*epitonium*] fitted with an iron handle [*manubrium*] is inserted in each of the channels. When these handles are turned they open the vents leading from the small container into the channels; leading off from the channels, the *kanon* has holes placed obliquely in it, corresponding to the vents in the upper board [*tabula*], which is called the πίναξ [*pinax*] in Greek. Between this board and the *kanon*, sliding bars [*regulae*] are inserted provided with matching holes; they are rubbed over with oil so that they can be pushed easily and then be taken back to their original positions; the bars which close these holes are called *plinthides* [little blocks] and their motion back and forth opens some holes and closes others.

4. These sliding bars have iron hooks [*coracia*] fixed to them connected to the keys [*pinnae*]: the keys, when pressed, cause the continual movement of the bars. Above the apertures of the board through which the air escapes from the channels, rings [*anuli*] are glued on into which the 'little tongues' [*lingulae*] of all the organ pipes are inserted. There is a set of pipes [*fistulae*] leading from the cylinders which are attached to the neck of the regulator and arrive at the vents in the small chamber. Valves turned on a lathe are fitted in these pipes preventing the blast of air from returning by blocking the holes when the compartment takes in the air.

5. So when the levers are raised, the piston-rods pull the pistons [*fundi*] down in the cylinders as far as they will go, and the dolphins mounted on pivots let the cymbal-shaped valves drop in the cylinders, filling up the space in the cylinders with air; the piston-shafts, pushing up the bottoms of the pistons in the cylinders with continuous vigorous thrusts and blocking the openings above by means of the cymbal-shaped valves, compress the air enclosed in the cylinders and force it into the pipes, through which it flows into the regulator and then into the small compartment through its neck. If the piston-rods are moved with greater force, the compressed air expands even further and flows through the apertures of the stopcocks and fills the channels with air.

6. So, when the keys, operated by hand, drive the bars forward and back continuously by alternately closing and opening the holes, they produce sounds emitted in a great variety of rhythms in accordance with the laws of music [Plate 43; Fig. 32].

I have done my utmost to set out an arcane subject clearly in writing, but its mechanism is not simple or immediately understood by everybody apart from those who already have practical experience of these kinds of instruments. But if anyone has understood little from my account, he will certainly find that I really have laid out everything carefully and precisely when he becomes familiar with the device itself.

Plate 43. The water-organ. This wonderful illustration, useless or nearly so from the point of reconstructing the instrument (see Fig. 32), includes a number of the components put together haphazardly, the bronze dolphins, the pistons in the cylinders operated manually, and others practically invisible under the keyboard.

CHAPTER IX

The Hodometer

1. Now the attention of the treatise moves on to a far from useless device of the greatest ingenuity handed down to us by the ancients by means of which we can discover the number of miles we have covered whilst sitting in a four-wheel carriage or sailing at sea. This is how it will work: the wheels [*rotae*] of the wagon should have a diameter of four feet so that if one has a reference point marked on it from which it begins to rotate as it proceeds along the road surface, it will have completed the distance of exactly twelve and a half feet when it arrives back at the point from which it began rotating.

2. Once the equipment has been prepared like this, then a disc [*tympanum*], with a single small tooth projecting from its circumference, should be fitted securely onto the hub [*modiolus*] of the wheel on the inside. Above this a frame [*loculamentum*] enclosing a revolving disc set on its edge and mounted on a small axle should be attached firmly to the chassis [*capsus*] of the four-wheeler;[18] on the circumference of the disc four hundred little teeth should be made, distributed equally and designed to engage with the little tooth of the lower disc. As well as this, another small tooth projecting further than the others should be fixed into the side of the upper disc.

3. Then above this should be located a horizontal disc housed in another frame with matching teeth which engage with the tooth that will be fixed in the side of the second disc; in this last disc a number of holes should be made corresponding to the number of miles that a four-wheeler can travel in the course of a day – or more or less, it is not important. Round pebbles should be put in all these holes, and in the housing [*theca*], or rather the frame of the disc, a single hole should be made connecting to a small tube: on arrival at the appropriate place, the pebbles previously placed in this disc can fall through the tube one by one into a bronze receptacle [*vas aeneum*] fixed below in the body of the four-wheeler.

4. In this way the wheel, as it goes forward, moves the lowest disc simultaneously with itself, and with each rotation its small tooth impels the teeth of the upper disc to move along; the result will be that, when the lower disc has turned four hundred times, the upper disc will have gone round once and the tooth fixed at the side of it will push forward the single small tooth of the third, horizontal disc. Accordingly when the upper disc has revolved once because of the four hundred revolutions of the lowest disc, the distance travelled will be five thousand feet, that is, a mile. Consequently the number of pebbles that will drop will alert us, by the noise they make, to each mile that has passed by; and the total number of stones collected from the bottom will tell us the number of miles travelled in a day [Fig. 33].

5. With respect to voyages on water the same techniques are applied with only a few adjustments. An axle is passed through the sides of the hull of a ship with the ends sticking out, and wheels four and a half feet in diameter are fixed to its ends, with projecting paddles touching the water fastened around their circumference. Again, the centre of the axle located in the middle of the ship carries a disc with one small tooth projecting from its circumference. A frame is set in place in that position housing a disc with four hundred regularly distributed teeth which engage with the tooth of the disc mounted on the axle. Besides that, it has a single tooth fixed at the side and projecting beyond its circumference.

6. Above this, in another frame attached to the first one, is mounted a horizontal disc provided with matching teeth which engage with the tooth fixed to the side of the disc set on its side, in such a way that the tooth turns the teeth of the horizontal disc during each of its revolutions and makes it complete a rotation. Holes should be made in the horizontal disc and round pebbles placed in them. In the casing, or rather the frame, of this disc, a single hole attached to a small tube should be drilled through which a pebble, free from obstruction, can drop into a bronze vessel and alert us by the sound it makes.

7. So, when a ship is being driven along either by oars or gusts of wind, the paddles attached to the wheels will strike the

Plate 44. Hodometers used on a four-wheeled carriage and on a ship. Another spectacular Renaissance visualization of which the value is romantic anachronism rather than technical accuracy.

water and, forced back by the vigorous impact, will turn them; the revolving wheels set the axle in motion, and the axle drives the disc; the tooth of the disc, turned by each revolution, will drive one of the teeth of the second [large, vertical] disc and make it revolve a little at a time. So, when the wheels have been made to revolve four hundred times by the paddles, the [large, vertical] disc, which has revolved only once, will drive forward the tooth of the horizontal disc with the tooth fixed on its side. Consequently, every time the revolution of the horizontal disc brings a stone to the hole, it should release it through the small tube, so indicating by the sound and the number of stones the length of the voyage in miles across the water [Plate 44].

I have now described carefully how machines that are of practical use and sources of amusement should be constructed in peaceful and tranquil times.

CHAPTER X

Scorpiones

1. Now I will explain the modular systems with which devices invented for our protection from danger and essential for our self-preservation, that is, *scorpiones* and *ballistae*, can be assembled.

All the proportions of these devices are worked out from a given length, that of the arrow which the instrument must shoot; the diameters of the holes in the frame [*capitulum*], through which the twisted cords [*nervi torti*] that hold back the arms [*bracchia*] are stretched, should be a ninth of that length.

2. The height itself and the breadth of the frame are then established in conformity with the size of these holes. The boards [*tabulae*] at the top and bottom of the frame are called *peritreti* [perforated], and should equal the diameter of one hole in thickness and one and three-quarters in breadth, and at their extremities, one and a half.[19] The supports [*parastaticae*] at right and left, excluding the tenons, should be four modules high and five-eighths of a module thick; the tenons, half a module. The distance from the support to the hole is a quarter of a module, as is that between the hole and the upright in the middle. The breadth of the post [*parastas*] in the middle is equal to one and three-quarter modules, the thickness, one module.

3. The hole [*intervallum*] in the central upright where the arrow is placed equals a quarter of a module. The four corners of the frame should be reinforced at the sides and fronts with iron plates or bronze pins and nails. The length of the groove [*canaliculus*], called σῦριγξ [*syrinx*] in Greek, will be eighteen modules long. The strips [*regulae*], which some call 'cheeks' [*bucculae*], nailed to right and left of the groove, will be nineteen modules long and one module in height and thickness. Two other strips, three modules long and half a module wide, are fixed to these and the windlass is fitted into them. The 'cheek' nailed onto them, called the 'little bench' [*scamillum*], or by some the 'little box' [*loculamentum*], and secured with dove-tailed tenons, is one module thick and half a module in height. The length of the windlass is four modules, its thickness five-twelfths.

4. The length of the retaining-hook [*epitoxis*] will be three-quarters of a module, the thickness a quarter; the same applies to its socket-piece. The trigger [*chele*], or 'handle' [*manucla*] as it is also called, is three modules long and a quarter of a module in breadth and thickness. The length of the bottom of the groove [*canalis fundus*] is sixteen modules, a quarter of a module in thickness and three-quarters in height. The base of the column [*columella*] at ground level equals eight modules, its breadth at the level of the plinth on which it is mounted is three-quarters of a module, its thickness five-eighths; the length of the column up to the tenon is twelve modules, its breadth and thickness three-quarters. The three struts [*capreoli*] of the column are nine modules long, half a module wide and seven-sixteenths thick. The tenon will be one and a half modules long and the top of the column two modules; the antefix, three-quarters of a module wide and one module thick.

5. The smaller strut at the back [*posterior minor columna*], called ἀντίβασις [*antibasis*; counter-strut] in Greek, is eight

modules long, three-quarters of a module broad and five-eighths thick. The strut [*subiectio*] below it is twelve modules long, with the same breadth and thickness as the smaller strut just mentioned. Above the smaller strut is a socket-piece [*chelonium*], otherwise known as the 'cushion' [*pulvinus*], two and a half modules long, one and a half high and three-quarters broad. The diameter of the drum housing the handles [*scutulae*] of the windlass will be two and a half modules, its thickness half a module. The handles of the windlass, including the side-pins, are ten modules long, and half a module in breadth and depth. The length of the arm [*bracchium*] is seven modules, its thickness at its base nine-sixteenths of a module, seven-sixteenths at the top, and its curvature, eight modules [Fig. 34].

6. Such devices are constructed according to these numerical relationships with additions or subtractions made to them. For if the frames are higher than they are wide, in which case they are called *anatona* [high-tensioned], the length of the arms should be reduced in such a way that the more the tension is reduced depending on the height of the main frame, the greater the velocity of the shot generated by the shortening of the arm. But if the frame were to be less high, in which case it is called *catatonum* [low-tensioned], the arms should be made a little longer on account of their high tension so that they may be pulled back easily. Just as a lever five feet long when operated by four men lifts the same load as a ten-foot lever operated by two men, so the longer the arms the easier they are to pull back, and the shorter they are, the harder it is.

I have now spoken of the mechanisms of catapults and of the elements and components of which they are constructed.

CHAPTER XI

Ballistae

1. The mechanisms used for *ballistae* are very varied, though they are constructed to fulfil one single function. Some are brought under tension with hand-spikes and windlasses, some by blocks with multiple pulleys, others with capstans, yet others with a system of discs with teeth. But no *ballista* is constructed without regard to the magnitude of the weight of stone which the device must throw. Accordingly the relevant principles are not immediately obvious to everybody but only to those who have a familiarity with numbers and their multiples using geometric methods.

2. For in fact the dimensions of the holes to be made in the frames through which the cords, made mainly of women's hair or sinew, are stretched, are based on the size and weight of the mass of stone which the *ballista* in question is intended to throw, just as the dimensions of catapults depend on the lengths of the arrows. So, in order that this subject will be comprehensible even for those who do not understand geometry and so that they will not waste precious time in the heat of battle doing the calculations, I will explain what I myself am certain of from personal experience and, in part, what I have learnt from my teachers, and will describe how in Greece weights are proportional to modules so that such relationships may also apply to our weights.

3. If, then, a *ballista* is intended to throw a 2-pound stone, it will have a hole of five digits in the frame; for a 4 pounder, six digits; for a 6 pounder, seven digits; for a 10 pounder, eight digits; for a 20 pounder, ten digits; for a 40 pounder, twelve and three-quarter digits; for a 60 pounder, thirteen and an eighth digits; for an 80 pounder, fifteen digits; for a 120 pounder, a foot and one and a half digits; for a 160 pounder, one and a quarter feet; for a 180 pounder, a foot and five digits; for a 200 pounder, a foot and six digits; for a 240 pounder, a foot and seven digits; for a 360 pounder, one and a half feet.

4. Once the diameter of the hole has been established [as the module], a lozenge [*scutula*], called περίτρητος [*peritretos*; perforated] in Greek, must be designed, two and three-quarter modules long and two and a half modules in breadth. The lozenge should be divided in the middle by a diagonal; this done, the edges of the figure should be pulled together so that it assumes an oblique configuration of a sixth in length and a quarter in width at the angle. The holes should be directed to the part where the curvature is and the points of the angles converge, and the breadth should be reduced towards the inside by a sixth; the aperture will then be more elliptical depending on the thickness of the tension-bolt [*epizygis*]. Having completed the design [of the *scutula*], its edges should be carefully finished off all round so that the curvature around it is smooth.

5. The thickness of the lozenge will be set at one module. The cylinders [*modioli*] of the holes will be two modules long, one module and five-twelfths wide and three-quarters of a module thick, excluding the part inserted in the aperture, and half a module wide on the outside. The length of the struts [*parastatae*] will be five and three-sixteenths, their internal slot half a module and their thickness eleven-eighteenths. An amount equal to that near the opening in the design is added to their breadth in the centre; the strip is five modules wide and deep, a quarter of a module high.

6. The wooden strip [*regula*] on the 'table' [*mensa*] will be eight modules long, with a breadth and thickness of half a module. Its tenons will be two modules long and a quarter of a module thick: the curvature of the wooden strip is three-quarters of a module. The outer strip [*regula exterior*] has the same breadth and thickness as the inner, but the length will be determined by the same point of the curvature of the scheme traced out and by the breadth of the strut [*parastatica*] at the point of curvature. As regards the upper strips, they will have the same dimensions as the lower; the cross-pieces of the 'table' will be a quarter of a module.

7. The shafts [*scapi*] of the 'ladder' [*climacis*] will be nineteen modules long and a quarter of a module wide; the space in the middle will be one and a quarter modules broad, and one and

one-eighth in height. The upper part of the ladder which touches the arms and is fastened to the table should be divided along all its length into five units, of which two should be allocated to the component which the Greeks call the χελώνιον [*chelonion*; socket-piece of a windlass]; its breadth is a module and three-sixteenths, its thickness a quarter of a module and its length eleven and a half modules; the projection of the trigger [*chele*] will be half a module and the thickness of the mortise [*pterygoma*] a quarter. The part resting on the axle [*axon*], called the 'transverse face' [*frons transversarius*], will be three modules.

8. The inner wooden strips will be five-sixteenths of a module wide and their thickness three-sixteenths. The housing [*replum*], or cladding [*operimentum*] of the socket-piece, is dovetailed into the shaft of the ladder [*scapus climacidos*] and is a quarter of a module in breadth and a twelfth in thickness. The square block [*quadratum*] attached to the ladder will be a quarter of a module wide at the edges, while the diameter of the circular pivot [*axis*] will be at the same level as the trigger except that at the height of the catches [*claviculae*], it will be seven-sixteenths.

9. The length of the struts [*anterides*] will be three and a quarter modules, their width at the bottom half a module and their thickness at the top three-sixteenths. The base [*basis*], which is called the ἐσχάρα [*eschara*; hearth], will be eight modules long, and the 'counterbase' [*antibasis*] four; both will be a module thick and wide. The struts are set halfway up the column and will be half a module broad and thick, while their height is not related to the module but will be adapted to practical exigencies. The length of the arm is six modules, its thickness at the base five-eighths, and at the end, three-eighths [Fig. 35].

I have now explained the relevant modular systems of *ballistae* and catapults which I thought would be most useful. But I will not neglect to discuss, in so far as I can express it in written form, how these devices are tuned by putting them under tension using twined bundles of sinew or hair.

CHAPTER XII

The Stringing and Tuning of Catapults

1. Beams of considerable length are selected; socket-pieces into which the windlasses are inserted are nailed onto them. The beams are cut down their centres and slots excavated in them in which the frames of the catapults are inserted and held in place with wedges, lest they slip when put under tension. Small bronze drums are inserted in these frames and little iron pins, which the Greeks call ἐπιζυγίδες [*epizygides*], are set in them.

2. Next, the ends of ropes are threaded through the holes in the frames and passed through to the other side; then they are thrown round the windlasses and wound round so that the ropes, stretched tight by means of hand-spikes, will make the same sound on both sides when struck with the hands. Then they are locked into the holes with wedges so that they cannot go slack. Then they are passed through to the other side and tightened up in the same way by means of the hand-spikes and windlasses until they sound the same. In this way, by keeping the device taut with wedges, the catapults are 'tuned' to the proper pitch by musical testing.

I have said what I could about these subjects. Now it remains for me to explain, with respect to sieges, how mechanical devices can contribute to the victories of military leaders and cities can be defended by them.

CHAPTER XIII

Siege Engines

1. First: it is reported that the battering-ram [*aries*] used in sieges was originally invented as follows. The Carthaginians pitched camp in order to besiege Cadiz.[20] After they had first captured one of their forts, they tried to destroy it: but since they had no iron implements with which to do so, they took a

wooden beam, and, holding it up with their hands and smashing the end of it repeatedly against the top of the wall, they gradually knocked down the highest courses of stones, and so demolished the entire fortification layer by layer.

2. Afterwards, an engineer from Tyre, called Pephrasmenos [Thoughtful], was inspired by the invention of this technique to set up a mast on which he hung another at right angles, like the beam of a balance, and so, by pulling it back and forth, knocked down the wall of Cadiz with a series of violent blows. But Geras of Carthage first made a wooden platform with wheels fixed under it on which he constructed a protective framework of uprights and cross-pieces; he suspended the ram inside and covered it with oxhides so that the men stationed in the machine to batter the wall would be better protected. Since it proceeded by fits and starts, he was the first to call it the 'tortoise-ram'.

3. Later on, after these first steps were taken in the development of this kind of machinery, and when Philip, the son of Amyntas, was besieging Byzantium,[21] Polyidos the Thessalian developed many other types that were more manageable. Diades and Charias, who served under Alexander, acquired the knowledge of all this from him.

So it is that in his writings Diades shows us that he invented mobile towers, which, in fact, he used to disassemble and carry around with the army on the march; similarly, the drill and the scaling-machine by which one could cross over to a wall at the same level, as well as the 'destructive raven', or 'the crane', as it is called by others.

4. He also used a ram mounted on wheels and has left a written account of how it worked. He says that the smallest tower should not be less than sixty cubits high and seventeen broad, but that it should diminish at the top by a fifth in relation to the bottom, while the uprights of the tower should measure three-quarters of a foot at the base and half a foot at the top. He also says that such a tower should be ten storeys high with apertures on all sides.

5. A larger tower, in fact, should be a hundred and twenty cubits high and twenty-three and a half cubits broad, diminishing like the other one to one-fifth less, with uprights of a foot at the bottom and half a foot at the top. He used to build a tower this size with twenty storeys, each with a gallery round it three cubits wide; he used to cover the towers with rawhide to protect them from any attack.

6. The tortoise-ram was constructed in the same way. It had a breadth [*intervallum*] of thirty cubits and its height, excluding the gable [*fastigium*], was thirteen cubits, while the height of the gable from the level of the platform [*stratum*] to the top was sixteen cubits. The gable emerged above the roof [*tectum*] in the middle by at least two cubits in height, and above that a small tower [*turricula*] of four storeys was erected: *scorpiones* and catapults were stationed on the upper floor [*tabulatum*] of this tower, and on the lower floors a vast quantity of water was stored to extinguish any fire that might be applied to the structure. The machinery of the ram [*arietaria machina*], called κριοδόχη [*kriodoche*; ram-holder] in Greek, was housed inside it; this comprised a roller, turned on a lathe, on which the ram was positioned and produced very effective results when pulled back and forth by means of ropes. This too, like the tower, was protected with rawhide [Fig. 36, A and B].

7. Diades also explained the mechanisms of the drill [*terebra*] in his writings as follows: the machine, like the tortoise, had a channel in the middle fifty cubits long and a cubit high supported by uprights like that usually built in catapults and *ballistae*, in which a windlass was set transversely.[22] At the front to right and left there were two pulley-blocks which moved a beam with an iron point along the channel. Below the beam, numerous rollers were inserted in the channel, which made its movements more rapid and powerful. Numerous arches placed close together along the channel were set up above the beam to support the rawhide with which the machine was covered [Fig. 37].

8. Diades thought it unnecessary to write anything about the 'raven' because he was well aware that the machine was not effective at all. With regard to the scaling-machine, called ἐπιβάθρα [*epibathra*; ladder] in Greek, and the machines used at sea which, as he wrote, could be used in boarding ships,

I noticed with particular regret that he merely promised to explain their operating principles, but did not do so.

I have reported what Diades wrote about machines and their construction. I shall now explain what I have learnt from my teachers and what seems to me to be of practical use.

CHAPTER XIV

The Tortoise

1. The tortoise built for filling ditches and thereby enabling access to the wall should be constructed as follows. A platform [*basis*], called ἐσχάρα [*eschara*; hearth] in Greek, twenty-one feet square with four transverse beams [*tranversarii*], should be assembled. These should be held together by two others, one and a half feet thick and half a foot wide. The transverse beams should be about three and a half feet apart, and the axle-bearings [axle-blocks, *arbusculae*], called ἁμαξόποδες [*hamaxopodes*; wagon-feet] in Greek, should be placed below the gaps between them; the axles [*axes*] of the wheels, enclosed in iron cylinders, turn in these bearings. These bearings should be so arranged that they have pivots [*cardines*] and apertures through which hand-spikes can be passed to set them turning, so that, thanks to the rotating bearings, the tortoise will be able to move forward or back or right or left, or if necessary, obliquely off at an angle.

2. Two wooden beams [*tigna*] should be put in place on the base, each projecting six feet on either side; two other beams should be nailed onto these projections, sticking out seven feet in front and as thick and broad as the platform mentioned above. On this framework a series of compound struts [*postes compactiles*], nine feet high, a foot and a quarter square and a foot and a half apart, should be set up, excluding their tenons. All these elements should be tied together at the top by mortised beams [*intercardinatae trabes*]. Above the beams, rafters [*capreoli*] tied together with tenons should be positioned and taken up nine feet high. Above the rafters a square beam [*tignum* *quadratum*] should be fixed by which they should be tied together.

3. Moreover, these rafters should be held together by purlins [*lateraria*] at all points and covered with boards [*tabulae*], preferably of palm-wood, or failing this, of any other wood with exceptional durability, except pine or alder, which lack density and catch fire easily. Wicker lattices made of small, very recently cut twigs spliced tightly together, should be placed over all the boarding, and the whole machine should be completely covered all around with two layers of absolutely raw hides sewn together and lined with seaweed or straw soaked in vinegar. In this way the outer coverings will resist the blows of *ballistae* and incendiary attacks [Fig. 38].

CHAPTER XV

Hegetor's Tortoise

1. There is another kind of tortoise as well which includes all the components as described above, except for the rafters; it is surrounded by a parapet [*pluteus*] and crenelations [*pinnae*] made of wooden boards, and, above, includes inclined eaves [*subgrundae*] securely held together with boards and hides. A coat of clay kneaded with hair should be spread over them to such a thickness that fire cannot damage the machine in the slightest. These machines can also be made with eight wheels when necessary, but they have to be adapted to the nature of the terrain. However, the tortoises made for digging mines, called ὄρυγες [*oryges*; pickaxes] in Greek, include all the components described above, but their fronts are constructed like the angles of triangles so that when missiles are shot at them from the wall, they will not suffer direct impacts on their surfaces but ricochet off to the sides, in this way protecting those digging inside from any danger.

2. It does not seem to me irrelevant to illustrate the principles on which Hegetor of Byzantium constructed his tortoise.[23] The

length of its platform was sixty-three feet, the breadth forty-two. Four corner-posts [*arrectaria*] made of two beams joined together were placed on this structure and each one was thirty-six feet high, a foot and a quarter thick and a foot and a half broad. The platform was manoeuvred by its eight wheels. The wheels were six and three-quarter feet high, three feet thick, and were constructed of three layers of wood clamped together with alternating dovetail tenons and bound together by sheets of iron worked cold.

3. The wheels revolved in the axle-blocks or *hamaxopodes*. Again, on the flat surface formed by the cross-beams above the base, posts [*postes*] eighteen feet high, three-quarters of a foot broad, five-eighths of a foot thick and a foot and three-quarters apart were erected; above these, beams a foot broad and three-quarters of a foot thick were built around the whole structure and held it together; above this, rafters were taken up to a height of twelve feet; above the rafters a beam of wood was fixed which ensured that they were all connected together. Again, the rafters had beams fixed to them obliquely on which flooring was laid all round covering the parts below.

4. It had, moreover, a floor [*contabulatio*] in the middle on smaller beams [*trabiculae*] where the *scorpiones* and catapults were housed. Two compound beams were set up there, both forty-five feet tall, a foot and a half wide and two feet broad, the ends of which were connected by a mortised cross-beam and, in the middle, by another one mortised between the two shafts and tied in place with iron strips. Above this, a piece of wood pierced by sockets and securely locked in place with clamps was set vertically between the shafts and the cross-beams. Inserted in it were two small axles, turned on a lathe, to which ropes holding the ram in position were fastened.

5. At the top of the structure holding the ram in position a parapet fitted out like a small tower was located so that a couple of soldiers standing there in safety could look out without risk and report what the enemy might be up to. The ram was a hundred and four feet long, a foot and a quarter broad at the base and a foot thick, tapering at the end to a foot broad and three-quarters of a foot thick.

6. This ram, moreover, had a point of hardened iron like those which warships usually have, and, leading back from the point, four iron bands about fifteen feet long were fastened into the wood. Four cables eight digits thick were stretched from the head to the back end of the beam and fastened in the same way as those on a ship are tied from stem to stern; the cables had other ropes wound round them transversely at intervals of a foot and a quarter. The whole ram was covered with rawhide on the outside; the ends of the ropes from which the ram hung were made of fourfold chains of iron, which in turn were wrapped in rawhide.

7. Again, the part of the ram which projected forward had a shelter constructed of boards nailed together in which there was a network of large ropes pulled tight whose rough surfaces prevented the feet from slipping and made it easy to arrive at the wall. This machine could be moved in six directions, forward and back, also to right or left, and likewise it could be raised by being extended upwards and sent down by being pointed downwards. The machine could be elevated to a height sufficient to demolish a wall of about a hundred feet, and similarly it could range over at least a hundred feet moving from right to left. A hundred men manoeuvred the machine, whose weight was four thousand talents, which is four hundred and eighty thousand pounds [Fig. 39, A and B].

CHAPTER XVI

Defence

1. I have explained, with regard to *scorpiones*, catapults and *ballistae*, but also tortoises and towers, what seemed to me to be particularly relevant, by whom they were invented and how they should be built. But I have not considered it necessary to describe ladders, derricks and other devices of which the mechanisms are much simpler, because soldiers can usually build them on their own initiative. But these machines cannot

be effective on all terrains using the same procedures because fortifications as well as the capacity of various nations to resist differ from each other. For mechanical devices must be built in one way to combat brave and unpredictable opponents, in another way to fight cautious ones and in still another against cowards.

2. And so whoever wishes to follow my instructions and, by selecting from a range of them, put them together to make a particular type of machine will not be left without the means, but will be able to face without hesitation whatever exigency arises because of events or the nature of the terrain.

Defence tactics do not lend themselves to treatment in writing because our enemies do not organize their defences following the letter of our treatise, but their siege engines are frequently destroyed, unexpectedly and without the use of machines, by some rapidly improvised plan; this is said to have been the experience of the Rhodians.

3. Diognetus was a Rhodian architect who was granted an honorific fixed annual payment from the public purse commensurate with his standing in the profession. In this period a certain architect from Aradus, Callias by name, came to Rhodes and gave a public lecture at which he presented the model of a wall on which he located a machine fixed on a revolving platform with which he seized hold of an *helepolis* [city-destroyer] as it was approaching the fortifications, and carried it inside the wall. When the Rhodians had seen this model, they were filled with admiration and took away the yearly stipend settled on Diognetus and transferred the privilege to Callias.

4. Meanwhile, when King Demetrius, known as Poliorcetes [Besieger of Cities] because of his tenacity,[24] was preparing for war on Rhodes, he took with him a distinguished Athenian architect named Epimachus.[25] With enormous expense and immense effort, Epimachus constructed an *helepolis* one hundred and twenty-five feet high and sixty feet wide. He reinforced it with goatskins and rawhide so that it could withstand the impact of a stone weighing three hundred and sixty pounds shot from a *ballista*; the machine itself weighed three hundred and sixty thousand pounds. Now when Callias was asked by

the Rhodians to get his machine ready to resist the *helepolis* and bring it inside the wall as he had promised, he said that he could not do it.

5. For not all machines can be made following the same principles: there are some which are just as effective on a large scale as they are in the form of small models; others for which models cannot be constructed and are made full-scale from scratch; and yet others which seem plausible when one sees them in the form of models, but which are not viable when they begin to be scaled up, as we can see from this example: holes of half an inch, an inch or an inch and a half can be made with a drill, but if we wanted to make a hole of a palm in the same way, it could not be done, and a hole of half a foot or more seems inconceivable.[26]

6. All the same, it is the case with some models that, once they are made on a large scale, they look exactly as they did on a small scale. So the Rhodians, fooled by the same reasoning, piled injury and insult on Diognetus. Consequently, when they saw that the enemy was relentlessly hostile, that there was the danger of enslavement, and noticed the machinery built to take the town as well as the devastation of the city awaiting them, they prostrated themselves before Diognetus, begging him to come to the aid of his fatherland.

7. At first he refused to do it. But after young girls and boys of noble birth came with the priests to implore him, he then promised to do it on condition that, if he captured the siege-machine, it would be his. Once agreement had been reached, he breached the wall in the area that the machine was aiming for and ordered everybody – people in public service and private citizens[27] – to pour whatever water, excrement and mud they had through pipes leading through the breach to the area in front of the wall. The next day, after a vast amount of water, mud and sewage had been poured out during the night, the *helepolis*, while making its approach, stuck fast in the liquid swamp created there before it could reach the wall, and afterwards could not go forwards or backwards. So Demetrius, when he saw that he had been foiled by the cunning of Diognetus, withdrew with his fleet.

8. Then the Rhodians, freed from conflict by the wiliness of Diognetus, thanked him publicly and loaded him with all kinds of honours and distinctions. Diognetus brought the *helepolis* into the city and set it up in a public place with an inscription: 'Diognetus dedicated this gift to the people out of his spoils of war.' Therefore, in matters of defence one must equip oneself not just with machines but also, above all, with good tactics.

9. Something similar happened at Chios; when the enemy had prepared machines [*sambucae*] on their ships for storming the walls, the Chians piled up earth, sand and stones in the sea in front of their walls during the night. So the next day, when the enemy tried to make their advance, their ships ran aground on this mass of submerged material and could not get near the walls or pull back, but were nailed to the spot with incendiary projectiles and burnt. So, too, when Apollonia was under siege and the enemy thought they could penetrate inside the walls without exciting suspicion by digging tunnels, this was reported by spies to the inhabitants of the town; they were terrified by the news, left clueless with fear, and started to lose all their morale because they could not find out exactly when or where the enemy would burst in.

10. But at that point Trypho the Alexandrian was the architect there. He traced out a number of tunnels inside the wall and, continuing the excavations, moved outside the wall as far as an arrow could be shot, and hung bronze vessels in all the tunnels. The bronze vessels hanging in one of the excavations in front of the enemy mines began to sound because of the blows of their iron tools. So this told the Apollonians where the enemy was thinking of penetrating the walls by digging tunnels. Once they had worked out the enemy's line of attack, Trypho prepared bronze vessels containing a mixture of boiling water, pitch, sewage and red-hot sand at a level above the enemies' heads. Then, at night, he drilled a great number of holes through which he poured all of it at one go, killing all the enemy in the mine.

11. When Marseilles was being besieged and the enemy was digging more than thirty tunnels, the inhabitants of the city became suspicious and lowered the level of the whole of the

moat in front of the walls by digging deeper;[28] so all the enemy tunnels would have come up into the moat. But in places where the moat could not be dug out, they made a pit of enormous length and breadth inside the walls, like a fishpond, in front of the area where the enemy tunnels were being dug, and filled it with water from wells and from the port. The result was that as soon as the ends of the tunnel were opened up, an irresistible deluge of water burst into it, knocking out the props, and all the men inside were overwhelmed both by the mass of water and the collapsing tunnel.

12. Again, when a rampart was being built against them near their walls, and the enemy, after felling trees and transporting them to it, were labouring to raise its level, the Marsilians fired red-hot iron bolts from their *ballistae* at it, setting the whole structure on fire. When a tortoise-ram had arrived at the wall to batter it down, they let down a noose with which they snared it, and, pulling the head up by turning the noose around a drum with a capstan, they stopped the walls from being touched, and eventually destroyed the entire machine with incendiary missiles and shots from *ballistae*.

So it was that these victorious cities were liberated not with the help of machines but by the tactical skill of architects deployed against mechanical devices.

In this book I have dealt in as much detail as I could with the mechanical systems which I thought most efficacious in times of war and peace. In the nine preceding books I have composed an account organized around single subjects and their subdivisions so that the entire work would explain all aspects of architecture in ten books.

Appendix
Barbaro's Illustrations: From Vitruvius to the Renaissance

The Plates in this book are reproduced from the second edition of Barbaro's translation and commentary on Vitruvius entitled *I dieci libri dell'architettura di M. Vitruvio tradotti e commentati da mons. Daniele Barbaro eletto Patriarcha d'Aquileia, da lui riveduti & ampliati hora in più commoda forma ridotti*, published by Francesco de'Franceschi and Giovanni Chrieger in Venice in 1567 with a number of woodcuts designed by Andrea Palladio particularly for the architectural content of Books I–VI. We have omitted practically all those cases in which Barbaro and his illustrator departed drastically from the text of Vitruvius, notoriously, for example, in the illustration of the temple *in antis*, where, following Fra Giocondo (1511), they place two columns at the side of the *antae*, of which they misunderstand the form, as well as in front of them. We include those Barbaro illustrations, often thoroughly labelled with letters, which, though departing in some details and dimensions from what Vitruvius requires, are useful, and extremely handsome in many cases, for illustrating the many technical terms in the treatise, particularly those relating to the orders.

The captions are intended to fulfil two functions: (i) to illustrate what Vitruvius said and to introduce readers to the relevant terminology; (ii) to show rapidly, using only the example of Barbaro and Palladio, how Renaissance interpreters tackled the problem of illustrating Vitruvius' text, given the immense difficulties they faced when trying to reconstruct the often elliptical and opaque words of the ancient author. The single example of Barbaro and Palladio can serve to illustrate the procedure of innumerable commentators on the text for the next three centuries; in the absence of Greek and Roman monuments they could study which exemplified exactly what Vitruvius said, they inserted, often with spectacular and creative anachronisms, elements from architecture they knew to fill out his account. It is hoped, then, that the captions give at least a glimpse of the glorious afterlife of Vitruvius.

Practically all questions relating to the Barbaro editions of Vitruvius – authorship of the drawings, the copyists, the sources, architectural theory, relationship with Palladio – are dealt with in a series of important articles by Louis Cellauro:

'Palladio e le illustrazioni delle edizioni del 1556 e del 1567 di Vitruvio', *Saggi e memorie di storia dell'arte*, 22, 1998, pp. 55–128, which contains a complete catalogue and commentary.

'Daniele Barbaro and his Venetian Editions of Vitruvius of 1556 and 1567', *Studi veneziani*, NS 40, 2000, pp. 87–134.

'The Architectural Theory of Daniele Barbaro', *Studi veneziani*, NS 42, 2001, pp. 43–56.

'Disegni di Palladio e di Daniele Barbaro nei manoscritti preparatori delle edizioni del 1556 e del 1567 di Vitruvio', *Arte veneta*, 56, 2000, pp. 52–63.

'Daniele Barbaro and Vitruvius: The Architectural Theory of a Renaissance Humanist and Patron', *Papers of the British School at Rome*, 72, 2004, pp. 293–329.

'Les Éditions de Vitruve par Daniele Barbaro, à Venise chez Marcolini en 1556 et chez de' Franceschi en 1567', in S. Deswarte-Rosa (ed.), *Sebastiano Serlio à Lyon: architecture et imprimerie*, vol. 1: *Le Traité d'architecture de Sebastiano Serlio: une grande entreprise éditoriale au XVIe siècle*, Lyons, 2004, pp. 392–6.

Of particular interest is the recent contribution of P. Gros, *Palladio e l'antico*, Venice, 2006, in which a number of celebrated Renaissance misunderstandings of Vitruvius' text are discussed (*stereobate/stylobate*, *scamilli impares*, the theatre, houses, etc.).

Other recent contributions include:

J. S. Ackerman, 'Daniele Barbaro and Vitruvius', in C. L. Striker and J. S. Ackerman (eds.), *Architectural Studies in Memory of Richard Krautheimer*, Mainz, 1996, pp. 1–5.

M. M. D'Evelyn, 'Varietà and the Caryatid Portico in Daniele Barbaro's Commentaries on Vitruvius', *Annali di architettura*, 10/11, 1998/9, pp. 157–74.

R. Tavernor, '"Brevity without Obscurity": Text and Image in the Architectural Treatises of Daniele Barbaro and Andrea Palladio', in R. Palmer (ed.), *The Rise of the Image: Essays on the History of the Illustrated Art Book*, Aldershot, 2003, pp. 105–33.

Hans-Christoph Dittscheid, 'Triumph der Architectura: zwei konkurrierende Frontispize in Daniele Barbaros Vitruv-Edition von 1556', in N. Riegel (ed.), *Architektur und Figur: das Zusammenspiel der Künste*, Munich, 2007, pp. 184–201.

Figures

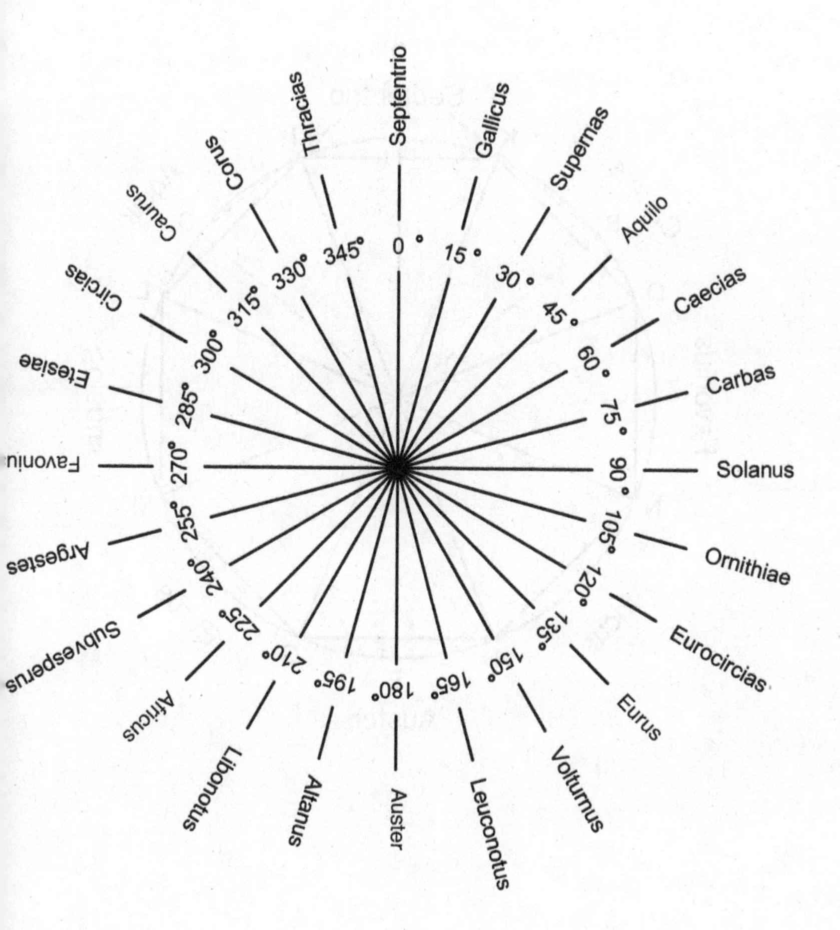

Fig. 1. The twenty-four winds described by Vitruvius (see p. 30)

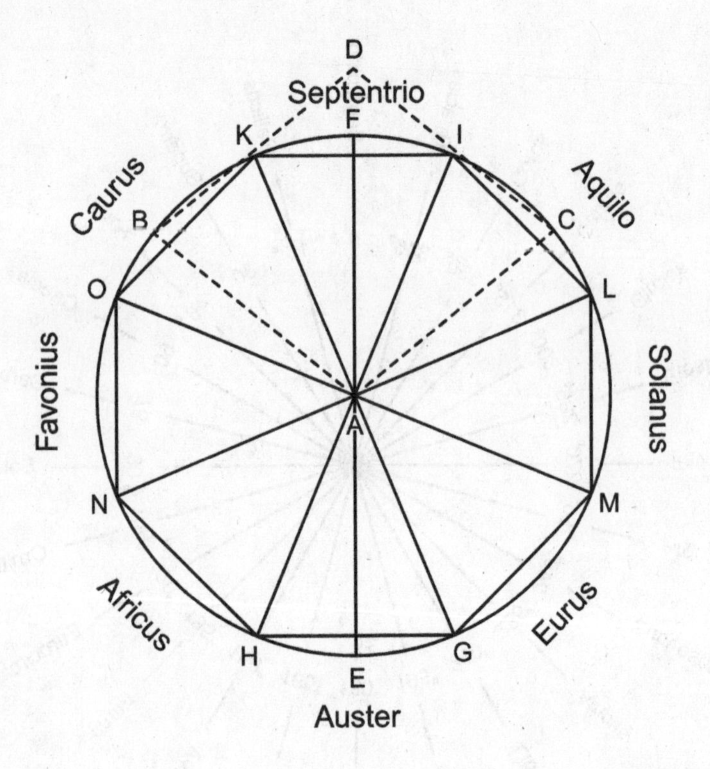

Fig. 2. The construction of the octagon of the wind-rose with which to site the city with the eight principal winds mentioned by Vitruvius (see p. 33)

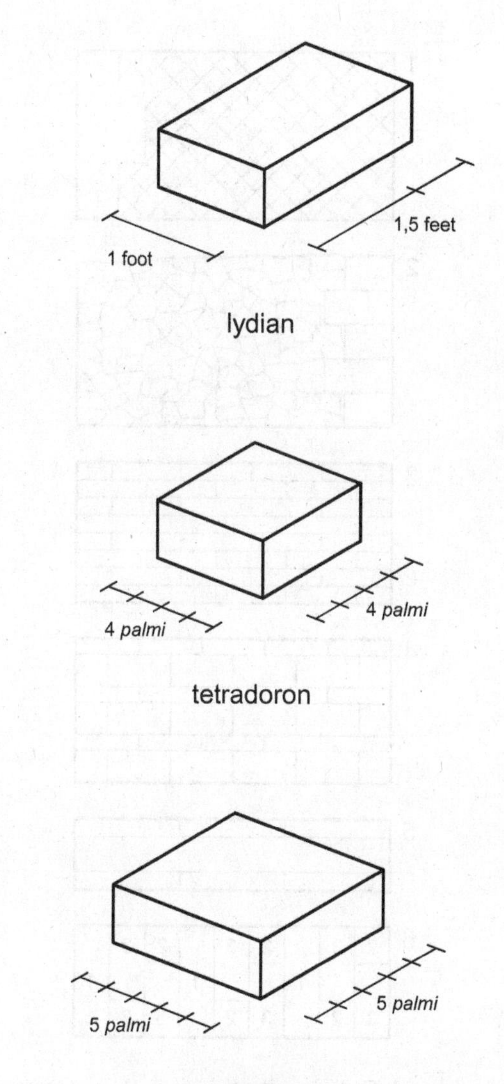

Fig. 3. Roman and Greek bricks (see p. 43)

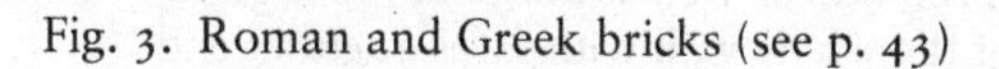

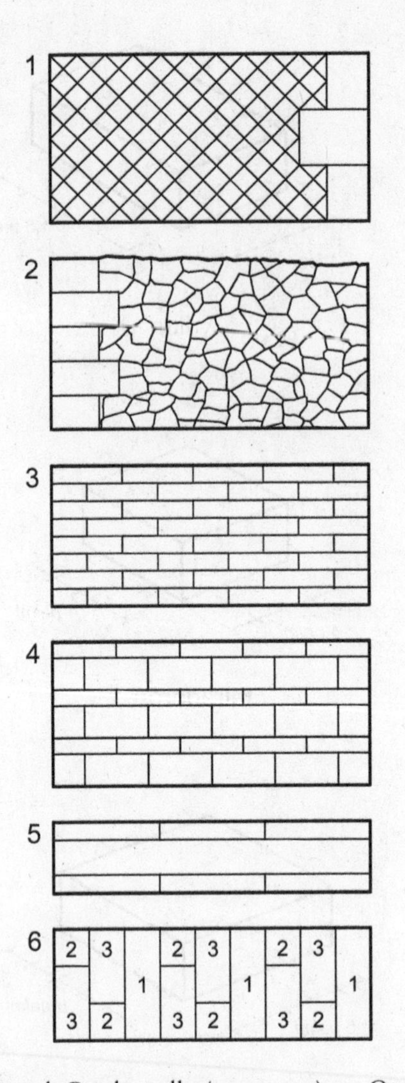

Fig. 4. Roman and Greek walls (see p. 52). 1 *Opus reticulatum*. 2 *Opus incertum*. 3 *Opus isodomum*. 4 *Opus pseudisodomum*. 5 Roman *emplekton* wall with infill between stone outer surfaces. 6 Greek wall with stones of alternating lengths (2, 3) making up the thickness of the wall as well as *diatonoi* (1) or stones running through the width of the whole wall with faces at either end

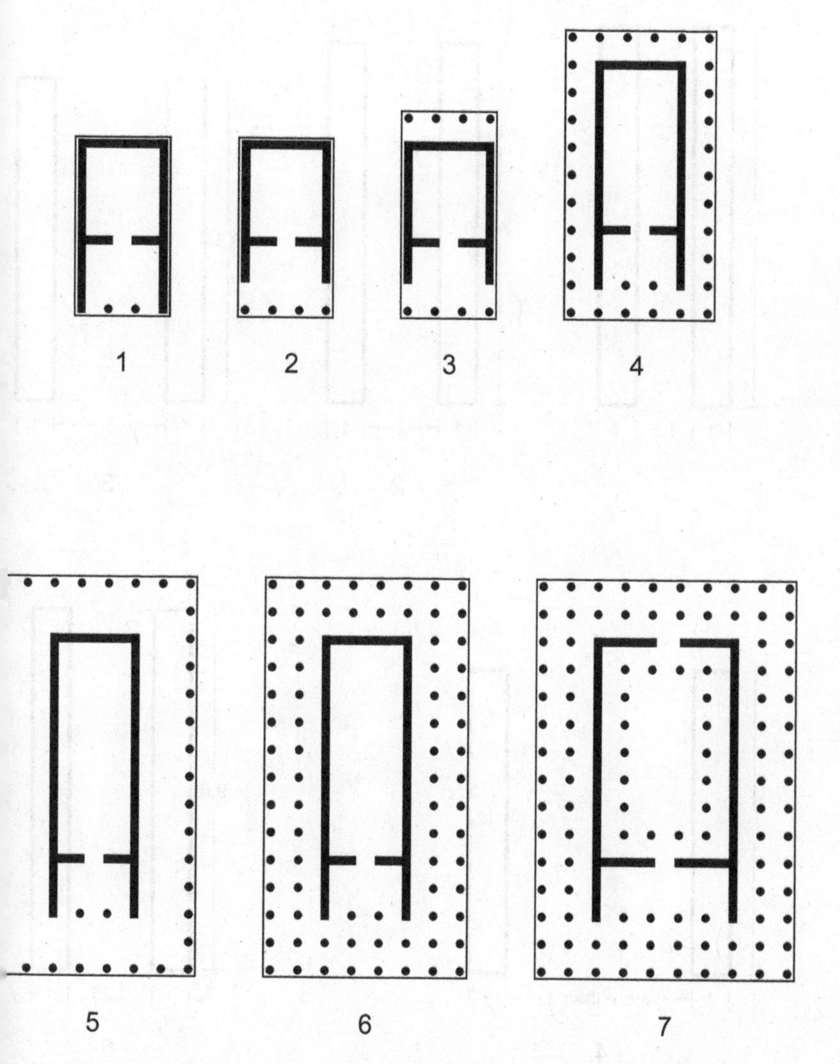

Fig. 5. Temple ground-plans described by Vitruvius (see p. 69). 1 Distyle *in antis*. 2 Prostyle. 3 Amphiprostyle. 4 Peripteral. 5 Pseudodipteral. 6 Dipteral. 7 Hypaethral

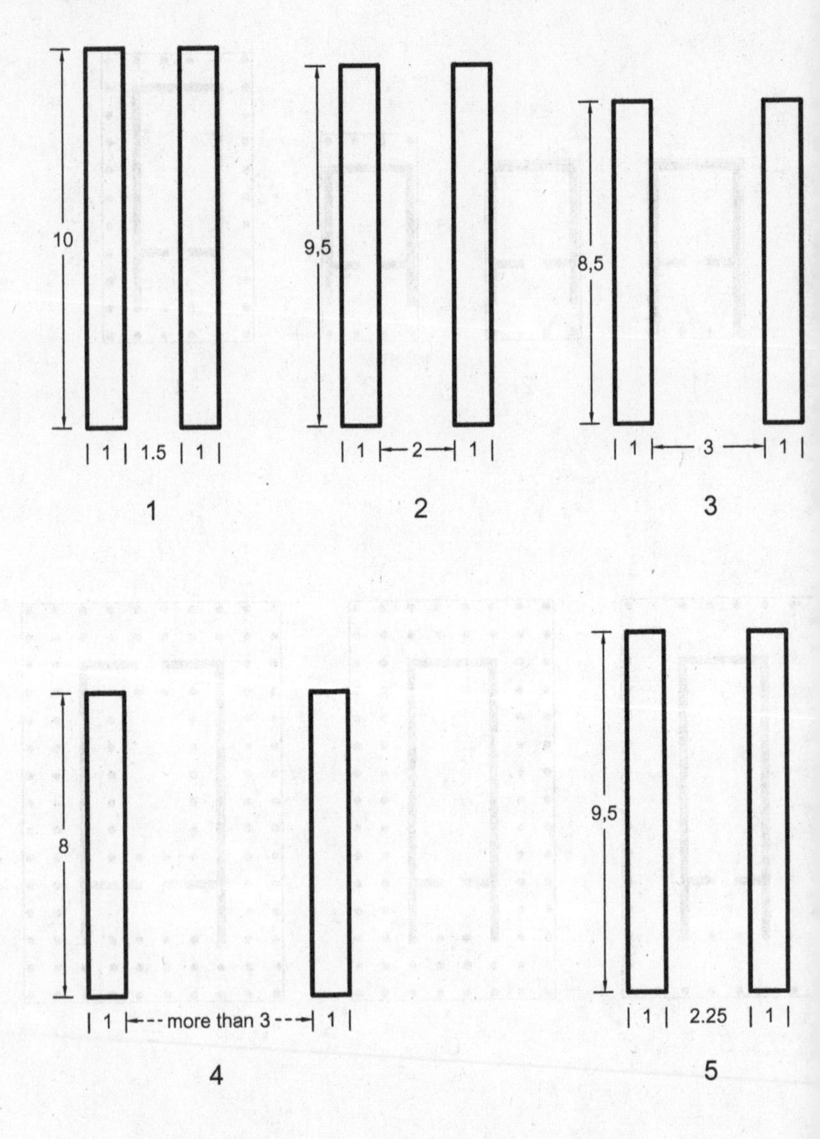

Fig. 6. Intercolumniations according to Vitruvius (see p. 73). 1 Pycnostyle. 2 Systyle. 3 Diastyle. 4 Araeostyle. 5 Eustyle

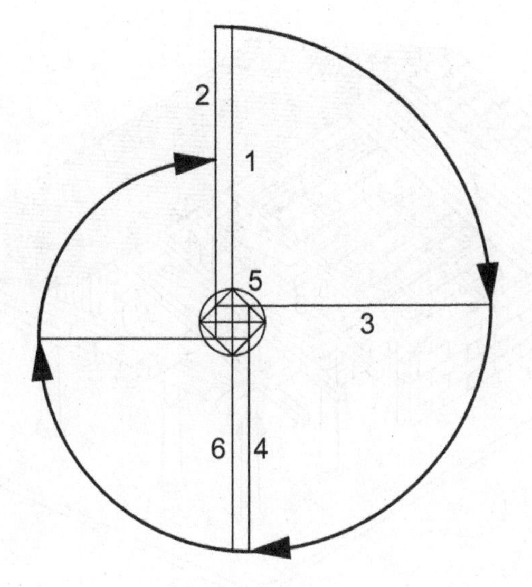

Fig. 7. The construction of the Ionic volute (see p. 84). 1 Upper segment of *cathetus*, 4.5 units. 2 Radius of first quadrant of 4 units (the height of the *cathetus* minus half the diameter of the eye of the volute) delimiting the outer edge of the rim of the volute. 3 Radius of second quadrant, reduced by half the diameter of the eye to 3.5 units. 4 Radius of third quadrant, reduced by another half of the diameter of the eye to 3 units, and so on. 5 Half the diameter of the eye of the volute. 6 Lower segment of the *cathetus* of 3.5 units

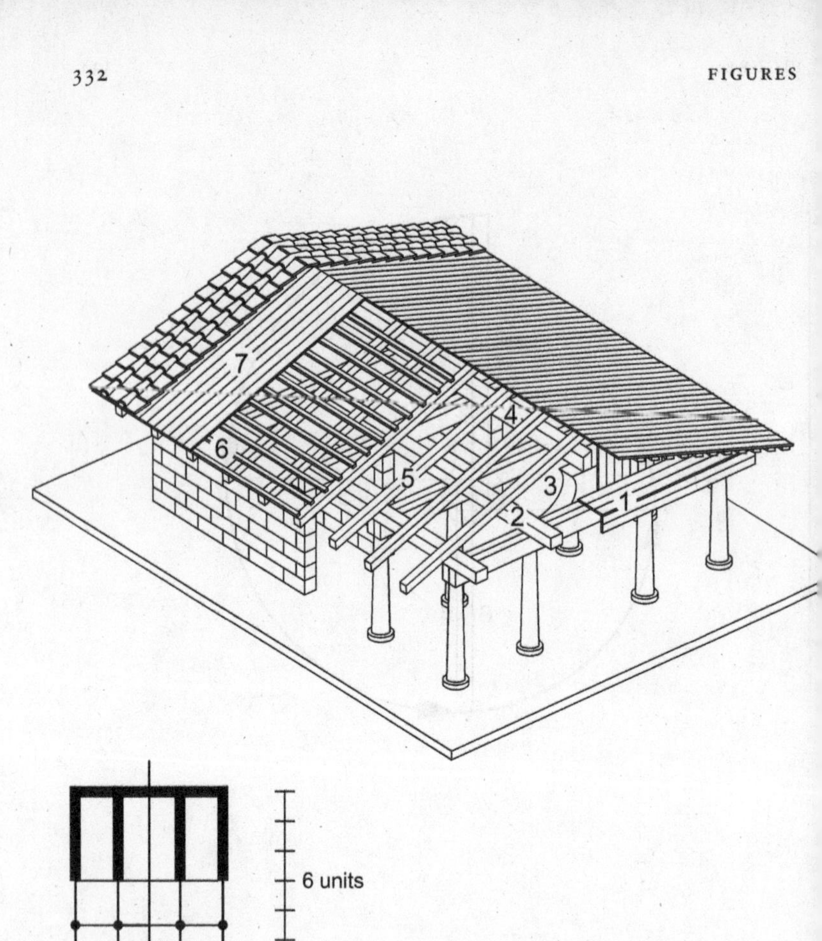

Fig. 8. Plan and axonometric of the Tuscan temple (see p. 112). 1 Compound beams. 2 Mutules. 3 Support for the ridge-piece. 4 Ridge-piece (*columen*). 5 Principal rafters (*cantherii*). 6 Purlins (*templa*). 7 Roof and eaves (*stillicidia*)

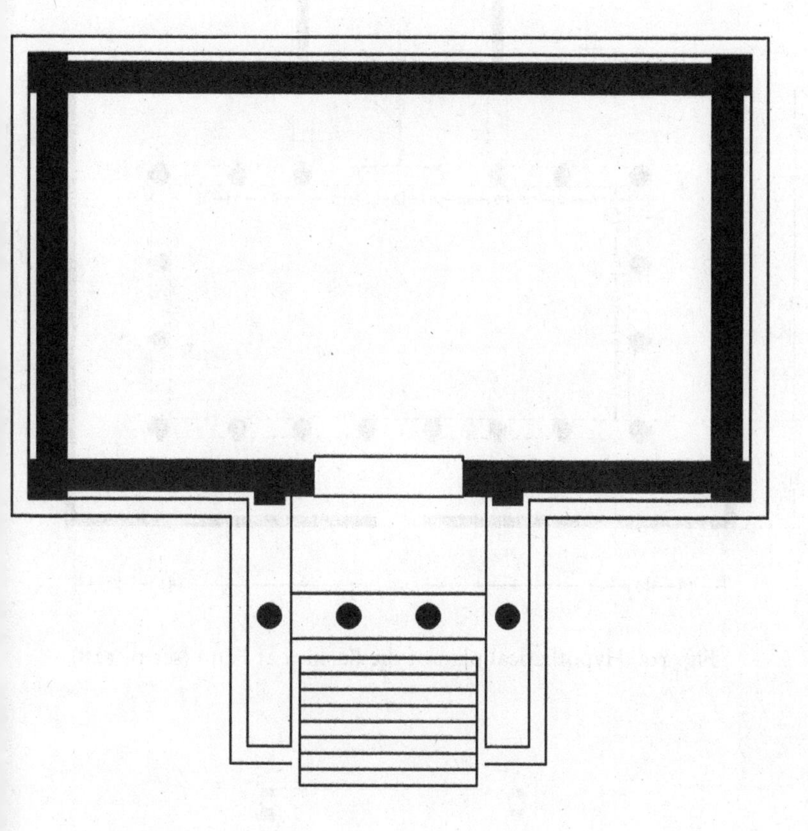

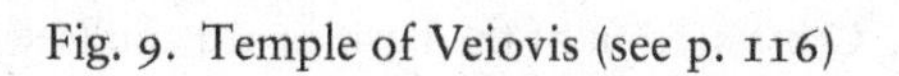

Fig. 9. Temple of Veiovis (see p. 116)

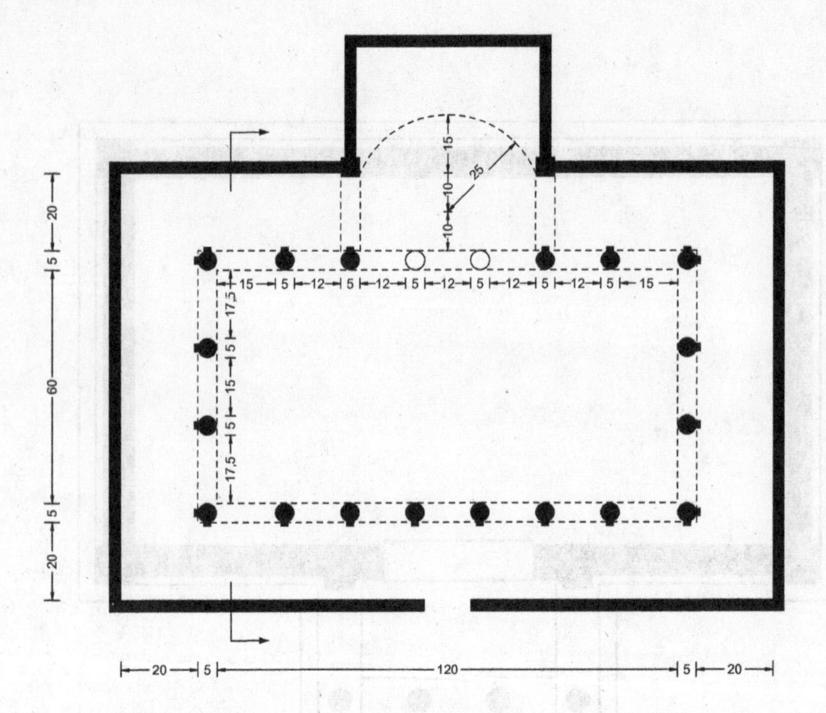

Fig. 10. Hypothetical plan of the Basilica at Fano (see p. 128)

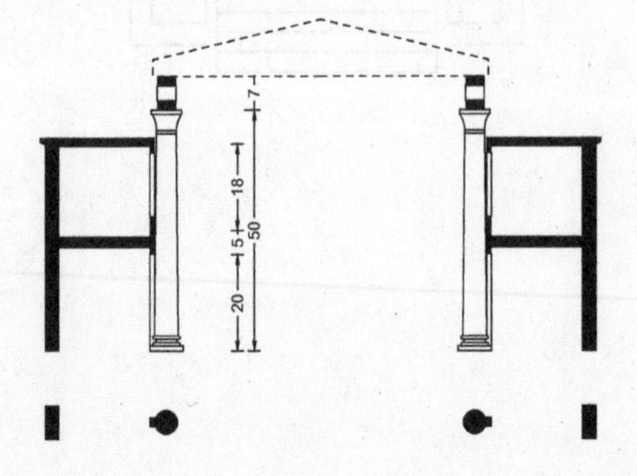

Fig. 11. Cross-section of the Basilica at Fano (see p. 128)

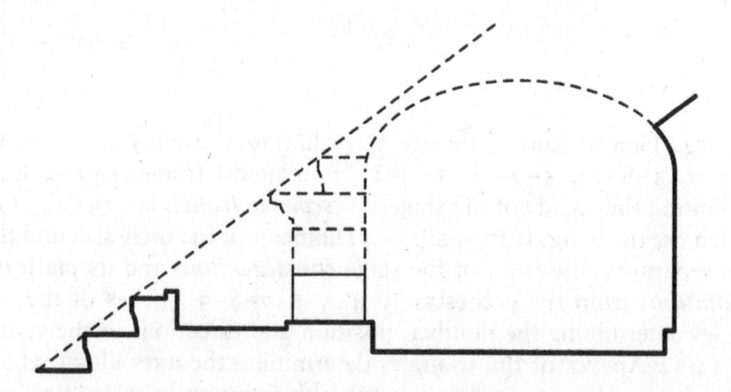

Fig. 12. Section of theatre at Scythopolis (ancient Bethsan, Palestine) showing the *cella* housing the acoustic vases with the aperture at the front facing the seats lower down; the theatre conforms to what Vitruvius recommends in the sense that a straight line drawn up the ranks of seats touches their edges, though there is no certainty if there is any direct connection between Vitruvius' text and the building (see p. 135)

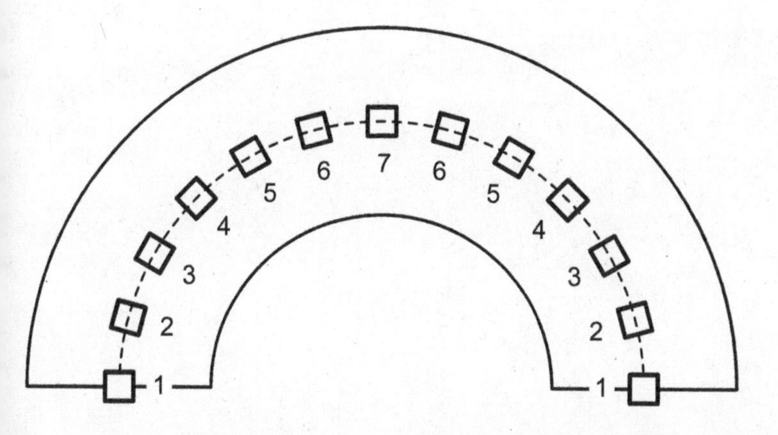

Fig. 13. Layout of acoustic cavities in small- to medium-sized theatres (see p. 135). 1 *Nete hyperbolaion*. 2 *Nete diezeugmenon*. 3 *Paramese*. 4 *Nete synemmenon*. 5 *Mese*. 6 *Hypate meson*. 7 *Hypate hypaton* (in the centre)

Fig. 14. Plan of Roman theatre according to Vitruvius (see p. 143). 1–2–7, 3–8–12, 5–9–10, 6–4–11 Equilateral triangles. 1–2 Line delimiting the façade of the stage-set (*scaenae frons*), left and right of which are the wings (*versurae*). 3–4 Diameter of the orchestra and the line separating the front of the stage (*proscaenium*) and its platform (*pulpitum*) from the orchestra. 3–4–5–6–7–8–9 Apexes of the triangles determining the number, position and directions of the stairs. 10-11-12 Apexes of the triangles determining the axes along which the doors in the stage-set are set (10 *valvae regiae* in the centre, 11, 12 *hospitalia*). 13–14 Entrances formed by cutting back the lowest seating to the height of a sixth of the diameter of the orchestra (3–4). 15 Locations of the *periaktoi*. 16 Blocks of seating (*cuneus*) in the auditorium. 17 Ramps of steps (*scalaria, gradus*). 18 Transverse aisle (*praecinctio, diazoma*). 19 Upper ramps of stairs on axis with blocks of seating below, but not with the ramps of stairs below. 20 Portico matching the height of the stage-set

Fig. 15. Section of Roman theatre according to Vitruvius (see p. 143). DO = Diameter of the orchestra. A Podium of the façade of the stage-set (*scaenae frons*) with its cornice and gola. B Column, base and capital. C Entablature. D Parapet. E Upper columns. F Entablature

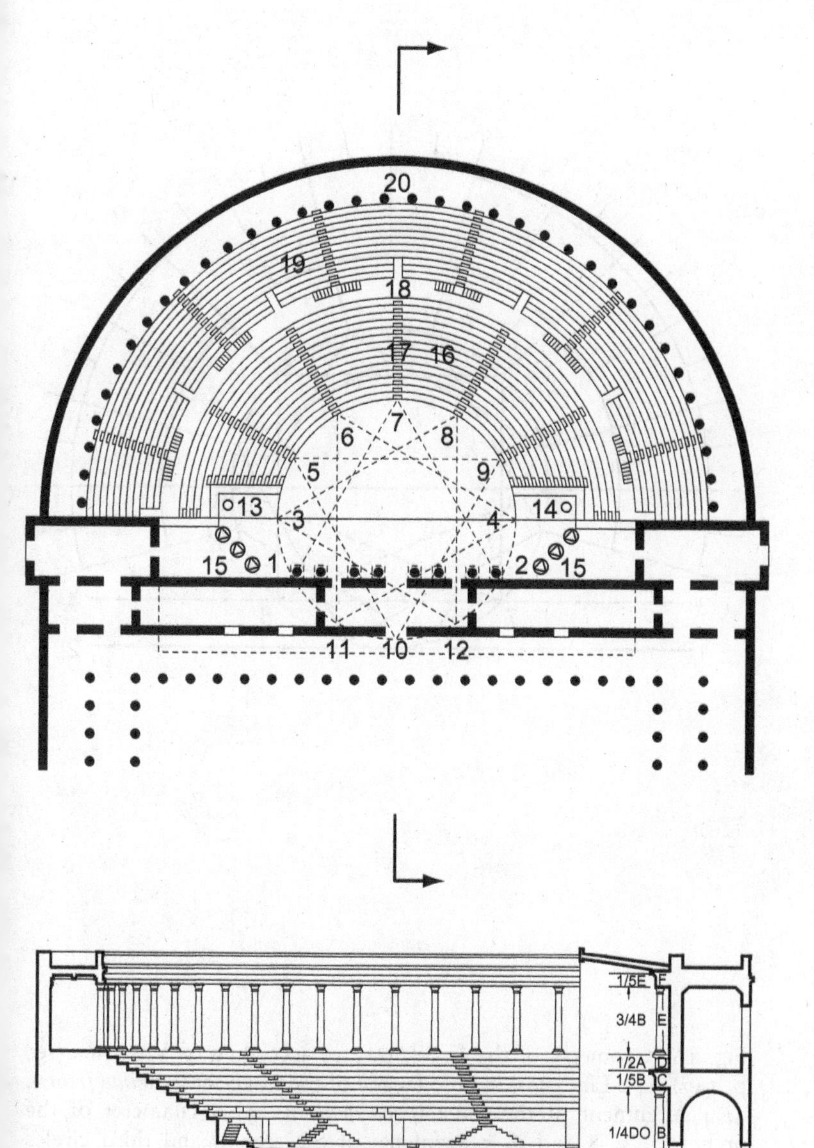
20
19
18
17
16
7
6
8
5
9
13
14
3
4
15
1
2
15
11
10
12
1/5E
F
3/4B
E
1/2A
D
1/5B
C
1/4DO
B
1/12DO
A
1/6DO
scaled up

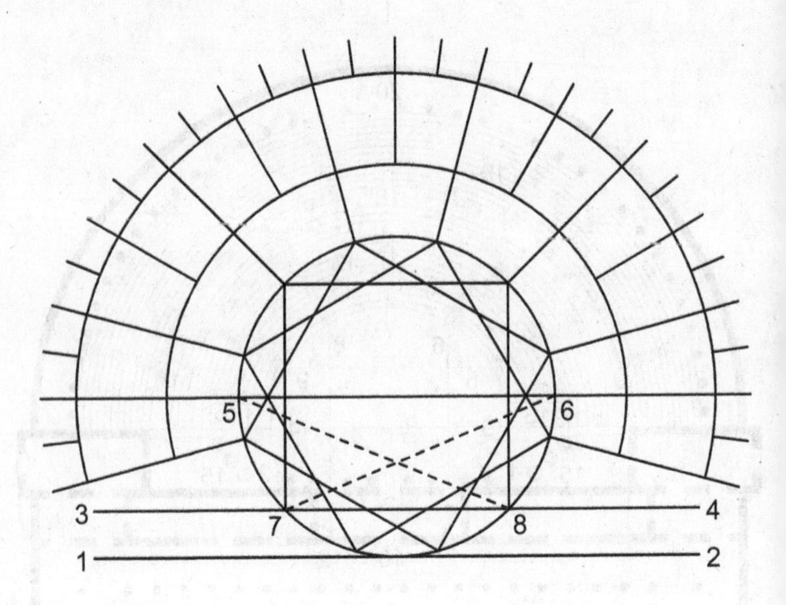

Fig. 16. Geometry of the Greek theatre according to Vitruvius (see p. 147). 1, 2 Line marking the façade of the stage-set (*scaenae frons*). 3, 4 Alignment of the stage (*proscaenium*). 5, 6 Diameter of the orchestra. 5, 8 and 6, 7 Segments of the second and third circles centred at left and right of the orchestra with diameters equalling that of the orchestra and extended to intersect with the stage-wall at 7 and 8

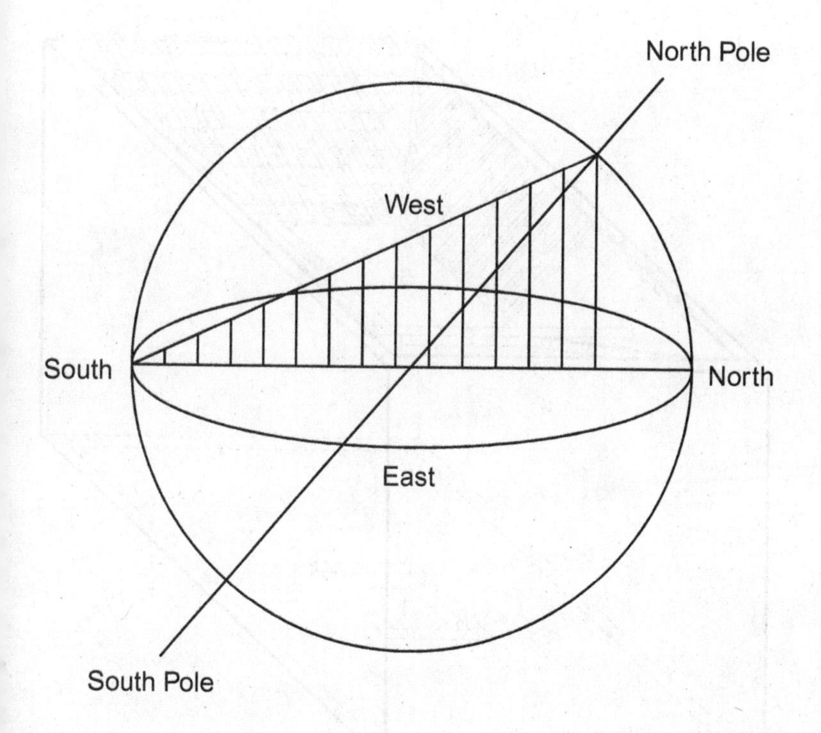

Fig. 17. The triangular configuration of the world according to Vitruvius (see p. 167)

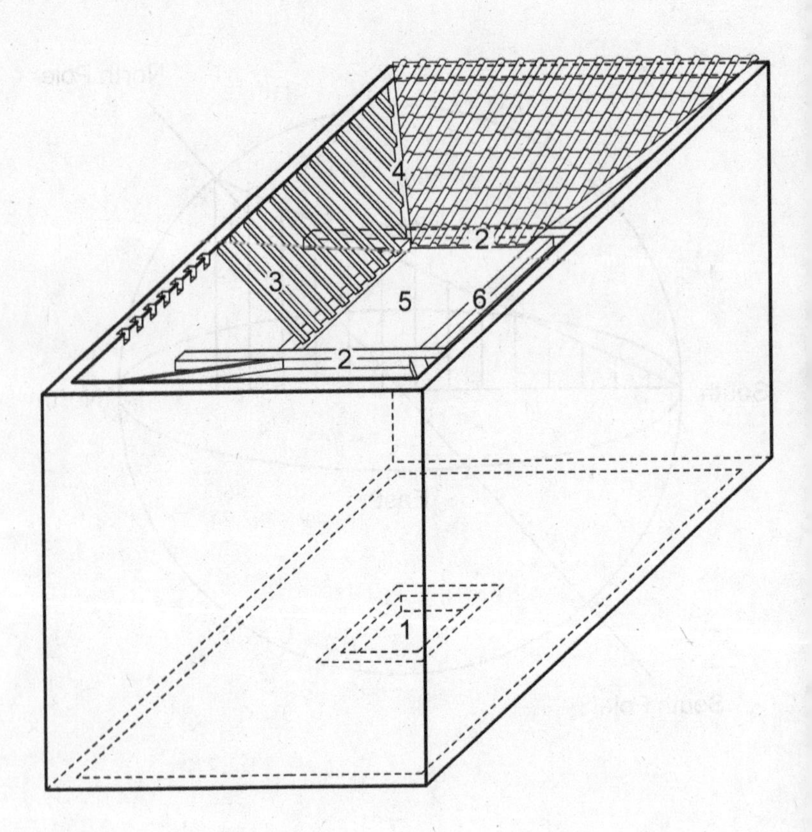

Fig. 18. The Tuscan courtyard (see p. 172). 1 *Impluvium*. 2 Main joists or beams (*trabes*). 3 Common rafters (*asseres*). 4. Water-channels (*colliciae*). 5. *Compluvium*. 6 Suspended cross-beams, or stringers (*interpensiva*)

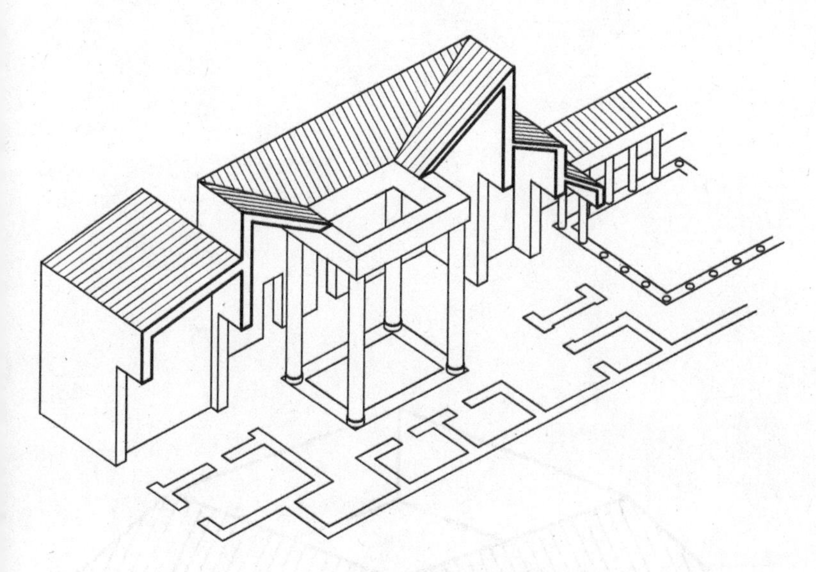

Fig. 19. The tetrastyle courtyard (see p. 172)

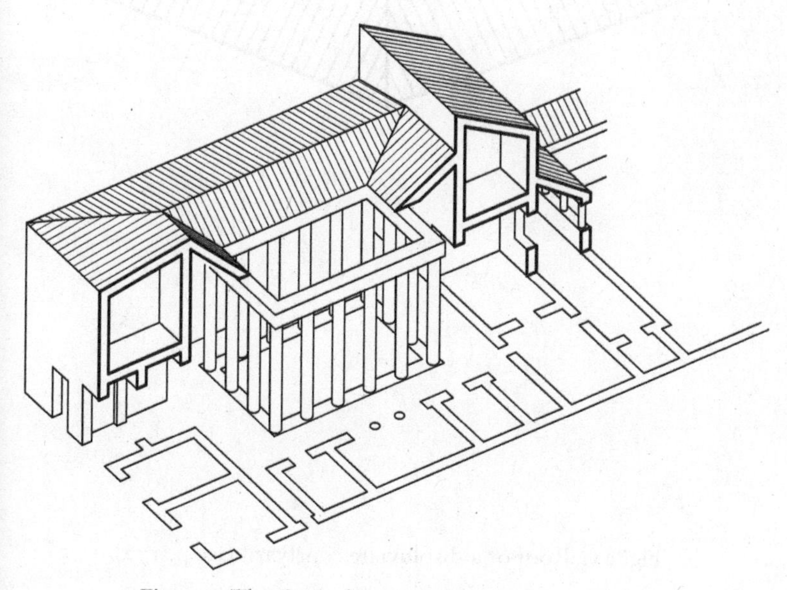

Fig. 20. The Corinthian courtyard (see p. 172)

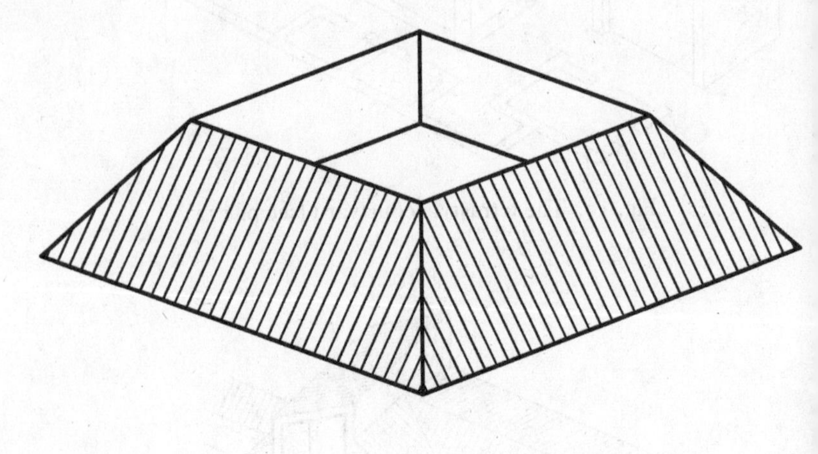

Fig. 21. Roof of a displuviate courtyard (see p. 172)

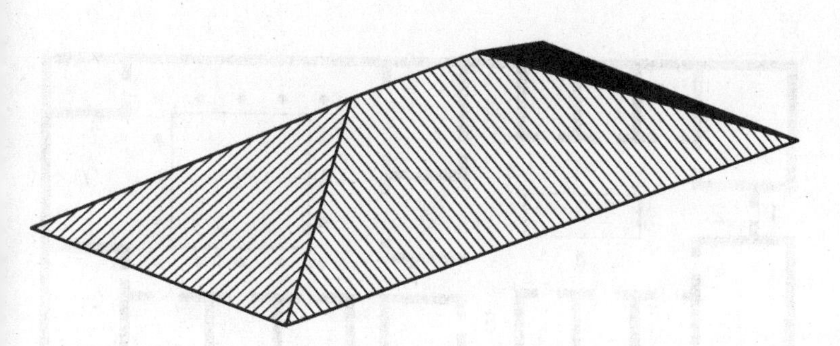

Fig. 22. Roof of a testudinate courtyard (see p. 172)

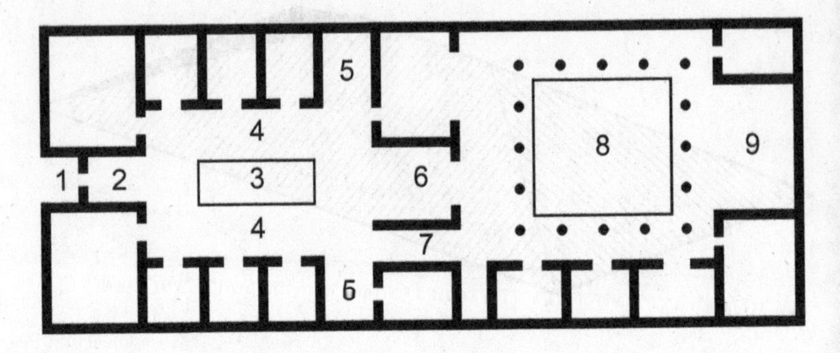

Fig. 23. Plan of a Vitruvian house (see p. 178). 1 *Vestibulum*. 2 Entrance corridor (*fauces*). 3 Rectangular catchment area or *impluvium* below the *compluvium*. 4 *Atrium*. 5 Side-rooms (*alae*). 6 *Tablinum*. 7 Corridor. 8 Colonnaded courtyard (*peristylium*). 9 *Exedra*

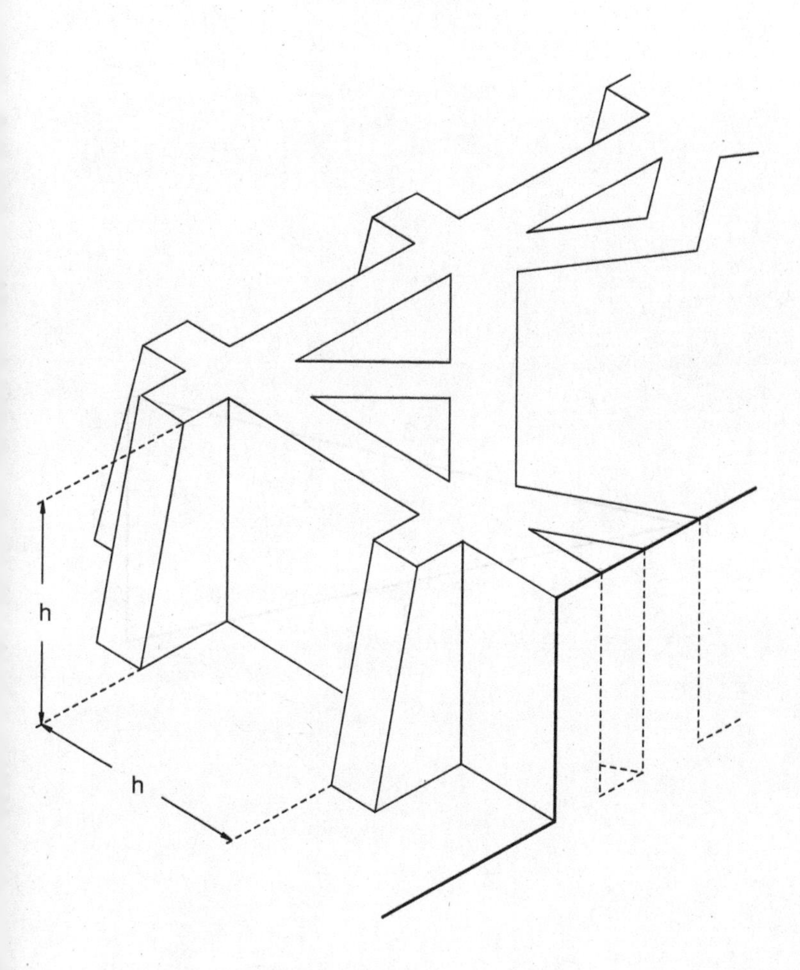

Fig. 24. Foundation walls with zigzag reinforcing walls inside, additional buttressing (*anterides*, *erismae*) on the outside and a wall-system in the corner forming a triangle with another wall let into the angle (see p. 190)

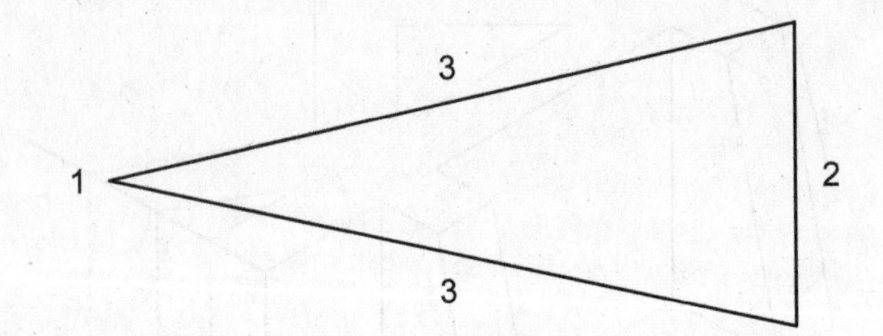

Fig. 25. Visualization of how the eye sees images on a stage-set (see p. 194). 1 Spectator's ideal viewpoint. 2 The stage-set (*locus certus*). 3 Visual rays emanating from the stage-set

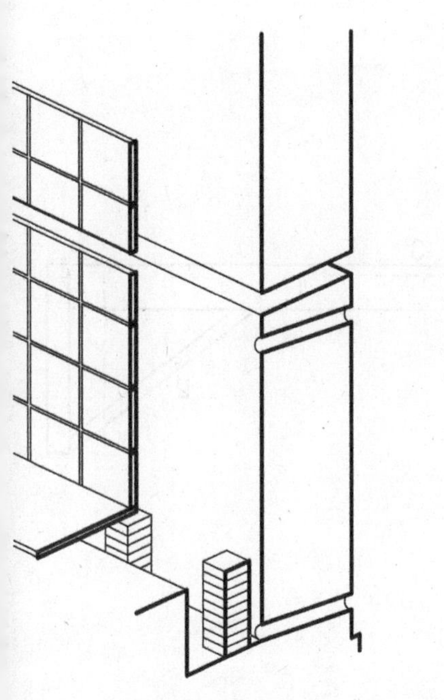

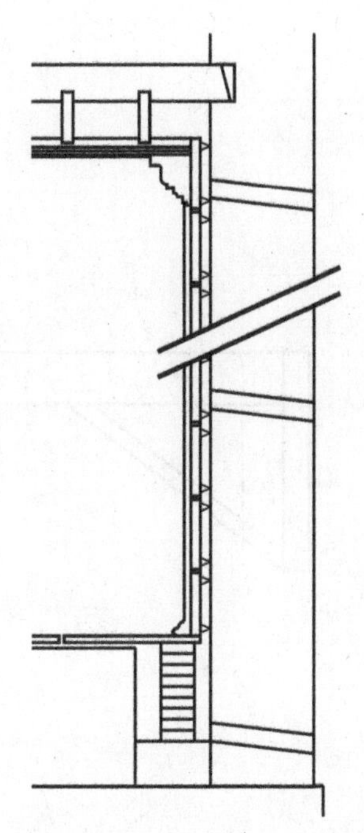

Fig. 26. Drainage channel below a damp room, with small piers supporting the floor and wall-tiles (see p. 205)

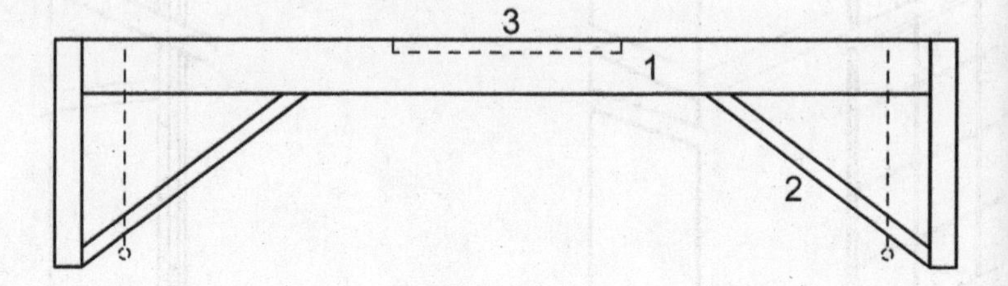

Fig. 27. The *chorobates* (see p. 236). 1 The main beam. 2 The struts calibrated with small vertical lines aligned with the plumb-bobs when the device is vertical (some make the struts connecting the legs run parallel with the beam). 3 Channel or groove five feet long in the top of the beam acting as a spirit-level

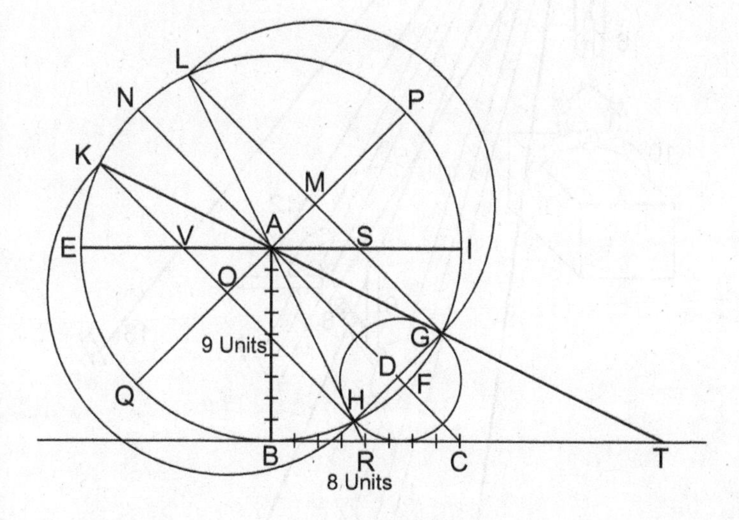

Fig. 28. The Vitruvian *analemma* (see p. 262). AB Gnomon. BC Equinoctial shadow. Large circle = meridian. CA Sunray at equinox. EI Horizon. PQ Axis. GH *Logotomus*. Smaller circle = *Menaeus*.

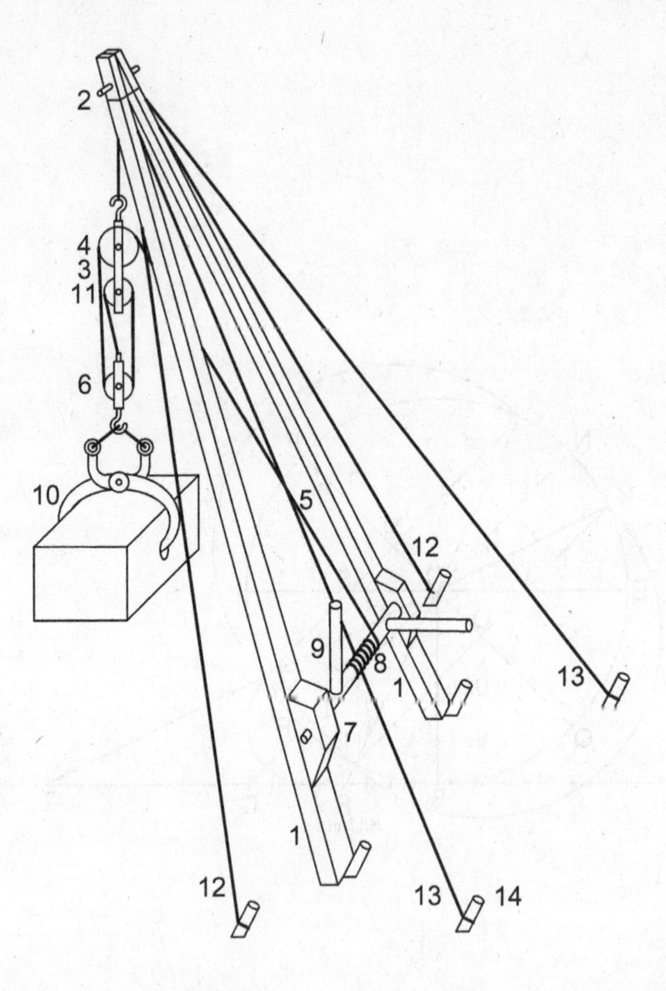

Fig. 29. Crane with three pulleys (*trispastos*) (see p. 280). 1 Beams (*tigna*). 2 Bolt (*fibula*). 3 Upper pulley-block (*troclea, rechamus*). 4 Pulley (*orbiculus*). 5 Traction-rope (*ductarius funis*). 6 Lower pulley (*troclea inferior*). 7 Socket-piece (*chelonium*). 8 Windlass (*sucula*). 9 Hand-spike or lever (*vectis*). 10 Iron pincers (*ferrei forfices*). 11 Little axle (*axiculus*). 12 Support-ropes (*antarii funes*). 13 Control or retaining rope (*retinaculum*). 14 Stakes for securing ropes (*pali*)

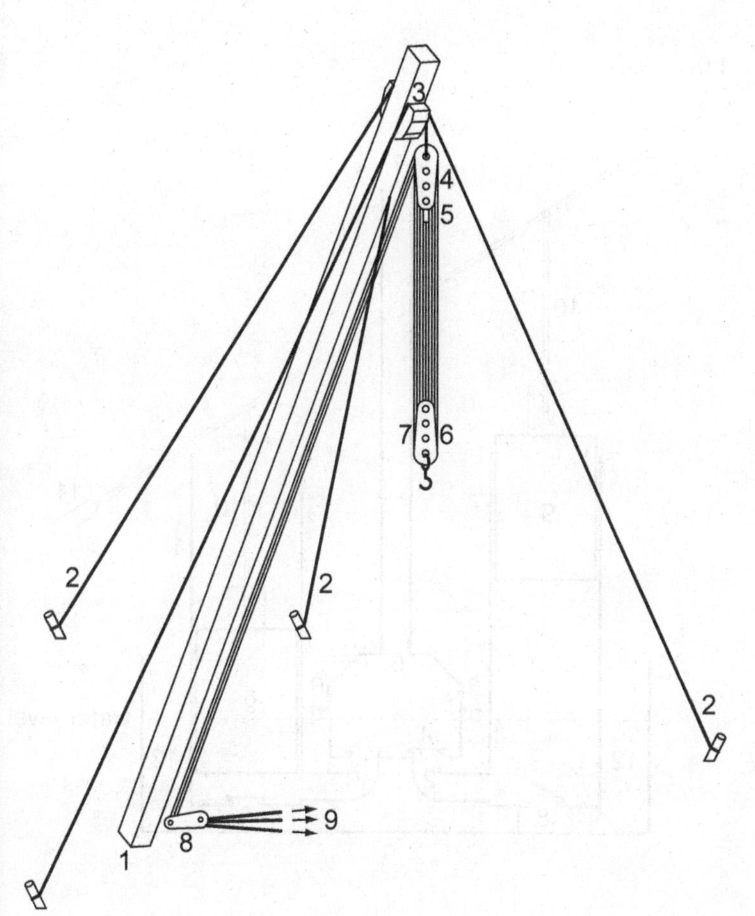

Fig. 30. Crane with a single beam and multiple pulleys (*polyspastos*) (see p. 283). 1 Beam (*tignum*). 2 Retaining ropes fixed in ground at four points (*retinacula*). 3 Socket-piece (*chelonium*). 4 Upper pulley-block (*superior troclea*). 5 Straight piece of wood (*regula*). 6 Lower pulley-block (*ima troclea*). 7 Lower set of pulleys (*ordines orbiculorum*). 8 Pulley-block (*artemon*). 9 Ropes for men to pull

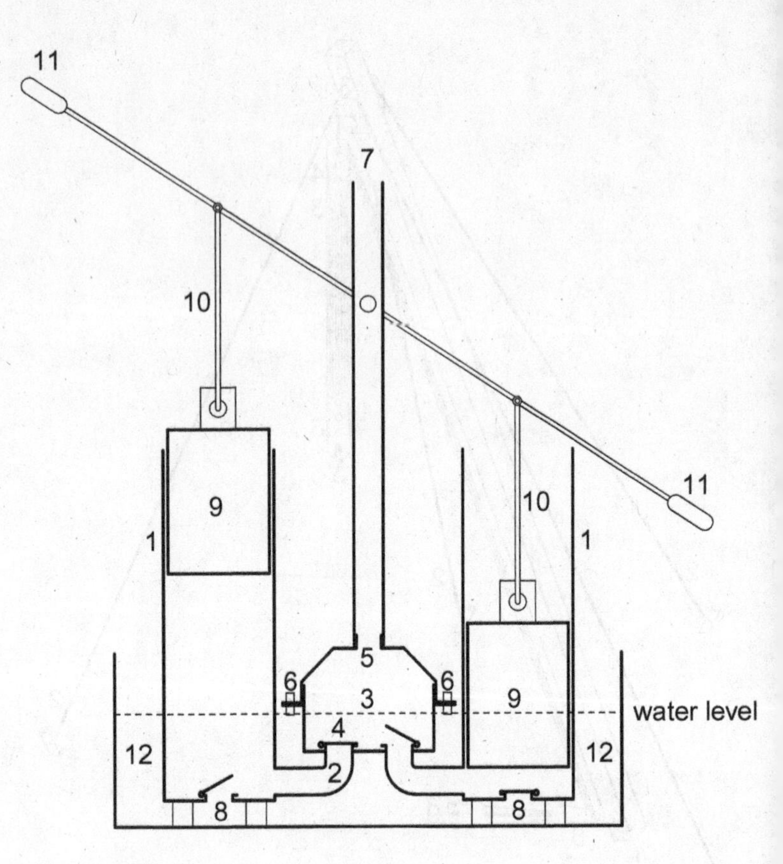

Fig. 31. Ctesibius' water-pump (see p. 296). 1 Cylinders (*modioli*). 2 Tubes converging like a fork (*fistulae*). 3 Tank (*catinus*). 4 Valves of tank (*asses catini*). 5 Hood like an inverted funnel (*paenula*). 6 Pins or bolts keeping the hood in place (*fibulae*). 7 Pipe, called a 'trumpet' (*tuba*). 8 Valves (*asses modioli*). 9 Pistons (*emboli masculi*). 10 Piston-rods (*regulae*). 11 Levers (*vectes*). 12 Tank (*castellum*)

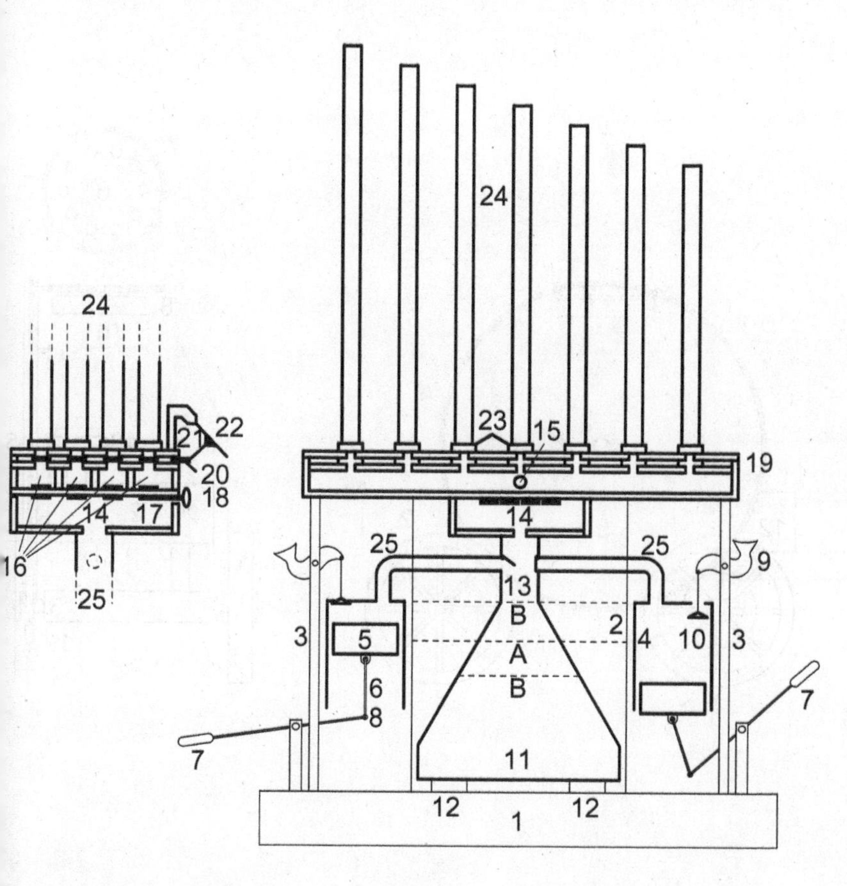

Fig. 32. The water-organ (see p. 298). A = water level when air-pressure is not applied. B = water level when air-pressure is applied. 1 Base (*basis*). 2 Altar-shaped container (*ara*). 3 Struts (*regulae*). 4 Cylinders (*modioli*). 5 Pistons (*funduli ambulatiles*). 6 Piston-rods (*ancones*). 7 Levers (*vectes*). 8 Hinges (*verticulae*). 9 Dolphins (*delphini*). 10 Cymbal-shaped valves (*cymbala*). 11 Regulator (*pnigeus*). 12 Little blocks (*taxilli*). 13 Neck of the regulator (*cervicula*). 14 Small chamber (*arcula*). 15 Main component (*caput machinae*). 16 Channels (*canales*). 17 Stopcock (*epitonium*). 18 Handle (*manubrium*). 19 Upper board (*tabula, pinax*). 20 Sliding bars (*regulae, plinthides*). 21 Iron hooks (*coracia*). 22 Keys (*pinnae*). 23 Rings (*anuli*). 24 Organ pipes (*organa*). 25 Pipes (*fistulae*)

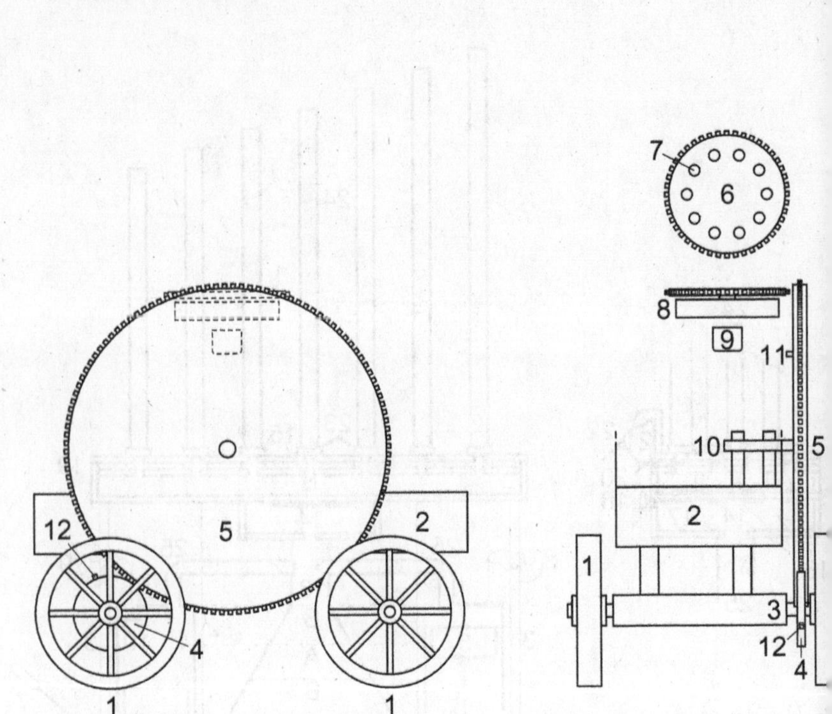

Fig. 33. The hodometer (see p. 301). 1 Wheel (*rota*). 2 Chassis of the cart (*capsus redae*). 3 Hub (*modiolus*). 4 Disc with one tooth. 5 Vertical disc with four hundred teeth. 6 Horizontal disc. 7 Holes (*foramina*). 8 Housing of the horizontal disc (*loculamentum, theca*). 9 Bronze receptacle (*vas aeneum*). 10 Housing of the second disc. 11 Tooth on the side of the vertical disc (*denticulus*). 12 Tooth on the lowest disc

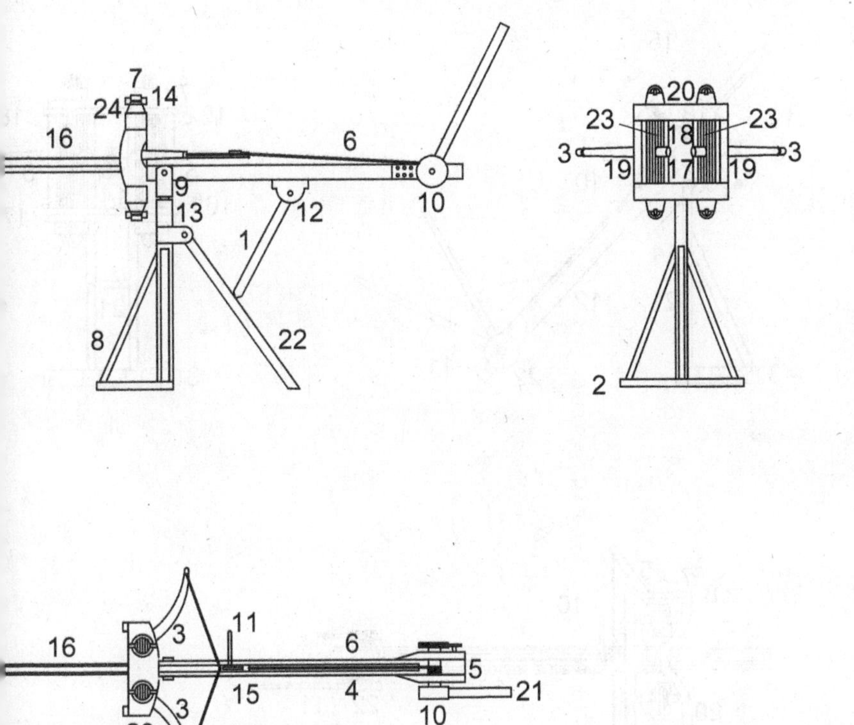

Fig. 34. The *scorpio* (see p. 305). 1 Small counter-strut at back (*minor columella, antibasis*). 2 Base of column (*basis*). 3 Arm of catapult (*bracchium*). 4 Reinforcing strips of wood (*regulae, bucculae*). 5 Box or block (*buccula, scamillum, loculamentum*). 6 Groove, channel (*canaliculus, canalis, syrinx*). 7 Head-piece, main frame of catapult (*capitulum*). 8 Strut (*capreolus*). 9 Top of the column (*caput columnae*). 10 Drum housing the levers of the windlass (*carchesium*). 11 Trigger (*manucla, chele*). 12 Socket (*pulvinus, chelonium*). 13 Column (*columella*). 14 Tension-bolt (*epizygis*). 15 Retaining hook (*epitoxis*). 16 Bottom of the channel or groove (*canalis fundus*). 17 Hole for the arrow (*intervallum*). 18 Central support, strut (*parastas media*). 19 Side supports, struts (*parastaticae*). 20 Boards (*tabulae, peritreti*). 21 Levers, handles of windlass (*scutulae*). 22 Back strut (*subiectio*). 23 Ropes under tension (*nervi torti*). 24 Cylinder (*modiolus*)

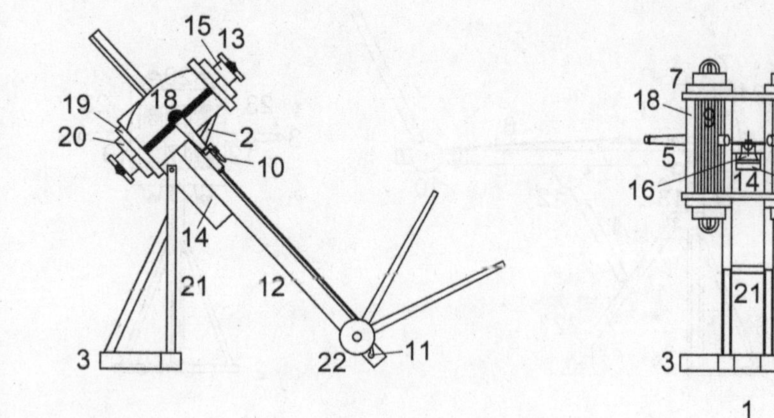

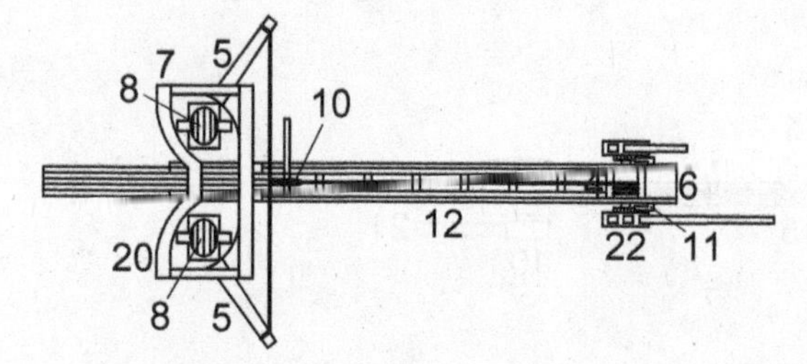

Fig. 35. The *ballista* (see p. 308). 1 Base (*basis*, *eschara*). 2 Front strut (*anteris*). 3 Counterbase (*antibasis*). 4 Pivot, axle (*axon*, *axis*). 5 Arm (*bracchium*). 6 Square block (*quadratum*). 7 Head-piece, main frame of machine (*caput*, *capitulum*). 8 Lozenge (*scutula*, *peritretos*). 9 Twisted ropes (*funes*). 10 Trigger (*chele*). 11 Catch (*clavicula*). 12 The 'ladder', the main beam of the device (*scapus climacidos*). 13 Tension-bolt (*cuneolus*, *epizygis*). 14 'Table' (*mensa*). 15 Cylinder (*modiolus*). 16 Mortise (*pterygoma*). 17 Cover, housing of the *chelonium* (*replum*, *operimentum*). 18 Strut (*parastata*). 19 Wooden strip (*regula*). 20 Outer wooden strip (*regula exterior*). 21 Columns, central supports (*columnae*). 22 Side face (*frons transversarius*)

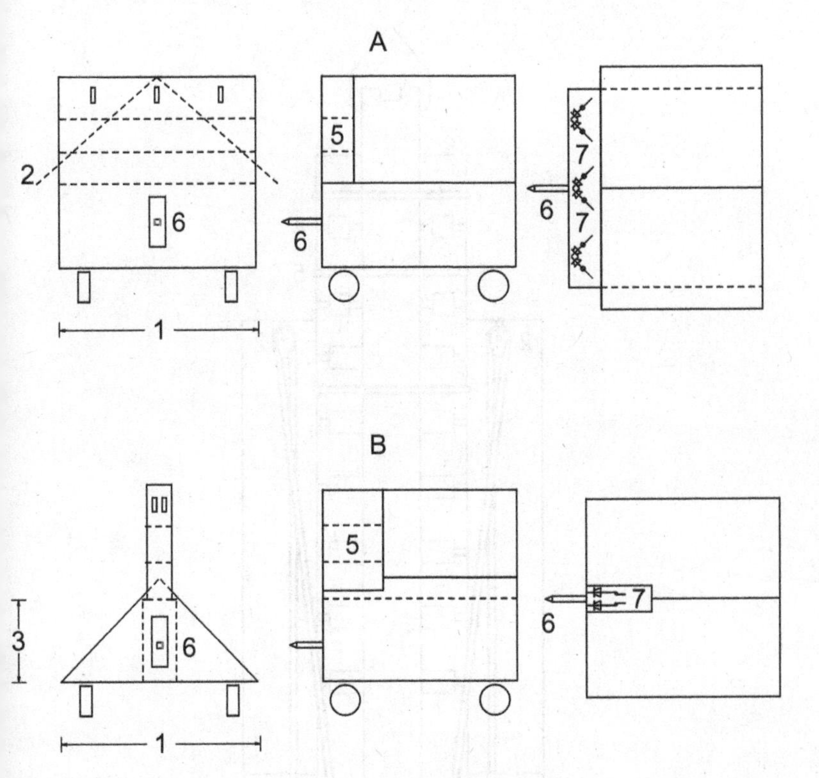

Fig. 36. A and B. The tortoise-ram (*testudo arietaria*) (see p. 311). There are two hypotheses. A: That the gable (*fastigium*) of sixteen cubits above the *stratum* of the machine, and at least two cubits above the roof of the machine, itself thirteen cubits in height, reached to at least thirty-one cubits, and enclosed the tower. B: That the gable rises sixteen cubits from the base of the machine (*stratum*), with the interior structures of thirteen cubits inside it, leaving a difference of at least two cubits between them, i.e. three. The main problem is the meaning of the word *stratum* (= platform of the machine or a floor at the top of the machine at thirteen cubits?). 1 Breadth of thirty cubits (*intervallum*). 2 Gable rises above the roof in the middle (*tectum*) by at least two cubits. 3 Height of the machine apart from the gable, thirteen cubits. 4 Height of the gable from the *stratum*, sixteen cubits. 5 Little tower (*turricula*). 6 Ram (*aries*). 7 *Scorpiones* and *catapultae*

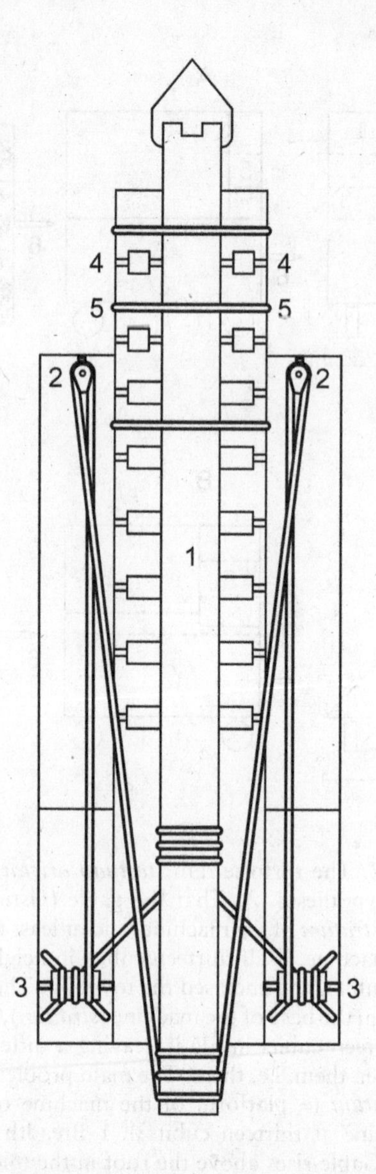

Fig. 37. Diades' drill (*terebra*) (see p. 311). 1 The beam with metal point set in the channel. 2 Pulleys. 3 Windlasses. 4 Rollers under the beam. 5 Arches covered with rawhide protecting the beam

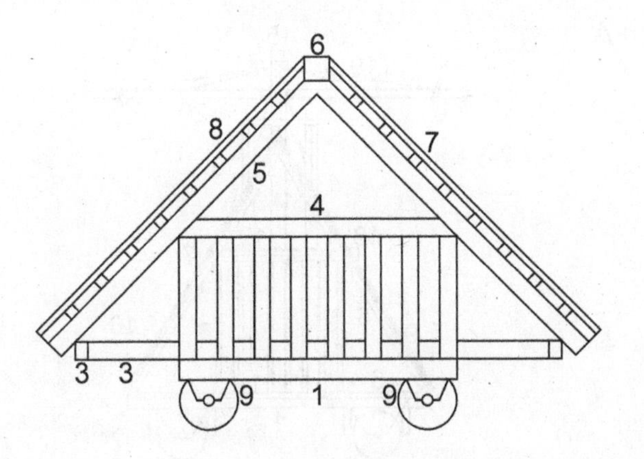

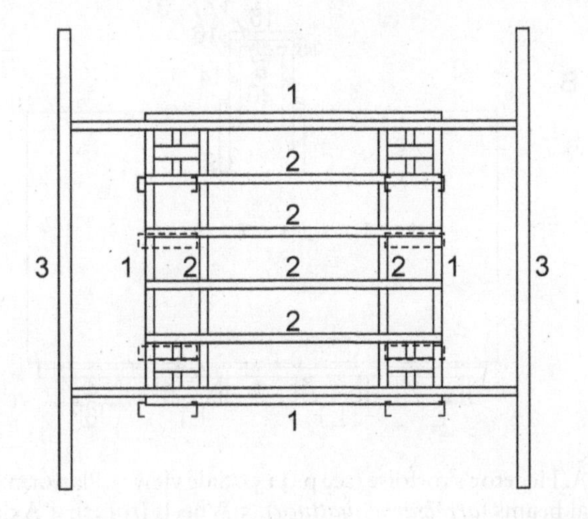

Fig. 38. The tortoise (*testudo*) (see p. 313). 1 Platform (*basis*, *eschara*). 2 Cross-beams laid at right angles to each other (*transversarii*). 3 Large beams connected at right angles to each other (*tigna*). 4. Horizontal morticed beam (*intercardinata trabs*) holding together the vertical compound posts (*postes compactiles*). 5 Rafters (*capreoli*). 6 Square ridge-piece (*quadratum tignum*). 7 Purlins (*lateraria*). 8 Wooden boards (*tabulae*). 9 Axle-block (*arbuscula*, *hamaxapous*)

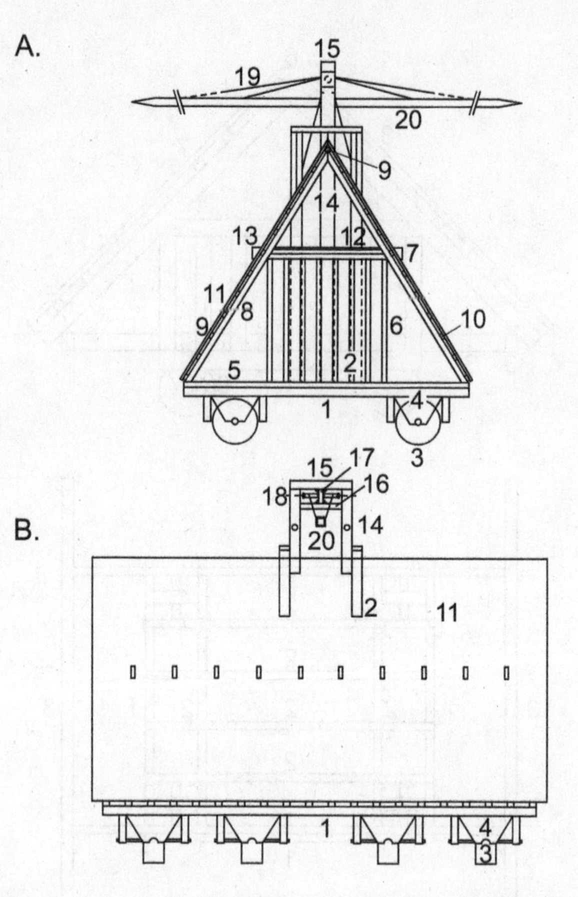

Fig. 39 A. Hegetor's tortoise (see p. 315). Side view. 1 Platform (*basis*). 2 Vertical beams (*arrectaria quattuor*). 3 Wheels (*rotae*). 4 Axle-bearings (*arbusculae, hamaxopodes*). 5 Floor (*transtrorum planities*). 6 Vertical posts (*postes*). 7 Horizontal bracing-beams (*trabes circumclusae*). 8 Rafters (*capreoli*). 9 Ridge-piece. 10 Purlins (*lateraria*). 11 Roof of wooden boards (*contabulationes circumdatae*). 12 Floor in the middle (*contabulatio*). 13 Little beams (*trabiculae*). 14 Two vertical beams (*arrectaria duo*). 15 Crowning beam with watchtower on top. 19 Ropes from which the ram was suspended. 20 Ram (*aries*).

B. Hegetor's tortoise. Front view. 16 Lower cross-beam. 17 Wooden vertical holding the axle in position (*materies inter scapos*). 18 Axle (*axiculus*)

Notes

BOOK I

1. *your far-ranging plans and decisions*: Vitruvius is addressing Augustus (born Gaius Octavius Thurinus), whose real father Gaius Octavius died in 59 BC: he was adopted by his great-uncle Julius Caesar in September 45 BC, and then preferred to be known as Gaius Julius Caesar rather than Octavian. The final composition of Vitruvius' treatise almost certainly comes after Actium (September 31 BC), with the decisive defeat of Antony and Cleopatra. Vitruvius addresses Augustus as 'Caesar' and Supreme Ruler ('imperator'): he does not use the honorific title Augustus, which Octavian received in January 27 BC after the close of the Civil Wars, but that does not mean the preface or the final compilation of the Ten Books predate that period, because the title was unusual and came into use gradually. The title 'imperator', adopted by Octavian as a *praenomen*, here does not yet mean 'emperor', and usually refers to a victorious commander of troops, but in this case covers much more than that since here and in the following paragraphs it alludes to the vast extent of Octavian's de facto dominion and civic building activities.
2. *your father ... found favour with you*: Refers to Augustus' adoptive father, C. Julius Caesar, deified in 42 BC.
3. *recommendation of your sister*: Octavia (63 BC–AD 11), Augustus' full sister. We do not know anything about the other persons mentioned here by Vitruvius, although it seems plausible that, like other contemporaries mentioned in the treatise, such as Fuficius (7.Introduction, 4) and M. Hostilius (1.4.12), they formed part of the forces of Julius Caesar.
4. *a more dependable record*: Probably meaning 'of his own work'.

5. *painted models*: Or 'painted/coloured designs'.
6. *modular systems . . . laws of geometry*: Vitruvius seems never to speak of 'symmetry' in the treatise, and the use of that word in some translations is particularly misleading. Depending on the context, I have translated 'symmetria' as 'modular system', 'modularity' or occasionally 'commensurability' in the singular, for want of a better abstract noun in English, and in the plural ('symmetriae', which Vitruvius often uses in a stock phrase with 'proportiones') as 'modular systems', since 'modularities' is surely inadmissible in English.
7. *war on the Carians*: Caria was a town in Laconia near Sparta of which the destruction in or after 479, when the Greeks decisively defeated the Persians at Plataea, is not otherwise recorded, but the story may be true. Otherwise Vitruvius may be misdating the destruction of Caria in 367/7 which took place after the city had taken sides with Thebes against Sparta at the battle of Leuctra (371).
8. *trophy of victory for future generations*: The decisive battle that ended the Persian War was Plataea (479 BC); Pausanias' father is otherwise recorded as Cleombrotus. The monument with Persian prisoners at Sparta was still extant in the second century AD when it was seen by Pausanias.
9. *architraves and their mouldings*: Vitruvius has no word for 'entablature'. For him, the most important horizontal element in the classical trilithic system is the architrave ('epistylium'). Although he has separate words for the frieze ('zophorus') and cornice ('corona'), he habitually refers instead to the architrave and its ornaments ('ornamenta'), which I usually translate as 'mouldings', meaning the frieze (if present) and cornice.
10. *Pytheos . . . Temple of Minerva at Priene so brilliantly*: Refers to the Temple of Athena Polias at Priene, funded by Alexander the Great after the battle of Granicus (334), but finished only in 154 BC. Pytheos, one of the greatest late-Ionic architects, also worked with Satyrus on the Mausoleum at Halicarnassus (see 1.1.15, 4.3.1, 7.Introduction, 2). His commentaries on these monuments do not survive: see Book IV note 6.
11. *though not incompetent in any of them*: A roll-call of celebrated intellectuals and artists. Aristarchus, the great grammarian, student of Homer and fifth head of the library at Alexandria (*c.* 216–144); Aristoxenus (born *c.* 370), a pupil of Aristotle, and the greatest musicologist of antiquity; Apelles, the most celebrated painter of antiquity, who lived at the court of Alex-

ander the Great (late fourth century BC); then come two of the most accomplished sculptors, Myron (*c.* 480–450) and Polyclitus (active 470–420), who devised a proportional scheme for the male body, the *canon*, and wrote a book on it (see 3.Introduction, 2), and Hippocrates (late fifth–early fourth century BC), the legendary codifier of medicinal theory and practice.

12. *laws of nature*: Another list of celebrated intellectuals: Aristarchus, astronomer and mathematician (*c.* 310–*c.* 230); Philolaus, a Pythagorean philosopher (*c.* 470–390) in contact with Democritus and Socrates; Archytas (*c.* 400–350), mathematician, also cited in 9.Introduction, 14 and 9.8.1 and mentioned by Horace, *Odes*, 1.28; Apollonius (*c.* 262–190), great geometrician, also mentioned by Vitruvius at 9.8.1; Eratosthenes (*c.* 284–192), celebrated polymath who worked at the library at Alexandria; Archimedes, mathematician, geometrician, astronomer and writer on statics (*c.* 285–194), mentioned by Vitruvius in 1.1.17, 7.Introduction, 14 and 9.Introduction, 9–10; Scopinas, otherwise unknown but mentioned again by Vitruvius at 9.8.1.

13. . . . *of the work*: At the end of this sentence the texts present the phrase 'cum qualitate', the sense of which is difficult to capture, and perhaps we are right to be suspicious that it may be an interpolation intended to expand on 'appropriate placement' ('apta conlocatio' or 'elegans effectus')? Is it supposed to mean 'in relation to quality', 'along with a certain quality or character', 'où apparait la qualité', all of which seem pretty meaningless?

14. *orthogonal*: I have used this word because (i) the drawing concerned is one of a flat plane or surface of a building evidently drawn to scale; if it was not orthogonal, the distinction Vitruvius makes between this type of drawing and the perspective drawing seems pointless; (ii) in any case Vitruvius tells us that such drawings are 'modice . . . picta', which surely means 'on a small scale, to scale' rather than 'correctly'; the latter seems unsatisfactory as a translation because the fact that the drawings should be made 'correctly' would have been obvious.

15. *the nature of its harmony is modular*: 'symmetros est eurythmiae qualitas'. I translate this as 'modular' rather than 'symmetrical', because (i) Vitruvius explains what he means immediately afterwards by listing the vertical and longitudinal dimensions of a man's body, not those relating to his breadth; but particularly because (ii) elsewhere Vitruvius never talks of the symmetry of buildings, only of their modularity, so the simile he is establishing here between men and buildings would break down if he now

refers to the symmetry, rather than the modularity, of human bodies.

16. *... [or from an embater]*: Meaning and reading are uncertain and much discussed. The context suggests that if after 'triglypho' there was indeed another word, it should have been a noun for another part of a building which was to be used as a module, rather than a word meaning 'module'. It is difficult not to suspect that the word *embater* was inserted here as a gloss by someone who had read Vitruvius 4.3.3, where he specifically says that 'embates' means module. See Glossary and Index of Technical Terms, s.v. module.

17. *order*: Vitruvius uses 'genus' ('category', 'style', 'type' when used in a general sense) in reference to the orders, but I can see no gain in not using the specific English word 'order' to translate it in the relevant contexts.

18. *pulvinate capitals*: That is, cushion-shaped, or Ionic capitals.

19. *system*: Here Vitruvius uses the word 'ordo': but in this context we should probably translate it with something less specific than the word 'order', for which, in any case, Vitruvius uses the word 'genus' many times; there seem to be only two other occasions in the treatise when the word 'ordo' could possibly be translated as a synonym for 'order': 1.7.2 and 4.8.7.

20. *heads of families*: Referring to ordinary households.

21. *Altinum, Ravenna, Aquileia*: Much of the coast along the north Adriatic was characterized by marshy land. Altinum was a city near Venice on the mouth of the river Silis; Aquileia was founded in 181 BC; and Augustus excavated a port at Ravenna, conquered by the Romans in 234 BC, as Vitruvius probably knew.

22. *Salpia ... was built in an area of this type*: The reference to the Pomptine marshes, south of Rome, was topical when Vitruvius was writing, since Julius Caesar intended to drain them and Augustus dug a canal there. The date of the foundation of Salpia is entirely uncertain (suggestions vary between the second century BC and after 89 BC). Marcus Hostilius is otherwise unknown. Another version of the foundation myth of the city says that Diomedes rather than Marcus Hostilius cut the canal.

23. *the walls corresponding ... not fixed down with nails*: If the walkways through the interiors of the towers really were as wide as the towers themselves (i.e. they crossed their diameters), as Vitruvius says, then the towers attached to the walls were presumably, in fact, semicircular.

24. *circular or polygonal*: Vitruvius says that towers are round or

polygonal; (i) he could be using 'or' ('aut') merely to indicate a synonym, in which case 'round' is the meaning, since polygons are round; or (ii) if 'or' is exclusive and not intended to signal an exact synonym, we can translate 'round' as circular. But see the previous note.

25. *antidotes*: That is, 'medicinae contrariae', medicines that produce an effect that is the opposite of that produced by the illness.

26. *Andronicus of Cirrhos . . . a demonstration of this fact*: Andronicus built his celebrated Tower of the Winds at Athens, which functioned as a wind-rose, water clock and sundial and still survives, at the end of the second century BC. It owes its survival to the fact that it was converted into a chapel in the Byzantine period. Each side is decorated with the figure of a wind in relief: Triton, the marine divinity, half-man, half-fish, was often regarded as the god of winds.

27. *Eratosthenes of Cyrene*: Eratosthenes (*c.* 284–194) was evidently the first to calculate the earth's circumference and the tilt of the earth's axis: given the primitive means at his disposal, his accuracy was astonishing; he calculated the equatorial circumference at 39,690 kilometres as against the real dimensions of 40,009 kilometres.

28. *two diagrams . . . at the end of this book*: Vitruvius' original illustrations are all lost, but he indicates that he placed them either at the end of books (3.3.13, 3.4.5, 3.5.8, 5.5.6, 8.5.3) or at the bottom of pages (5.4.1, 9.Introduction, 5).

29. *layouts*: *Ordines*; see note 19 above.

BOOK II

1. *Dinocrates . . . arrived at such a level of fame*: A celebrated story, repeated by a number of other ancient sources and with a long afterlife in the Renaissance. Dinocrates evidently joined Alexander's army in Thrace on the way to conquer Persia in 334 BC; he worked on the last stages of the construction of the Temple of Artemis at Ephesos and the laying-out of Alexandria in 332–331.

2. *roofs . . . made of earth worked with straw*: Vitruvius probably saw Marseilles when Julius Caesar besieged the city in 49 BC.

3. *an exemplar . . . roofed with mud*: 'The ancient exemplar on the hill of the Areopagus', or 'the ancient exemplar of the Areopagus', referring to the main buildings of the Areopagus in Athens.

4. *hut of Romulus . . . practices in ancient times*: The hut, which Vitruvius probably saw for himself, even if he includes information taken from Varro, stood on the site of the present apse of S. Maria in Ara Coeli on the Capitol.
5. *about the scope of the art*: Difficult not to wonder whether this is not a resumptive, interpolated marginal heading, since it repeats almost exactly what precedes it.
6. *For this book does not pronounce . . . as they were*: The contrast Vitruvius tries to establish here seems very opaque; he tells us that he will not discuss the origins of architecture in Book II (but surely does), but rather the origins of the construction of buildings (which of course he discusses at great length).
7. *separate entities . . . retain their integrity for all time*: Another list of celebrated intellectuals, including Thales (*c.* 620–546) for whom the main source is Aristotle; Heraclitus (*c.* 535–475), Epicurus (341–270) and Democritus (460/57–*c.* 370), based on information which Vitruvius presumably derived from Posidonius, Varro, Lucretius and others.
8. *decision of a magistrate*: Utica was near ancient Carthage in Tunisia. Perhaps Vitruvius visited it when Julius Caesar arrived there in 46 BC.
9. *courses of full-size bricks . . . on the other*: Vitruvius has in mind walls comprising courses with full bricks down one side of a given course mortared to half-bricks down the other side of the same course; then, in the next course, above or below, the full bricks and half-bricks reverse their positions.
10. *Further Spain . . . Pitane*: Further Spain (*Hispania ulterior*) included roughly Baetica and the Guadalquivir valley, as well as Lusitania (modern Portugal): perhaps Vitruvius' information here derives from Varro. Pitane was a coastal city of Aeolis, in Turkey (which Vitruvius calls Asia).
11. *assailed by them*: Reading 'tanguntur', not 'ducuntur'.
12. *quarry-stones*: That is, rough, smaller stones, as opposed to squared slabs.
13. *because of its strength*: The meaning remains obscure.
14. *Signian sand*: Used to make the mortar for impermeable revetments and pavements (see 7.6.14).
15. *ligneous*: The soil or earth to which Vitruvius refers is mysterious; he uses the word 'materia', his preferred word for wood, as against 'legnum'. Whatever the type of soil or earth he is referring to we can hardly call it 'woody', which makes no sense in the context, or 'rocky', which is too hard a substance (cf. Budé,

vol. 2, p. 98 n. 1). To maintain the idea that it is in some way woody or like wood or related to wood, I translate it as 'ligneous'.

16. *these quarries*: Vitruvius talks about the quarries as though they were types of stone in the next few paragraphs.
17. *the City*: Vitruvius often refers to Rome simply as 'the City'.
18. *it comprises . . . in all parts of the wall*: A curious way of expressing the fact that perfectly aligned *opus reticulatum* has no overlapping horizontal or vertical joints, so allowing long diagonal cracks to form. Cf. Book IV, note 23, and Book X, note 9.
19. *masonry*: Probably meaning masonry with mortar, i.e. concrete.
20. *they bind the other stones . . . vertical joints*: What Vitruvius means by saying that they bound these internal stones with alternating vertical joints (therefore, presumably, with random horizontal joints) is not easy to understand.
21. *Another type of wall . . . stability*: The passage on *emplecton* apparently refers to walls with stone faces and a rubble or concrete in-fill, but the passage about Greek walls that immediately follows in this paragraph does not seem to describe this kind of wall, but completely solid walls without infill constructed of alternate long and short blocks inserted through the whole thickness of the wall, punctuated at intervals with *diatonoi*, or blocks as long as the whole thickness of the wall. Pliny (*Natural History*, 36.172) produces a garbled version of this passage, in which he reports that 'when the space between the walls is packed with rubble, they call it "diatonicon" ', but Vitruvius specifically says that 'they do not put infill in the middle' of this kind of wall, and this does not mean that they left the middle empty, since he also says that from one face to another the thickness was uniform: see Glossary and Index of Technical Terms, s.v. facing; stone.
22. *aedileship of Varro and Murena*: The source for Vitruvius' allusion to this episode from 68 BC is probably Varro. Sparta was in economic crisis and a number of works seem to have been taken from the city at that point, including Myron's *Discobolus* from nearby Amyclae.
23. *since he ruled over the whole of Caria*: Mausolus of Halicarnassus, satrap of Caria from 377 to 353, made Halicarnassus his capital instead of Milasa; he rebuilt the town and his palace (from 375) and started his celebrated Mausoleum *c.* 353 with Pytheos as the architect. He was succeeded by his wife-and-sister Artemisia II (died 351), who almost completed the tomb and fended off an attack from the Rhodians, mentioned only by Vitruvius (2.8.14–15).

24. *accomplished master*: In view of what follows immediately, Budé, vol. 2, p. 29, is surely right to delete 'Leocharis' after 'accomplished master'.
25. *erotic fixations*: Not venereal disease; the point of the story is that the water caused the barbarians to become friendly with the Greeks, not their lovers.
26. *partitions of great practical use*: Reading 'disparationes', which provide for the partitioning of the floors; 'despectationes' or 'views-from-the-top' is surely too small a detail to be included in this list about the advantages of large-scale structures.
27. *city-walls*: Elsewhere in Vitruvius, the word *moenia* refers only to city-walls, so the walls here are not those of the buildings themselves.
28. *well-defended fort in the area, called Larignum*: Vitruvius was presumably an eyewitness to the siege of Larignum, which probably took place 59/8 BC when Caesar was in Cisalpine Gaul: it is not mentioned by any other source, not even by Julius Caesar.
29. *their fibres are long*: Following the interpretation given in Budé, vol. 2, p. 44.

BOOK III

1. *Socrates was the wisest of all men*: Chaerephon, who asked Apollo who was the wisest man of all, was a close friend of Socrates, was mentioned in Aristophanes' *Clouds*, *Wasps* and *Birds*, and Xenophon's *Memorabilia*, and appears in Plato's *Charmides*, *Gorgias* and particularly the *Apology*, the ultimate origin of the story here.
2. *artists may recommend their own skills . . . they lay claim to*: a complicated passage, much disputed and emended. By contrast, some want it to say that poor artists, even if they have a famous studio behind them and great eloquence, can never rise to the top – a sense which requires a drastic emendation of Fra Giocondo, 1511; but the emendation certainly has some weight, because Vitruvius gives the impression in the next paragraph that it is *only* with the help of great cities, kings and powerful patrons that an artist can succeed, and not otherwise.
3. *misfortune . . . prevented them from becoming famous*: In the first part of this paragraph Vitruvius presents a standard list of celebrity artists (Myron, Polyclitus, Phidias, Lysippus); in the second half another list, but this time of artists who did not get

as far as they should have: of these, Chion, Myager, Boedas (son and pupil of Lysippus), Aristomenes and Polycles are attested elsewhere.

4. *cubit*: The measurement from the elbow to the end of the middle finger.
5. *since they derived it from the number of fingers on the hand*: here I follow Budé, vol. 3, p. 8: 'numero [ab palmo pes] est inventus'.
6. *the Greeks called* μονάδες: the reference is to Plato, *Timaeus*, 35b–36a.
7. *for the [first four] single units are the components of that number*: 1 + 2 + 3 + 4 = 10.
8. *Further . . . of his height*: Reading 'Non minus etiam quod pes hominis altitudinis sextam habet partem [ita etiam ex eo quod perficitur pedum numero corporis sexis altitudinis terminatio] eum perfectum constituerunt.' I have no trust in the text here which seems, in its present form, to include a tortuous, tautological repetition, which I omit.
9. *above the architraves*: Or 'above the corner columns, architraves like those of the temple *in antis*'.
10. *opisthodomus*: The specific meaning of *posticum* in this context.
11. *Temple of the Deified Julius . . . and any others designed like them*: The Temple of the Deified Julius in the Roman Forum was voted for in 42 BC, and built between 36 and 29, when it was dedicated. The Temple of Venus Genetrix stood in the Forum of Julius Caesar, dedicated in 46: it was in 48, before the battle of Pharsalus, that Caesar voted to build a temple to Venus, his mythical ancestor, in the Forum.
12. *the Stone Theatre*: The debate about whether this is the Theatre of Pompey or that of Marcellus is still open.
13. *mothers . . . must go single file*: Cf. the famous complaint about Lord Burlington's York Assembly Room, that ladies with bustles could not pass between the columns (R. Tavernor, *Palladio and Palladianism*, London, 1991, p. 169).
14. *octastyle temple*: Vitruvius is evidently making a simple and perhaps surprising mistake when he says that Hermogenes invented the octastyle, pseudodipteral temple, since he mentions earlier octastyle temples not invented by Hermogenes and he does not think pseudodipteral and octastyle are synonyms. The reading 'exostylon', which should, or could, mean 'with an external colonnade, a colonnade on the outside', has been suggested to avoid these difficulties, but is not found elsewhere in Latin.

But clearly Vitruvius wants to make an architectural hero of Hermogenes, who codified the rules for eustyle and pseudodipteral Ionic temples and built, between 230 and 180 BC, those of Dionysus on Teos and of Zeus Sosipolis and Artemis Leucophryene at Magnesia ad Maeandrum and constituted, directly or indirectly, the major source for Vitruvius' discussion of those categories of Ionic temples and certain faults in Doric design (3.2.6, 3.3.8, 3.3.9, 7.Introduction, 12).

15. *those who have simply doubled . . . by one intercolumniation*: When the number of columns down the sides is double the number of columns on the façade (e.g. 6 × 12), then there are too many intercolumniations (5 × 11) in the sense that it makes the proportions of the temple greater than the desirable 2 : 1. But Vitruvius says that counting intercolumniations instead produces the correct proportions: e.g. a temple with 6 × 11 columns, such as Hermogenes' Temple of Dionysus at Teos, produces 5 × 10 intercolumniations, and one with 8 × 15 columns, such as the same architect's Temple of Artemis Leucophryene at Magnesia, produces 7 × 14 intercolumniations (ignoring the greater spacing in the centres of the short façades). However, while counting intercolumniations can be made to give a ratio of 2 : 1, it does not reflect the real dimensions of a given temple because it ignores the thicknesses of the columns; better still is to count the interaxial distances (for which Vitruvius does not have a word and does not talk about), which include both intercolumniations and columns, leaving only the half columns at the ends of each row not counted.

16. *varying height*: The meaning of 'scamilli impares' remains one of the most famous problems in Vitruvian studies from the Renaissance onwards. Perhaps the answer is something like this: the little projections may have been small blocks of stone left on the upper surface of the stylobate or small wooden blocks, of regularly varying height, used to establish the curvature of the stylobate. Perhaps a tight, horizontal cord was taken along the length of the stylobate; the piece of string would remain tangent with the centre of the stylobate; the little blocks on the stylobate (either small projecting bosses in stone, or small wooden blocks) were lined up with the string: the more the curvature of the stylobate required towards the corners of each side, the higher the *scamilli* had to be which were placed under the piece of string at regular intervals, one for each slab of the stylobate, as one

moved towards the corners. The surfaces of the slabs of the stylobate would then be cut or bedded down progressively more towards the corners depending on the depth below the piece of string indicated by the *scamilli* for each one.

17. *the other three divided equally in two units*: That is, $3/2 = 1.5$.
18. *so that the breadth of the base*: Text reads: 'uti latitudo spirae *quoque versus* sit columnae crassitudinis adiecta crassitudine quarta et octava'; since the base is circular it is difficult catch the sense of *quoque versus*: along all axes? The base should be each way equal in breadth (Morgan, 1960)? One suspects that at some time a reader who believed that the base being described was *square*, or that at this point Vitruvius was in fact talking about the plinth, inserted a gloss.
19. *pulvinate*: Cushion-shaped (*pulvinatus*), bulgy, referring to Ionic.
20. *the heights of the columns*: Probably meaning 'the heights of still taller columns'.
21. μετοχή: Reading *metoche*, instead of μετοπή: 'intersectio' seems to refer to the run of dentilation as a whole, evidently considered by Vitruvius as the element connecting the frieze to the cornice, and it is only with some difficulty that 'intersectio' can be taken as a synonym for μετοπή.

BOOK IV

1. *Dorus . . . ruled Achaea*: Some mythographers put the date of Dorus' domination of the Peloponnese near the end of the sixteenth century BC: at this point Vitruvius was presumably using now lost antiquarian histories of Argos, which are known to have been popular in the fourth and third centuries BC.
2. *apothesis*: The outward curve at the bottom or the top of the column shaft. The impression is that *apothesis* is identical in meaning to *apophysis*, which is used twice elsewhere by Vitruvius in 4.7.3.
3. *the central curve of the abacus*: 'Intra sinum medium', not 'intra suum medium'.
4. *entablatures*: 'ornamenta', usually used by Vitruvius to indicate the frieze and cornice above an architrave; but here perhaps the translation 'entablature' is permissible in view of the discussion that follows.
5. *mutules or dentils*: The addition of the word 'mutules' here is

required by the subsequent mention of principal and common rafters ('cantherii . . . asseres') to which the mutules and dentils correspond.

6. *Hermogenes . . . built an Ionic temple to Father Liber instead*: Tarchesius (spelling disputed) is known only from Vitruvius (cf. 7. Introduction, 12). Pytheos was a celebrated architect, writer and sculptor, active *c.* 360–330 in Ionia where he worked on the Mausoleum at Halicarnassus, on which he wrote a treatise, and built the Temple of Athena Polias at Priene (see also 1.1.12, 1.1.15, 7. Introduction, 12). Hermogenes: see Book III, note 14.
7. *the distribution of triglyphs and coffers*: Here, and at 4.3.5, 'lacunaria' evidently refers to the mutules in the soffits of cornices.
8. *where the columns are placed*: In effect, then, the length of the stylobate across the façade.
9. *module*: See Book I, note 16.
10. *The triglyphs . . . back façade*: Vitruvius counts one triglyph over the intercolumniation as two, and two as three, because he includes the half-triglyphs over the columns to left and right of the intercolumniation.
11. *rivers*: Reading 'flumina' not 'fulmina'.
12. *the structure is to be built as systyle and monotriglyphic*: That is, with one triglyph and therefore two metopes over the intercolumniation, or, as Vitruvius defines it at the beginning of the next chapter, one whole triglyph plus the two half-triglyphs at right and left over the nearest columns, with two metopes in between: see note 10 above.
13. *three triglyphs and three metopes*: Two whole triglyphs plus two half-triglyphs over the columns, making three, with three metopes between them.
14. *flat*: That is, as facets of the column, making it in effect polygonal.
15. *the diameters of the inner columns . . . a different method*: The diameters of the inner columns will be made to look equal in thickness to those of the outer columns, but using a different method.
16. *reliefs*: Meaning architectural forms, such as triglyphs, mutules, *taeniae*, *guttae*, etc.
17. *the rusticated panels . . . more picturesque*: A curiously backhanded way of describing the panels of rustication on blocks of stone separated by vertical and horizontal joints. Cf. Book II, note 18.
18. *corridors*: Accepting the noun 'alae' (wings, side corridor), a

famous emendation of Fra Giocondo (1511), not the adjective 'aliae' ('other'): grammatically 'aliae' would have to refer to units ('partes'), not to the *cellae*; in which case Vitruvius would be saying that 'the width should be divided into ten units, of which three on the right and three on the left (of the central *cella*) should be given to the smaller *cellae*, or to other units (of the building) if there are to be any', which seems incomprehensible, since the breadth of the building has already been decisively established as ten units by Vitruvius. The term 'wing' or 'corridor' refers to temples with central *cellae* with long rectangular spaces at the sides enclosed by the back wall and side-walls on the outside running forwards, sometimes as far as the front columns (e.g. temples at Fiesole and at Tarquinia, the Temples of Fortuna and Mater matuta in the S. Omobono complex in Rome).

19. *so that angle-columns ... exterior walls*: Referring to the two angle-columns at the front of the façade.
20. *the middle two columns ... in the centre*: Referring to the two columns of the façade between the corner columns.
21. *the other columns ... along the same axes*: Referring to the row of four columns behind, and on axis with those of the façade.
22. *... on the Acropolis in Athens*: The Erechtheum.
23. *Other architects ... exterior colonnade*: A curiously inverted way of expressing the fact that architects discarded the exterior colonnades, leaving temples as broad as the *cella* alone, to which they applied half-columns (cf. Book II, note 18).
24. *layouts*: 'Ordines': see Book I, note 19.

BOOK V

1. *the Julian colony at Fano*: Fano, or the Colonia Julia Fanestris, was an Augustan colony founded in 27 BC or a little later. No remains of Vitruvius' celebrated Basilica, which included a temple of Augustus, have yet been decisively identified: it was destroyed by the Goths in AD 635.
2. *on contact*: Reading 'tactu'; the reading 'actu', becoming activated, active, as opposed to staying merely potential, seems unconvincing because the waves of sound exist whether they are heard by anybody or not.
3. *writings of Aristoxenus*: Aristoxenus (born *c.* 370 BC), the peripatetic philosopher and writer on music and rhythm, learnt musical theory from the Pythagorean thinkers Lamprus of Erythrae

and Xenophilus and later studied under Aristotle. His *Elementa harmonica* has survived in incomplete form and a fragment of a treatise on metre may be his.

4. *Lucius Mummius . . . dedicated them . . . the Temple of the Moon*: Mummius, the destroyer of Corinth in 146 BC, also brought back to Rome a large number of valuable non-religious metal artefacts.
5. *the lowest part*: Referring to the ground level of the orchestra.
6. *The apexes . . . of the theatre*: Referring to the points of the triangles at ground level in the upper half of the orchestra.
7. *the other five . . . the stage-set*: Referring to the points of the triangles, still notionally at the level of the orchestra, in the lower half of the circle where the stage is.
8. *arranged*: Or 'planned'.
9. *the corners*: Meaning the corners of the seating of the auditorium to left and right of the diameter of the orchestra, which are about to be cut back.
10. *the innermost seats*: That is, the lowest seats, nearest the orchestra.
11. *equalling that sixth*: That is, to a height equalling a sixth of the diameter of the orchestra.
12. *the podium*: Of the *scaenae frons*.
13. *points should be marked . . . to right and left*: The points at either end of a diameter drawn horizontally across the orchestra.
14. *a radius based on that distance*: Between the points at left and right, or the diameter of the orchestra.
15. *with a radius based on that distance*: See previous note.
16. *aligned with the corners*: Or better, along the diagonals.
17. *the number of intermediate flights . . . cross-aisles*: That is, after each *praecinctio*, or cross-aisle, equidistant stairways should be inserted between the continuous stairways dividing the blocks of seats.
18. *restored by King Ariobarzanes*: Only Vitruvius says that Themistocles used the masts of Persian vessels captured at Salamis in 480 for the roof of the *odeum*, and another source adds that the building was shaped like the tent Xerxes used when he was in Attica in that year. It was burnt down by the Athenians themselves in 86 BC to prevent Sulla, who was besieging the city, from getting his hands on the timber it contained. Ariobarzanes I, King of Cappadocia (65–52 BC), restored it using three Roman architects, Gaius and Marcus Stallius and Melanippus: presumably Vitruvius saw the building just after its restoration.

19. *the diagram*: 'Descriptio' here must mean a drawing or diagram, not a description, because Vitruvius, famously, can hardly be said to have described 'scamilli impares' at 3.4.5, but did mention a diagram and explanation (*forma, demonstratio*).
20. *half-cylinders of the baths*: Semicircular metal containers for heating water, open at one end and closed at the other.
21. *exedrae*: 'Exhedrae'; there is no English word that corresponds to this space; 'alcove' or 'recess' imply spaces that are too small.
22. *anointed athletes doing their exercises*: To say that the *dressed* spectators are not interrupted, bothered by the anointed ('uncti') athletes, surely makes better sense than saying that the spectators are not interrupted by 'all' (*cuncti*) the athletes, since it gives a contrast between the clothed bystanders and the naked or semi-naked athletes; and why 'all' the athletes would interfere with the spectators is anything but clear.
23. *cleared out by men working on small cross-beams*: Or 'cleared out with small cross-beams'? The meaning remains obscure.
24. *immensely strong platform . . . constructed of masonry*: The verb used by Vitruvius here ('struere') probably means 'constructed with masonry' rather than just 'constructed', in view of the description of the platform as 'immensely strong': a similar problem with the interpretation of the verb occurs at 4.2.2, where 'inter tigna struxerunt' is translated as '[they] built masonry between the joists'.
25. *. . . with a horizontal surface*: Apparently meaning, to judge from what follows immediately afterwards, 'for less than half its length' rather than 'for less than half its breadth'.
26. *slope forward*: Vitruvius evidently envisages an artificial promontory designed to close the harbour, hence the reference to the part that faces the shore, meaning the shore opposite it across the expanse of water about to be enclosed.
27. *the centres of the blocks . . . [of the row above]*: Ineptly expressed; one would expect Vitruvius to say instead that the middles of the blocks above hold down the vertical joints below. Cf. Book II, note 18, and Book IV, notes 17, 23.

BOOK VI

1. *Aristippus . . . signs of men's presence*: Either Aristippus of Cyrene (end fifth century BC–first half fourth century BC), a pupil of Socrates, or his grandson; the story is told with variations by

ancient sources, and some have him arriving at Syracuse, not Rhodes, and his companions wishing to return to Cyrene.

2. *Theophrastus . . . trust in money*: Probably Theophrastus (*c*. 370–287), the peripatetic philosopher who succeeded Aristotle at the Lyceum (see also Vitruvius 8.3.27).

3. *foils the plots of southerners by force of arms*: 'forti manu [consiliis] meridianorum cogitationes' (Rose, 1867). I think it probably right to omit 'consiliis' because it obscures the clarity of the strong contrast Vitruvius is making about the Romans' virtues: they repel the strong but stupid northerners with their brains ('consiliis'), but foil the weak but wily southerners with their brawn ('forti manu').

4. *by means of additions or subtractions*: The phrase ('detractiones et adiectiones') is clearly a favourite of Vitruvius': but one wonders whether here it may have been added as a gloss by someone who had read forward to 6.2.4, 6.3.11 or 10.10.6, where it recurs, but not pleonastically as here where it uselessly anticipates and replicates the idea of adjusting the modular system by making additions or subtractions specifically described in the next clause.

5. *displuviate*: The roofs of displuviate courtyards sloped upwards and inwards from the walls of the courtyard, not downwards as in the cases of the Tuscan, Corinthian and tetrastyle courtyards.

6. *Tuscan courtyards . . . the aperture of the compluvium*: These were constructed using two large parallel beams built across the courtyard; two other lighter, smaller beams or scantlings (*interpensiva*) rested on them at right angles and formed the square or rectangular aperture in the middle (*compluvium*) through which rain fell. The roofs sloped inwards and downwards towards the *compluvium* from the walls of the courtyard; the *colliciae* were the diagonal water-channels running down from the angles of the walls of the courtyard to the *compluvium*.

7. *In Corinthian courtyards . . . cross-beams*: Corinthian and tetrastyle courtyards were the same in principle as the Tuscan: the roofs still slope down and inwards from the walls of the courtyard, but here there was no need for the large joists across the whole courtyard. In both cases columns took the weight of the beams forming the aperture of the *compluvium*, four in the case of the tetrastyle courtyard, no fixed number in the case of the Corinthian. In principle, making columns carry the weight of the roofs in Corinthian and tetrastyle courtyards rather than suspended beams meant much larger courtyards could be built.

8. *the first category . . . to the breadth*: Evidently referring to the Tuscan *atrium*.
9. *in the second . . . to the width*: Evidently referring to the Corinthian *atrium*.
10. *in the third . . . to the atrium*: Evidently referring to the tetrastyle *atrium*.
11. *side-rooms*: *Alae*; lateral *exedrae*: rectangular rooms with only three walls at the sides of *atria*.
12. *The intercolumniations . . . the diameter of the columns*: That is, no less than diastyle (three diameters), no more than four diameters (one of the options with araeostyle).
13. *single rows of columns*: That is, one storey of columns, not two.
14. *not ignorant*: Since, surely, no farmers were ignorant ('imperiti') of the seasons, it is essential to insert *non*, or perhaps read *periti*.
15. *culleus*: 525 litres (approx. 120 gallons).
16. *raised*: Probably 'sublimata', instead of 'sublinata' (coated), which explains better why these granaries are cooled by the air, as Vitruvius says immediately afterwards.
17. *their coats become rough*: Vitruvius tells us at 6.6.1 that cattle do not become shaggy when exposed to fire and light, but here, that draught-animals ('iumenta') do.
18. *lintels*: Reading 'limina' or lintels: others change the reading to mean some kind of roof-light or dormer-window (*lumina* = *lucernaria*). Vitruvius seems to be saying that if, on a given floor, there are obstructions to making windows, then they should be inserted higher up in the house, in which case the mention of roof-lights, alongside beams and wooden floors in the storey in question, would not seem to make much sense.
19. *in the manner of piers and antae*: It is difficult to catch the sense exactly. If the translation of the word 'secundum' as 'in the manner of, like' is correct, presumably the vertical supports would be like *antae* if they were at the sides of the windows, and like pilasters if they were in the space in the middle; but other translations such as 'in parallel with' or 'in line with piers and pilasters' cannot be ruled out.
20. *pavements and the revetments of walls*: The word 'expolitiones' refers to the important surfaces that must be applied once a structure has been finished, i.e. in Latin it covers the making of pavements, the revetments of walls and vaults, and cornices; the periphrasis is necessary because in English, as far as I am aware, the word 'revetments' cannot be applied to floors, only to vertical surfaces and soffits.

BOOK VII

1. *library at Alexandria . . . with just as much effort*: The library at Alexandria was founded under Ptolemy I Soter (321–283) and Ptolemy II Philadelphus (283–247), that at Pergamum by Eumenes II (197–159). Vitruvius inverts the chronology here probably because he has followed a Pergamonese source motivated by the famous rivalry between the two libraries.
2. *Aristophanes . . . commitment and assiduity*: Aristophanes of Byzantium (*c.* 257–180) succeeded Eratosthenes (Vitruvius 1.1.16) as librarian at Alexandria; a great textual scholar, he produced editions of Homer, as well as of Hesiod, Alcaeus, Alcman and others.
3. *attacking the Iliad and the Odyssey to the king*: Zoilus of Amphipolis (late fourth century BC), cynic philosopher, who famously attacked Isocrates, Plato and, in nine books, Homer.
4. *Agatharcus . . . left a commentary on it*: Agatharcus of Samos (*c.* 490–*c.* 415) worked with the great playwright Aeschylus (others say Sophocles), and, although the exact nature of his achievements is unclear, he clearly made great strides in the invention of something like perspectival scenery. It was apparently the same person who was kidnapped by Alcibiades in around 430 to paint his house.
5. *once a particular spot has been fixed . . . and others to project*: The passage, much discussed, seems to mean that once a particular spot on the stage-set has been chosen in relation to the spectator in his seat, then the images will be painted on the scenery, but distorted above and below that line to counteract the increasing distance from that optimum viewpoint.
6. *appraisal*: 'Probandum' surely refers to scrutiny or appraisal rather than refinement or perfection (the usual translation), as is indicated by the fact that Vitruvius says that the artists worked in competition with each other ('certatim'). This was presumably seen by Vitruvius, or his source, as a competition in which the artists appraised each other's works at the end, a little like that reported by Pliny, *National History*, 34.53, involving Polyclitus, Pheidias, Cresilas, Cydon and Phradmon and the Ephesian *Amazons*.
7. *Fuficius . . . Publius Septimius has two*: Fuficius may have been C. Fuficius Fangus, a senator, recorded as a veteran of Caesar's

campaigns; M. Terentius Varro was the great polymath (116–27), author of *De lingua latina*, *Rerum rusticarum libri III*, the *Disciplinarum libri IX* to which Vitruvius refers here, written before 27 BC, and many others. Publius Septimius is perhaps the person to whom Varro dedicated Books 2–4 of the *De lingua latina*.

8. *but also by experts*: Or perhaps 'but is regarded as one of the small number of masterpieces'.
9. *increasing the size of the vestibule*: Vitruvius says the vestibule was increased in size, not that it was added from scratch; and it seems unlikely that Ictinos had left no provision at all for a vestibule, even though the temple lacked a peristyle.
10. *walls that do not reach up to the top*: 'qui non exeat ad summum', to the top of the house presumably.
11. *another thin wall should be built*: The fact that the wall is thin may mean that it was not made of masonry, but the verb Vitruvius uses ('struere') seems to imply that it is.
12. *the pillars*: Or 'the tiles'.
13. *sextarii*: 2.2. litres (about 4 pints).
14. *in iron mortars*: Or 'with iron pestles'.
15. *Faberius the Secretary*: Secretary to Julius Caesar, also mentioned by Cicero and Appian.
16. *immediately*: Or 'he was therefore the first to make a contract'.
17. *just as marble statues of nudes are treated*: Refers either to statues of nudes and the treatment of the skin colour required for them, or to statues devoid of any decoration, unpainted, hence 'nuda', which could therefore include statues with clothing as well as bare arms, legs and head.
18. *they are broken up all round the edges . . . like tears*: The text here is difficult. I have made it say that the shells are cut around, then worked on in the mortar, thus producing the liquid, rather than that the liquid emerges from the shells which have been cut and then it, the liquid, is ground up in mortars.

BOOK VIII

1. *Philosopher of the Stage, chose air and earth*: it is not clear whether the title refers to the great playwright Euripides (*c.* 480–406), a pupil of Protagoras, Prodicus and Anaxagoras according to some ancient sources, or in fact to Anaxagoras

himself (*c.* 500–428), who wrote a treatise on theatrical scenery, mentioned by Vitruvius (7. Introduction, 11).

2. *creation*: 'Inventio'; either its creation by the Divine Beneficence or its discovery by men.
3. *it*: Probably the *caelum*, the upper zone of the hot bath.
4. *hot water has no distinctive quality*: Evidently the text as it stands has to be translated like this, but the meaning must be: 'It is not a natural characteristic of water that it is hot' (so Budé, vol. 8, p. 1; Morgan, 1960, p. 232), as is indicated by what Vitruvius says immediately afterwards in the same paragraph: 'if it were naturally hot, its heat would not diminish': perhaps emend to 'neque enim *calida* aquae est qualitas' or something like it.
5. *the gushing Aqua Marcia*: The aqueduct the Aqua Marcia was begun in 144 BC by the praetor Marcius Rex and completed in 140 BC, then restored by Agrippa *c.* 33 BC. The water was famous for its clarity and pleasant taste according to Pliny, Frontinus, Propertius and others.
6. *picnics*: Derives from a plausible emendation by Fra Giocondo, 1511 ('pransitare' for the manuscript's 'transitare'), which Rowland captures perfectly with 'picnics' (Rowland, 1999, p. 101).
7. *the king was murdered by Antipater using this water*: Many sources rehearse this famous story, though some attribute the plot to kill Alexander to Aristotle: Alexander in fact died of malaria, and the story of Antipater's plot may be due to the fact that at that point he was in disgrace. Iollas was one of Alexander's cup-bearers.
8. *conversation inevitably turned to erudite matters*: Vitruvius says that Gaius Julius, son of a certain Masinissa, served with Augustus' 'father', meaning Julius Caesar; the name, Gaius Julius, suggests that the person concerned gained Roman citizenship from Caesar. Most unfortunately Vitruvius does not say where he put Gaius up.
9. *bought handsome men and nubile girls from abroad*: So the text; even against the objections of Romano, in Gros, 1997, vol. 2, p. 1134 and n. 196, that we should understand that the girls were local and hence have good voices, and that the young foreign men supply the good looks. The text hardly supports that view and Vitruvius specifically says the resulting children will have both good voices and good looks from such marriages; so the good looks of the children were supplied by both foreign-born parents and their good voices by the local water.

10. *plumb-lines hang vertically from the wooden beam, one on each side*: Presumably meaning that four plumb-bobs are let down from the beam, two at each end of it, lining up with each of the four cross-pieces or struts.
11. *sicilicus*: 0.6 cm. (quarter of an inch).
12. *three equally spaced pipes*: The meaning remains unclear.
13. *valves*: The meaning required seems moderately clear but the word in Latin is not.
14. *two hundred actus*: 6 km. (20,000 feet).

BOOK IX

1. *Milo of Croton*: Proverbial strong-man, allegedly won six victories at the Olympic games, the first in 540 BC, six at the Pythian games, ten at the Isthmian and nine at the Nemean. He was able to carry a heifer, kill it with one blow, then eat it; most sadly when trying to rip a tree apart he was caught in the trunk then eaten alive by wolves.
2. *as he formulated it*: The story derives from Plato, *Meno*, 82b–85b.
3. *Pythagoras ... without the intervention of workmen*: This discovery was not in fact Pythagoras', but was already known in the Middle East; the story of the 'discovery', including the sacrifice, was very famous in antiquity.
4. *truly infinite ingenuity*: Hiero II was Tyrant of Syracuse between *c.* 271 and 216 BC. Archimedes (*c.* 287–212/11), mathematician and inventor, was an extraordinary genius who spent most of his life at Syracuse, of whom a number of works survive (*On the Sphere and Cylinder*, *Measurements of the Circle*, *On Spirals*, etc.) as well as other works known in Arabic translations.
5. *sextarius jug*: A jug with the capacity of 0.54 litre (1 pint).
6. *theorizing*: 'Cogitationes'; but 'concertiones' (disputes) is also attractive (Budé, vol. 9, p. 7).
7. *to mark in soft wax . . . to the test*: The text remains incurable.
8. *teachings of writers*: Changing 'scriptorum' to 'scriptoribus' would yield the attractive thought, 'Since honours for outstanding writers are not provided either by social customs or institutions . . .'
9. *living disputants*: In this and the previous paragraphs Vitruvius lists Ennius (239–169 BC), Accius (170–*c.* 86), Lucretius

(*c.* 94–55), Varro (116–27) and Cicero (106–43), the last three therefore contemporaries of Vitruvius, all of whom influenced him strongly.

10. *in his tragedy Phaethon*: what remains of the play is so fragmentary that one cannot tell whether this was part of an extensive scientific and philosophical discourse or a single observation about Phaethon.
11. *Berosus . . . propounded the following*: Vitruvius was particularly interested in Berosus, a priest of Baal who lived in the third century BC and wrote a history of Babylon in Greek up to the time of Alexander the Great; but this person may not be the astrologer Berosus. Vitruvius' knowledge of Berosus' account of Chaldean astronomy may come indirectly through Nigidius Figulus (*c.* 98–45 BC) or Bolos of Mendes (second century BC).
12. *Aristarchus of Samos*: Aristarchus (*c.* 310–230), the great astronomer, who proposed that the planets revolved around the sun, calculated its diameter and distance from the earth and fixed the duration of the solar year at 365⅛ days: Vitruvius attributes to him the theory of lunar phases which, according to other sources, was developed by Thales, Parmenides or Anaxagoras.
13. *the winter solstice . . . their brevity*: Assuming the etymology *bruma = brevissima.*
14. *brandishing a club in the direction of the Twins*: Some want him to hold a club in each hand.
15. *not because I was put off by laziness*: Or perhaps: 'not because I was put off by the fear of boredom', i.e. his own or the readers'. Vitruvius is not saying that he has left out a discussion of the *analemma*, although he does omit a number of details, but that he omits a discussion of all the other different types just mentioned.
16. *Ctesibius . . . was born in Alexandria*: Ctesibius of Alexandria lived in the third century BC, probably under Ptolemy II (283–247) and was a favourite with Vitruvius, being mentioned seven times altogether.

BOOK X

1. *would practise the profession of architecture without diffidence*: The phrase may mean: 'without any equivocation in the mind of the public that they were the genuine architects'.
2. *praetors and aediles must provide machinery for the games*: Under the Republic the aediles looked after the games, but in 22 BC Augustus transferred this obligation to the praetors: the conjunction of aediles and praetors here suggests that there was a transitional period after 22 when they worked together at the games.
3. *beams arranged vertically and bound together by cross-bars*: That is, ladders.
4. *instruments*: It is difficult to stick to Vitruvius' distinction between *machinae* and *organa* in English; for example, later on he calls water-screws and *scorpiones organa*, which we cannot very well translate as 'instruments'; so occasionally *organum* has been translated as 'device'.
5. *at the point where they spread apart*: A very curious expression, since the beams of the derrick separate immediately from the top where they are connected by a bolt.
6. *the capstan*: The word 'is' ('this') is sometimes taken to refer to the 'funis' (rope) rather than to the 'ergata' (capstan).
7. *revolving platforms*: Or a boom that turns around a mast, according to some scholars.
8. *devised by Chersiphron*: Chersiphron of Cnossus and his son Metagenes designed this version of the Temple of Artemis at Ephesus as dipteral and Ionic and wrote a treatise on it *c.* 560–550 BC. It was burnt down in 356 BC but reconstructed with essentially the same plan, which Vitruvius probably saw when he went with Caesar to Asia.
9. *the pivots . . . made the wheels go round*: Another of Vitruvius' curiously back-to-front expressions: cf. Book II, note 18.
10. *confident that it would make his name*: Or 'proud of his reputation' or 'with confident pride'.
11. *the burning question of the day*: The question was where the Ephesians might find the right marble for their temple.
12. *the prow slices easily through the water*: Literally, 'the prow cutting through the lightness (low-density) of the water'.
13. *congius*: 3.5 litres (6 pints).
14. *as low as possible*: That is, below the surface of the water.

15. *at either end*: At the upper end on land and at the lower end in the water.
16. *procedure which should be followed*: Cf. Plate 34, where the same process is illustrated for the construction of a spiral staircase.
17. *angobatae*: Meaning very obscure, but perhaps refers to figurines that emerge from vases and pour out water.
18. *a frame . . . attached firmly to the chassis [capsus] of the four-wheeler*: Scholars have different views about the relative sizes of these components; the size of the large disc and its housing ('loculamentum') is anything but clear.
19. *should equal the diameter of one hole . . . one and a half*: The diameter of this hole constitutes the module according to Vitruvius, and is so translated hereafter.
20. *in order to besiege Cadiz*: The date of the siege is unknown but may have been around 500 BC.
21. *besieging Byzantium*: Philip II unsuccessfully besieged Byzantium, allied with Athens, in 340–339 BC.
22. *a windlass was set transversely*: Vitruvius speaks of one transverse windlass; in the reconstruction in Fig. 37 we have left two windlasses, because otherwise it is difficult to see how a single windlass could operate ropes passing through two pulley-blocks at the other end to left and right of the beam and its point.
23. *Hegetor . . . constructed his tortoise*: Hegetor was perhaps the engineer of Demetrius Poliorcetes (336–283).
24. *Demetrius, known as Poliorcetes . . . because of his tenacity*: Demetrius I of Macedon, son of Antigonus Monophthalmus ('one-eyed'), was heavily involved in the wars of the Successors to Alexander's empire and sought to conquer Greece: he was proclaimed King of Macedon in 294 BC but ended up defeated and imprisoned by Seleucus I for three years and died in 283. Apparently his famous nickname, Poliorcetes, was given him on the occasion of the unsuccessful siege of Rhodes in 305–304, which Vitruvius describes.
25. *Athenian architect named Epimachus*: Recorded only here and by Athenaeus Mechanicus, who flourished during the reign of Augustus (30 BC–AD 14).
26. *holes of half an inch . . . seems inconceivable*: Vitruvius' description of the difficulty of making drill-holes over a certain size is a very bad illustration of the point that he is trying to make about the relationship between models and scaled-up versions of them.
27. *people in public service and private citizens*: The meaning of the

phrase used by Vitruvius ('publice et privatim') is not clear and other possible translations are 'by public proclamation and in private' and 'from public and private sources'.

28. *the inhabitants of the city . . . by digging deeper*: Rather than 'the inhabitants were on guard, and made a deeper ditch than the one in front of the ramparts' (Granger, 1931–4).

Glossary and Index of Technical Terms

The Glossary includes a necessarily limited selection of terms used by Vitruvius for architectural theory, for temples and houses and their decorations, components and spaces, for floors and interior and exterior walls and their surfaces, materials and related processes and for most of the machines and military equipment which he describes in great detail. Interested readers should consult the exhaustive *Dictionnaire des termes techniques du* De architectura *de Vitruve*, ed.L. Callebat and Ph. Fleury, Zurich and New York, 1995, which runs to 415 pages, with some 347 columns of definitions. Comparisons with Renaissance terminology may be made by consulting the diagrams and glossary in R. Tavernor and R. Schofield, *Andrea Palladio: The Four Books of Architecture*, Cambridge, Mass., 1997, pp. xxxii–xxxv and 379–419. Words in bold have separate entries. Latin words are italicized. V. = Vitruvius.

abacus (*abacus*): the low square slab between the **echinus** of a **capital** and the **architrave**; in the Corinthian order it takes the form of a square with concave sides with the points of the corners sliced off (Pll. 9.Ah; 11; 13.H; 17.E); at 4.3.4 and 4.7.3, V. uses the word *plinthus* to describe the abacuses of Doric and Tuscan capitals.
acanthus: plant of which the leaves are said to have been copied to decorate the Corinthian **capital** (2.7.4; 4.1.9; Pl. 11)
acoustics: of sites for the **theatre** (5.8.1; Figs. 12–13); circumsonant location (περιηχῶν, *circumsonans locus*, repelling sound all round); consonant location (συνηχῶν, *consonans locus*, amplifying sound); dissonant locations (κατηχῶν, *dissonans locus*, pushing sound back); resonant locations (ἀντηχῶν, *resonans locus*, creating echoes)
acroterium (*acroteria*): pedestal at the ends and centres of pediments (3.5.12–13; Pll. 10.I; 14.S): see **gable**
addition: to the modular system (*adiectio modulorum*), the opposite

of **subtraction** (*detractio*). V. cites two examples: (i) the **entasis** of columns (3.3.13; 3.5.14; 4.3.10); (ii) the convex curvature added to the **stylobate** creating a *stylobata alveolatus* matching the **entablature** (3.4.5; 5.9.4): *see* **diminution; module; *scamillus***

aeolipila [reading disputed]: a bronze sphere representing Aeolus' head with a small hole in it through which water was poured and which released steam when heated over a fire (1.6.2)

aggregate: *see* **concrete**

agnus castus (*vitex*): 2.9.9; 8.1.3; 10.6.2: see **timber**

aisle: transverse semicircular aisle in **theatre** (*praecinctio*, 5.3.4; 5.6.2; 5.7.2); also *diazoma* (5.6.7; Fig. 14.18)

akrobatikon (ἀκροβατικόν = *scansorium genus*): of a device with which to climb or ascend, i.e. a ladder (10.1.1): *see* **machines and mechanical systems (I)**

ala: *see* **corridor; side-room**

albarium opus, album opus: stucco revetment with powdered marble sometimes worked in relief (5.2.2; 5.10.3; 6.3.9; 6.7.3; 7.3.4)

alder (*alnus*): 2.9.10; 8.1.3; 10.14.3: see **timber**

altar (*ara*): orientation of (4.5.1); heights of (4.8.7; 4.9; 9.Introd.13); the altar-shaped container of the water-organ (10.8.1–2; Fig. 32.2)

amphiprostyle (*amphiprostylos*): of a temple with identical porticoes (and tympanum) at either end (3.2.1, 4; Fig. 5.3): *see* **dipteral; façade; monopteral; peripteral; prostyle; pseudodipteral; pseudoperipteral; temple**

amphithalamos: one of the bedrooms (of disputed function) along with the *thalamus* situated at right and left of the *prostas* of the *gynaeconitis* in the Greek household; *see also* **bedroom; houses**

amphitheatre (*amphiteatrum*): 1.7.1

analemma (ἀνάλημμα): a diagram, not an instrument, with which to construct a **sundial** (9.1.1; 9.6.1; 9.7.6–7; 9.8.1, 7, 8; Fig. 28)

analysis (*cogitatio*): 1.2.2; 2.Introd.2; 2.1.2,7; 7.Introd.16; 9.Introd. 10; *see* **architecture; invention; projection (II)**

anatonum: *see* **tension**

andron: (I) in Latin, a passageway between two courtyards in a house, synonymous with *mesauloe* (6.7.5): *see* **corridor**; (II) ἀνδρών: in Greek, an all-male dining room, a type of *oecus* (6.7.5): *see* **hall**

andronitis: a colonnaded **courtyard** (*peristylium*) reserved for men in the Greek house (6.7.4): *see* **houses; portico**

angobata: perhaps a toy operated by water: reading and meaning very obscure (10.7.4)

anta (παραστάς): pilaster, or better, spur-wall; in Books III–V the *antae* are those of temples, referring to the projecting ends of the front two walls of the *cella*; a temple *in antis* (ναὸς ἐν παραστάσιν, *naos in parastasin* = *aedes in antis*) includes columns between the ends of the spur-walls of the *cella* (3.2.1, 3; Fig. 5.1). In Book VI, the *antae* are those in houses (6.7.1; 6.8.2)

aperture (*hypaethrum*): of a door (4.6.1–2; Pl. 15C, E)

apophysis: the curve of a column of a quarter-circle surmounting the fillet above the upper **torus** of the **base** and below the *listello* under the **capital** (used twice at 4.7.3; Pll. 9.f, e; 13.F, N; 17.D, L); for the word *apothesis* at 4.1.11, see p. 371 n. 2: *see* **fillet, *listello***

apothesis: *see* ***apophysis***

appropriateness (*decor*): 1.2.1, 5–9 and *passim*: one of the six requisites of **architecture**: appropriateness is achieved (i) by following a rule (θεματισμός, *thematismos*) such as building **hypaethral** temples to Jupiter, Caelus and the Sun and Moon, that is, without roofs over the ***cella*s** (cf. Pl. 8; Fig. 7.7); (ii) by following a convention (*consuetudo*), such as not putting dentils in Doric entablatures; (iii) by following nature, for example, by choosing healthy sites for temples of Aesculapius

araeostyle (*araeostylos*): of a colonnade with an intercolumniation which is too wide, or rather of more than three diameters or six modules (3.3.1, 5, 10–11; 3.4.3; Fig. 6.4)

arch: *fornicatio* (6.8.3) is an arch comprising voussoirs (*cunei*); *fornix* (6.8.4) is an arch with voussoirs built between pillars: also *arcus* (6.8.3): *see* **vault**

architect (*architectus*): education of, 1.1: *see* **patron**

architecture (*architectura*): principles and subdivisions (1.2.1–9). Architecture consists of (i) **planning** (*ordinatio*, τάξις, *taxis*); (ii) **projection** (*dispositio*, διάθεσις, *diathesis*); (iii) **harmony** (*eurythmia*); (iv) **modularity, commensurability** (*symmetria*); (v) **appropriateness** (*decor*) and (vi) **distribution** (*distributio*, οἰκονομία, *oikonomia*). Architecture is concerned with (i) building (*aedificatio*), which is of two types: public and private buildings; public buildings are divisible into three types, for defence, religion and other public use; (ii) **clocks and sundials** (*gnomonice*); (iii) **machines and mechanical devices** (*machinatio*, 1.3.1): *see* **gnomon; instrument; projection (II)**

architrave (*epistylium*): Doric Pl. 14.A; Ionic Pl. 10.A; V. uses the word *epistylium* 47 times, often in conjunction with *ornamenta* (**ornaments**) meaning the **frieze** and **cornice,** so collectively the **entablature**; at 1.2.6, however, one must translate *epistylium* as 'entablature', and perhaps also at 5.1.1 and 7.5.5; *see* p. 362 n. 9

arm (*bracchium*): firing arm of *ballista* and *scorpio* (1.1.8; 10.10.1, 5–7, 9; Figs. 34.3; 35.5); *see* **tension**

arris: *see* **flute**
arsenal (*armamentarium*): 7.Introd.12
arsenic sulphate: *see* **orpiment**
ash (*farnus, fraxinus*): 2.9.11; 7.1.2: *see* **timber**
ashlar: *see* **stone; walls**
astragal (*astragalus*): small circular moulding of convex profile between the **scotias** of the **Ionic base** (3.5.3; Pl. 9.1 1); convex moulding at top of column-shaft below Ionic and Corinthian capitals (3.5.7; 4.1.11; Pl. 9); used with the crowning moulding of the lintels and frames of Doric and Ionic doors (*cymatium cum astragalo*, 4.6.2–3; Pl. 15.K)
astronomy (*astrologia*): 1.13, 10, 17; zodiac and planets (9.1); phases of the moon (9.2); the sun and constellations (9.3); northern constellations (9.4); southern constellations (9.5); astrology and astronomy (9.6); the ***analemma*** (9.7); types of sundials (9.8.1)
atlas (ἄτλας): *see* ***telamon***
atrium: 6.3.1, 3–6; 6.5.1–3; 6.7.1; 7.5.1; *see* **courtyards; side-room**; Fig. 23.4; cf. Pll. 31–2
Attic (*atticurges*): of a base with two **toruses** separated by a ***scotia*** (3.5.3; Pl. 13); of a type of **door** (*thyroma*, 4.6.1, 6)
auditorium (*cavea*): 5.6.7; perhaps the translation 'auditorium' is permissible for the word *theatrum* at 5.6.2
avenue (*platea*): avenue, a main road in a town (1.6.1, 7–8, 12; 1.7.1; 2.8.11; 3.Introd.4); *see* **street**; Pl. 5
axle, axis (*axis*): of the world (6.1.4–7; 9.1.2, 4; 9.5.4; 9.7.6); of the anaphoric clock (9.8.8–9); of the waterwheel (10.4.1); of the windlass of a crane (10.2.2, 5, 7); of Metagenes' device (10.2.12); of water-raising wheels (10.4.1–4); of watermills (10.5.2); of marine hodometer (10.9.5, 7); in ***ballista*** (10.14.1; Fig. 35.4). *Axon*: the axis of the celestial sphere in the *analemma* (9.7.5, 6; Fig. 28 PQ); axle of the windlass of the *ballista* (10.11.7 = *axis* 10.11.8; Fig. 35.4). *Small axle* (*axiculus*): in cranes (10.2.1; 10.3.2; Fig. 29.11), in the hodometer (10.9.2), in Hegetor's tortoise (10.15.4; Fig. 39B.18): *see* **flute; volute**
axle-block: *see* **wheel-housing**
azurite: *see* **blue**

bakery: *see* **farmhouse**
balcony (*maenianum*): upper balcony or loggia of the portico of a forum; projecting balcony of a house (5.1.2; 5.6.9)
ballista: machine for hurling rocks (1.Introd.2, 8; 1.2.4; 10.1.3, 9; 10.10.1; 10.11.1–3, 9; 10.13.7; 10.14.3; 10.16.1, 4, 12; Fig. 35)

balteus: band around the centre of the 'cushion' or bolster separating the volutes of Ionic capitals (3.5.7): *see* **pulvinate**

balustrade, parapet (*pluteus*): 4.1.1; 5.1.5, 10; 5.6.7; 5.10.4

bar, strip (*regula*): strip of willow used in water-screw (10.6.2); metal bar used in vault construction (5.10.3); element of the water clock (9.8.5–6); bar of uncertain function attached to the upper pulley of the *polyspastos* (10.2.8; Fig. 30.5); sliding bar of water-organ (10.8.1; Fig. 32.20); reinforcing strip of wood along the slide of the *scorpio,* also called 'cheek' (*buccula*, 10.10.3; Fig. 34.4–5); in the *ballista*, wooden strip below the cylinders (10.9.6; Fig. 35.19–20; cf. 10.11.6, 8): *see* ***gutta***; **piston-rod; ruler; strut**

barn: *see* **farmhouse**

base (I) Base of a column (*spira*): Attic (3.5.3; 4.1.6–7; Pl. 13); Ionic (3.5.1–3; Pl. 9); Tuscan (4.7.3; Pl. 17); at 3.4.5, *spira* refers to the base moulding of the podium of a temple. (II) Other types of base (*basis*): base, socle of water-organ (10.8.1; Fig. 32.1); of *ballista* (10.10.4; 10.11.9; 10.14.1; also called ἐσχάρα, *eschara*: Fig. 35.1); of *scorpio* (Fig. 34.2); of tortoise (also called ἐσχάρα, *eschara*; Fig. 38.1); of Hegetor's tortoise (Fig. 39A.1); at 10.2.13, *basis* refers to the pedestal of a large statue. *Antibasis* is the counter-base of the *ballista* at 10.10.5 (Fig. 35.3); *stratum* for the base or a platform in the siege engine at 10.13.6 (Fig. 36A and B)

basilica: law court (5.1.4–5; Pll. 19–21); V.'s Basilica at Fano (5.1.6–10; Pl. 22; Figs. 10–11); place for meetings in a wealthy man's house (6.5.2): *see* **Chalcidian portico; tribunal**

basin (*labrum*): could be very large, often circular and supported on a pedestal, made of metal, stone or marble, but English does not have an appropriate word for this except for basin, which usually implies a small receptacle used by one person at a time; the *labrum* was used for aspersion, the *alveus* (bath) for immersion (5.10.4; Pl. 28.L): *see* **bath**

bath: bathroom, domestic and public (*balneum*, 1.2.7; 1.3.1; 5.9.9; 5.10.4; 8.2.4; 8.6.2; 9.Introd.10); in private houses (*balnearius*, 6.4.1–2; 6.6.2); baths in the *palaestra* (5.10.1–5; Pll. 28–30); bath for immersion, of very variable size (*alveolus*, 5.10.1; *alveus*, 5.10.4); Archimedes' bath-tub (*solium*, 9.Introd.10); bathroom, hot or cold, in *palaestra* or farmhouse (*lavatio*, 5.11.2; 6.6.2); bathroom with cold-water baths (*frigidarium*, 5.11.2; Pll. 28.K; 29.g; 30.F); steam bathroom (*sudatio*, 2.6.2; 5.10.5; 5.11.2; Pl. 30.H); bathroom with underfloor heating and hot-water baths (*caldarium*, *calda lavatio*, 5.10.1–3; 8.2.4; Pl. 30.K); cold bathroom with basins (*frigida lavatio*, 5.11.2 = λουτρόν, *loutron*, Pl. 30.D); bathrooms with warm-water basins heated by braziers and, later, hypocausts (*tepidarium*, 5.10.1, 5; Pll. 28.H; 29.f); bathroom with warm-water basins (*propnigeum* = *tepidarium*, 5.11.2; Pl. 30.G): *see* **basin; farmhouse; furnace; gymnastic complex; Spartan hot room; water**

battering-ram (*aries*): 1.5.5; 10.2.5; tortoise-ram (*testudo arietaria*, 10.13.1–2, 4); 10.13.6 for the (*arietaria*) *machina* = κριοδόχη, *kriodoche*; 10.15.4–6; 10.16.12: Figs. 36B.6; 39A.20

beam: V. has a number of words for beams. *Arrectarium*: vertical beam (2.8.29; 7.3.13; 5.12.3–5, also *postis*, 10.14.2; 10.15.3; Fig. 39A.2, 14). *Interpensivum*: in Tuscan courtyards the light cross-beam or stringer used to form two sides of the aperture of the *compluvium* and supported on a pair of large joists crossing the whole of the courtyard which form the other two sides (6.3.1; Fig. 18.6). *Tignum*: the ridge-piece of the tortoise is a *tignum quadratum* (10.14.2; Fig. 38.6); elsewhere *tignum* means joist, a large beam used in cranes, floors and war machines (Figs. 29.1; 30.1; 38.3). *Trabicula*: small beam in Hegetor's tortoise (10.15.4; Fig. 39A.13). *Trabs*: large beam placed over columns, piers and pilasters (cf. Fig. 18.2); *trabs euerganea* for the inclined (?) beams in V.'s Basilica (5.1.9); mortised beam (*intercardinata trabs*, 10.14.2; Fig. 38.4). *Transtrum* for cross-beam, transverse beam (10.15.3; also *transversarius*, Fig. 38.2): *see* **woodwork in roofs**

bedroom: room for sleeping in but also a room for reclining in (*cubiculum*, 1.2.7; 6.4.1; 6.5.1; 6.7.2, 4); daughter's bedroom? (*amphithalamos*) and master bedroom (*thalamus*) placed left and right of the *prostas* in the *gynaeconitis* of the Greek house (6.7.2): *see* **women's quarters**

beech (*fagus*): 2.9.9; 7.1.2: *see* **timber**

bitumen: 1.5.8; 2.6.1; 8.2.8; 8.3.1, 5, 8–9; 8.6.12

black (*atramentum*): used in paint and for ink (7.10.1–4): *see* **pigments and colours**

block: small block used in the water-organ to maintain the distance between the regulator and the base (*taxillus*, 10.8.2; Fig. 32.12); block or box at the end of the shaft of the *ballista* (*quadratum*, 10.11.8; Fig. 35.6)

blue (*caeruleus*): 4.2.2; 7.11.1; 7.14.2; dark-blue, of the dark face of the moon (9.2.1); Armenian blue, azurite (*armenius*, 7.5.8; 7.9.6): *see* **pigments and colours**

board, boarding (*tabula*): upper board or πίναξ (*pinax*) of the water-organ (10.8.3; Fig. 32.19); board at top and bottom of the head-piece of a *scorpio* with holes in it (also called *peritretos*: 10.10.2; 10.11.4; cf. 1.2.4; Fig. 34.20); the *mensa* is the board below the head-piece of the *ballista* holding the two cylinders together (10.11.6, 7; Fig. 35.14)

boarding bridge (*sambuca*): from ship to land (10.16.9)

bolt, dowel, pin (*fibula*): for holding together the beams at the top of a crane (10.2.1; 10.7.2; Fig. 29.2); for keeping in place the hood of the tank of the water-pump (10.7.2; Fig. 31.6)

border (*cuneus*): apparently used of a raised yellow ochre or cinnabar border around a monochromatic black wall-panel (7.4.4; 7.5.1)

boxwood (*buxus*): 7.3.1

bracket: *see* **console**

brick (*later, laterculus*): fired brick (*coctus later*); half brick (*semilater*, 2.3.4); unfired brick (*crudus later*); bricks of three-quarters of a foot (*besalis laterculus*, 5.10.2; 7.4.2); brick four palms square (τετράδωρον, *tetradoron*) and brick five palms square (πεντάδωρον, *pentadoron*, 2.3.3; Fig. 3); Lydian, a brick of one and a half feet by one foot (*lydius*, 3.2.3; Fig. 3): *see* **walls**

broom, Spanish (*spartum hispanicum*): 7.3.2

bucket: square or rectangular aperture or 'bucket' (*quadratus modiolus*) around exterior of waterwheel (10.4.3; 10.5.1; Pl. 41 ii)

building: origins of (2.1–9); elements of building materials (2.2); **brick** (2.3); **sand** (2.4); **lime** (2.5); **pozzolana** (2.6); **stone** (2.7); **walls** (2.8); **timber** (2.9); highland and lowland **fir** (2.10)

buttress: in foundations (*anteris, erisma*, 6.8.6; Fig. 24): *see* ***anta***

caementicium, opus; caementum: rough stones, quarry-stones, hence sometimes rubble; it is not entirely clear whether the phrase *caementiciae structurae* (2.4.1; 2.7.5) refers to walls surfaced with rough stone or to walls built entirely of rough stone: *see* **concrete**; ***expolitio***; **pavement**; **walls**

candle (*candela*): used for the final polishing in the process of ***ganosis*** (7.9.3): *see* **linen**

cane: *see* **reed**

capital (*capitulum*): Corinthian (4.1.1, 11–12; Pl. 11); Doric (4.3.4; Pl. 13); Ionic (3.5.5–8; Pl. 9; Fig. 7); Tuscan (4.7.3; Pl. 17); of triglyph (Pl. 14.I)

capstan (*ergata*): used separately from the main structures of **cranes** (10.2.7, 9) and ***ballistae*** (10.11.1), whereas the windlass was mounted on them: *see* **windlass**

carpentry (*opus intestinum*): interior woodwork (2.9.7, 17; 4.4.1; 5.2.2; 6.3.2, 9; 6.7.3); also *materiatio* and *materiatura* (4.2.1, 2) of the carpentry or woodwork of wooden temples: *see* **timber**; **woodwork in roofs**

caryatid (*caryatis*): figure of draped woman substituted for a column: in V.'s account they support Doric mutules, not Ionic or Corinthian cornices (1.1.5; cf. Pl. 1)

catapult (*catapulta*): 1.1.8; 10.11.1, 2, 9; 10.12.1–2; 10.13.6, 7; 10.15.4; 10.16.1; machine for firing arrows, of which the *scorpio* was a small version (1.Introd.2; 1.1.8; 1.5.4; 10.1.3, 9; 10.10.1; 10.13.6; 10.15.4; 10.16.1; Figs. 34; 36B.7)

catatonum: *see* **tension**

catch (*clavicula*): catch or latch which prevented the windlass of the *ballista* from turning (10.9.8; Fig. 35.11)

cathetus: *see* **line**

cedar (*cedrus*): 2.9.13

cedar oil (*cedrium*): 2.9.13

ceiling, panelled or coffered (*lacunar*): 2.9.13; 4.6.1; 5.2.1; 6.3.4, 6; 6.7.3; 7.2.2; but at 9.8.1 *lacunar* means 'coffer'; for 4.3.1, see p. 372 n. 7; *see also* **vault**

cella: inner shrine or enclosure housing a cult statue in a temple: *see also* ***naos***

cellar, covered wine cellar: *see* **farmhouse**

Chalcidian portico (*chalcidicum*): a columniated hall at the end of a basilica (5.1.4)

chalk (*selinusia creta*): bead chalk or Selinusian chalk (7.14.2)

chamber (*arcula*): small chamber in the water-organ below the main component (*caput machinae*) and above the regulator (Fig. 32.14)

channel, groove: V. has various terms. *Canalis*: the concave horizontal moulding connecting the two volutes on the face of the Ionic capital (3.5.7; Pl. 9.m); water-channel (8.6.1, 3, 5; 10.16.7); channel in main board of water-organ (Fig. 32.16); channel housing the ram in Diades' drill (10.13.7; Fig. 37); channel or groove of the slide of the *scorpio* (also *canaliculus*, σῦριγξ, *syrinx*, 10.10.3–4; Fig. 34.6, 16). *Canaliculus*: the small vertical grooves in triglyphs (4.3.5; Pl. 14.F). *Colliciae*: the diagonal water-channels of the roof of the Tuscan courtyard carrying water down to the *compluvium* (6.3.1; Fig. 18.4). *Semicanaliculus*: the half-channels or half-flutes at either side of triglyphs (4.3.5; Pl. 14). *Specus*: an underground water-channel (8.1.6; 8.6.3)

chassis (*capsus*): of four-wheel cart (10.9.2–3; Fig. 33.2)

chorobates: wooden device for levelling (8.5.1–3; Fig. 27; Lewis, 2001, pp. 31–5)

cinnabar, vermilion (*minium*): 7.4.4; 7.5.8; 7.8.1; 7.9.1–3, 5–6: *see* **pigments and colours**

circuit, double (δίαυλος): part of the sports facilities next to the *palaestra* (5.11.1)
cistern, reservoir: *see* **chamber; tank**
city (*civitas*): city planning (Pll. 5; 19; Fig. 2); sites (1.4); walls, ramparts, towers (1.5; Pl. 5); **winds** (1.6; Figs. 1–2); sites for public buildings and **temples** (1.7); **forum, basilica** (5.1; Pl. 19–22; Figs. 10–11); prison, senate house, treasury (5.2)
clay: V. has three words for clay: *argilla*, sometimes mixed with hair (5.10.2–3; 10.15.1); *creta*, for making bricks and also used in revetments (2.8.19; 5.12.5; 7.3.3; 7.7.4; 7.14.1–2; 8.1.2, 5); *lutum*, which evidently means mud in some contexts (2.1.2–5; 2.3.2), clay mortar in others (7.3.11)
climacis: the main beam, 'ladder', of the *ballista* which is wider than that of the *scorpio* because it has to fire rocks rather than arrows (10.11.7, 8; Fig. 35.12)
clock and sundial (*horologium*): types of sundial (9.8.1); anaphoric winter clock indicating the rising of the stars (*anaphoricum horologium*, 9.8.8; Pll. 38–9): *see* **gnomon**
coffer: *see* **ceiling**
cold bath: *see* **bath**
colonnade: *see* **portico**
colours: *see* **pigments and colours**
column (*columna*): column in basilicas, houses, porticoes, temples, etc.; column of water clock on which the hours are marked (9.8.7; Pl. 35); column supporting ***scorpio*** and ***ballista*** (Figs. 34.13; 35.21): *see* **shaft; strut**
comitium: gathering-place for citizens (2.8.9)
commensurability: *see* **modularity**
commentary, body of writing, treatise (*commentarium*): 1.1.4, 12; 2.8.8; 4.Introd.1; 6.Introd.4; 7.Introd.1, 2, 11, 14, 17; 9.Introd.14; 10.7.5
compluvium: rectangular or square aperture in roofs of Corinthian, displuviate, **tetrastyle** and **Tuscan** courtyards, directly above the *impluvium* (which V. does not mention), the catchment area for rainwater on the ground (6.3.1, 2, 6; Figs. 18.1, 5; 23.3)
conclave: room in the private sector of the house, such as a bathroom, bedroom or dining room, which can be locked and so isolated (6.3.8; 6.6.7; 7.2.2; 7.3.4; 7.4.1; 7.5.1; 7.9.2)
concrete (aggregate with mortar): the word *rudus* refers both to the aggregate, hard core or **rubble** used in concrete and the concrete itself, *ruderatio* to concrete and the process of laying it (2.8.20; 5.12.6; 7.1.1, 3, 5–7; 7.4.5): *see* ***expolitio***; **pavement**; ***signinum***; **walls**

console (*ancon*): of Ionic door, also called 'ear-piece' (*parotis*, 4.6.4; Pl. 16.f)
copper: 7.8.4; 7.11.1 (Cypriot)
cord: *see* **rope**
Corinthian, Corinthian order: for the Corinthian **courtyard** (6.3.1), *see* p. 376 n. 7; Fig. 20; *see also* **base** (I); **capital**; **hall**
cornice (*corona*): crowning element of an entablature, sometimes referring to the whole of a cornice of a temple including the ***simae***, sometimes exclusive of them (Pll. 10.G; 14.L); cornices inside buildings, sometimes made of wood or stucco (5.2.2; 6.3.9; 6.7.6; 7.3.3–5); cornice of a podium (3.4.5)
corridor: *ala* is a corridor at the side of the central *cella* of a Tuscan temple, according to a famous emendation of Fra Giocondo (1511) accepted here (4.7.2; p. 372 n. 18); *fauces* is the corridor from the vestibule to the **atrium** (6.3.6; Fig. 23.2); *mesauloe* is a corridor connecting two courtyards in the Greek house and is synonymous with the Latin ***andron***, according to V. (6.7.5); *iter* is a walkway or passage, often narrow, on a fortified wall, in a house, temple or theatre (1.5.4; 4.4.1; 5.3.4; 5.6.2–3, 5, 7; 5.11.2)
courtyards (*cava aedium*): 6.2.5; 6.3.1; 6.5.1. V.'s types are: **Corinthian; displuviate; roofed; tetrastyle; Tuscan** (6.3.1–2; Figs. 18–22); colonnaded courtyard (*peristylium*, Fig. 23.8); *see* **atrium; farmhouse; Rhodian courtyard; side-room**
cover-joint, cover-stile (*replum*): vertical plank or strip of a door covering the gaps between the leaves (4.6.5; Pl. 16.h)
crane, derrick: with three pulleys (*trispastos*, 10.2.1–3; Fig. 29); with five pulleys (*pentaspastos*, 10.2.3); heavy duty cranes have a drum on the **axle** and a separate **capstan**; the drum is called a *tympanum* or wheel (*rota*), but also a περιθήκιον (*perithekion*) or ἀμφίεσις (*amphiesis*, 10.2.5); V. uses the term *polyspastos* to refer to cranes with more than five pulleys (10.2.10; 10.11.1; Fig. 30): cf. Pl. 40; *see* **disc; 'raven'**
craticium, opus: wattle-and-daub, half-timber (2.8.20; 7.3.11)
crenelation (*pinna*): of Hegetor's tortoise (10.15.1)
crepido: *see* **plinth**
cross-bar (*impages*): of a wooden panel of a door (4.6.5; Pll. 15.T; 16.N)
cross-beam: *see* **beam**
cubit (*cubitus*): the distance between the elbow and the tip of the middle finger, about a foot and a half (444 cm. or 17.5 inches: 3.1.7, 8; 10.13.4–7)
curia: *see* **senate house**

curvature, of columns, of stylobate: *see* **addition; entasis; subtraction**
'cushion' (*pulvinus*): bench or socle below the **bath** (*alveus*) in bathroom (5.10.4); a platform, pier, mole in the sea (5.12.5–6): *see* **pulvinate; socket-piece**
cylinder (*modiolus*): in water-pump and water-organ (10.7.1–3; 10.8.1, 4, 5; Figs. 31.1; 32.4); in head-pieces of *scorpiones* and *ballistae* enclosing the ropes under tension (Figs. 34.23–4; 35.15)
cymbal: *see* **valve**
cypress (*cupressus*): 1.2.8; 2.9.5, 12–13; 5.1.3; 7.3.1: *see* **timber.**
Cyzicene hall: *see* **hall**

deal (*sap(p)inus, sap(p)ineus*): plank, strip of deal (fir, 1.2.8; 2.9.7, 17): *see* **knotwood**
decastyle (*decastylos*): of a temple with ten columns across the front and back façades and therefore so wide as to be **hypaethral** (3.2.8; Pl. 8; Fig. 5.7)
***decor*:** *see* **appropriateness; architecture**
defence strategies: 10.16.1–12
dentil (*denticulus*): small block or modillion, an element in the continuous horizontal moulding or dentilation between the frieze and cornice of an Ionic entablature (1.2.6; 3.5.11; 4.1.2, 4, 5); V. regards dentilation (for which he evidently does not have a collective noun) as the connecting element (*intersectio*) or μετοχή (*metoche*: reading disputed) between the frieze and cornice (3.5.11; Pl. 10.E, O); *see* p. 371 n. 21
diagram, drawing, representation: V. uses a number of terms: *adumbratio* for drawing (1.2.2); *conformatio* for plan (5.6.1); *deformatio* for diagram (1.1.1; 3.Introd.4; 4.9.1; 10.11.4, 6); *descriptio* for diagram, drawing; *designatio* for figure, configuration; *diagramma* for musical diagram (5.4.1; 5.5.6; 6.1.7); *exemplar* (1.1.4, where it may mean model; 8.5.3); *forma* = σχῆμα, *schema* for figure or diagram (1.6.12); *imago* for image, representation; *subiectio* for projection (9.7.7; 9.8.1): *see* ***orthographia*; projection (II); *scaenographia*; scale**
diapegma (διάπηγμα, *interscalmium*): the distance between thole-pins on a boat, used as a module (1.2.4): *see* **module**
diastyle (*diastylos*): of a colonnade with an intercolumniation of three diameters or six modules (3.3.4, 10; 3.4.3; 4.3.7; Fig. 6.3)
diathyron (διάθυρον): the space between the doors at the either end of the entrance corridor in a Greek house; known as the *prothyron* in Latin according to V. (6.7.5)
diatonos (διάτονος): bond-stone set through the whole thickness of a

wall with facings at either end (2.8.7; Fig. 4.6); διάτονος is translated by V. as *frontatus* (p. 367 n. 21): *see* **facing; walls**
diaulos (δίαυλος): double circuit (5.11.1): *see* **gymnastic complex**
digging machine or tortoise (*testudo ad fodiendum*): also called 'pick-axe', digging device (ὄρυξ, *oryx*, 10.15.1); *see* **machines and mechanical systems (III)**
diminution of columns, contraction (*contractura*): 3.3.12; 3.5.4; 4.6.3: *see* **addition; entasis; subtraction**
dining room (***triclinium***): 6.3.2, 8, 9–10; 6.4.1, 2; 6.5.1; 6.6.7; 6.7.2–4; 7.4.4–5
dioptra: levelling instrument (8.5.1; *see* Lewis, 2001, pp. 51–105)
dipteral (*dipteros*): of a temple with a *cella* surrounded by two rows of columns (3.2.1, 7–8; 3.3.8; 7.Introd.15; Pl. 8; Fig. 5.6, 7): *see* **amphiprostyle; façade; monopteral; peripteral; prostyle; pseudodipteral; pseudoperipteral**
disc, drum (*tympanum*): a disc, sometimes with teeth, or a drum; of waterwheels, hodometers, watermills (5.12.4; 10.3.9; 10.4.1–2; 10.5.2; Pll. 41–2; Fig. 33.4–6); of water clock (9.8.5; 9.8.8–14); at 10.10.5, *carchesium* refers to the drum housing the levers of the windlass of the *scorpio* (Fig. 34.10): *see* **crane**
displuviate (*displuviatus*): 6.3.1; Fig. 21; p. 376 n. 5; *see* **courtyards; houses**
dispositio: *see* **architecture; projection (II)**
distribution (*distributio*, οἰκονομία, *oikonomia*): one of the six requisites of **architecture**
dolphin (*delphinus*): decorative bronze dolphin attached to a pivot for raising and lowering the cymbal-shaped valve in the water-organ (10.8.1, 5; Fig. 32.9–10): *see* **valve**
dome, roof: of the conical roof of a circular peripteral temple (*tholus*, 4.8.3; cf. Pl. 18); of the **Spartan hot room** (*hemisphaerium*, 5.10.5; Pll. 28.I; 29.e; 30.I)
door: large door of temple or house (*ostium*, 4.6.1; 6.3.6; also *thyroma*, 4.6.1; 6.3.6); of private house (*ianua*, 6.5.3; 6.7.1, 3–5); door with two leaves (*foris*, 4.4.1; 4.6.4–5); door with folding panels of theatre, temple, house (*valvae, valvatus*, 3.2.8; 3.3.3; 4.4.1; 4.6.1; 4.8.2; 5.5.7; 5.6.3, 8; 6.3.10; Pll. 6; 7; Fig. 14.10–12). *Types*: Attic (4.6.6); Doric (4.6.1–2; Pl. 15); Ionic (4.6.3–5; Pl. 16); royal doors (5.6.8); *see* **cover-joint; fascia; hinge-stile; stage**
door frame, architrave of door (*antepagmentum*): 4.6.1, 2–6; Pll. 15.M; 16
Doric order: wooden origins (4.2); columns, entablature, intercolumniations, plans of Doric temples (4.3; Pll. 13; 14); ***cella*** and

pronaus (4.4); *see* **abacus; base; capital; door; echinus; entablature**
doron (δῶρον): *see* **brick**
drill (*terebra*): as well as modern meaning, also refers to a type of siege engine (10.13.3, 7; Fig. 37)
drum (*tympanum*): *see* **disc**
dyer's weed (*luteum*): 7.14.2

earthwork (*opus terrenum*): 1.5.6
eaves: of Tuscan temple (*stillicidium*, 4.7.5; Fig. 8.7); also *suggrundatio*, referring to the ends of the rafters of a roof (4.2.1); the *suggrundium* made of larch of an *insula* or housing-block (2.9.16); small inclined roof of Hegeter's tortoise (*suggrunda*, 10.15.1)
echinus (*echinus*): circular element between the top of the column-shaft and the **abacus** of a **capital** with a profile varying between, roughly, a truncated cone and quarter-circle depending on the period (4.3.4; 4.7.3); of Doric and Tuscan capitals (Pll. 13I; 17.F). V. also uses *cymatium* (moulding) for the echinus of the Ionic capital (3.5.7; 4.1.7; Pl. 9.I)
Egyptian hall: *see* **hall**
ekklesiasterion (ἐκκλησιαστήριον): little theatre or place of assembly at Tralles (7.5.5)
element (στοιχεῖον, *stoicheion*): 1.4.5
elevation drawing (*orthographia*): *see* **architecture; diagram; projection (II)**
elm (*ulmus*): 1.2.8; 2.9.5; 2.9.11: *see* **timber**
embater (?) **and ἐμβάτης** (*embates*): *see* **module**
emplekton (ἔμπλεκτον): 2.8.7; Fig. 4.5–6: meaning 'woven' or 'spliced'; but 2.8.7 is very puzzling. V. seems to describe *emplekton* walls as those with stone faces filled with concrete (= *fartura*) made by (i) the Greeks and (ii) by Roman peasants, though in slightly different ways: he then (iii) says the Greeks do not build like this ('non ita') but, evidently, make solid walls without infill which include *diatonoi* (p. 367 n. 21). Is type iii still *emplekton*? If so, the sense of the word, referring now to stonework without infill, is different from that used in (i) and (ii). One suspects either that the word *emplekton* refers only to (i) and (ii) and not to (iii), or that *emplekton* does not refer to walls with facings plus infill but to some other quality all three types of walls share, a 'spliced' or 'woven' look of the outside facings, perhaps; but then how would that make them different in principle from ashlar walls or the isodomic and pseudisodomic walls just described in 2.8.6 which are also 'woven' in a similar sense? *See* **stone; walls**

entablature: Corinthian 4.1.2; Doric 4.3.5–6; Pl. 14; Ionic 3.5.8–13; Pl. 10; Tuscan 4.7.4–5; Pl. 17; *see also* **architrave**; **cornice**; **frieze**; **ornament**

entasis (ἔντασις): the curvature added to columns (3.3.13; Pl. 9.e): *see* **addition**; **diminution**; **module**; **subtraction**

epaietis (ἐπαιετίς): *see* **sima**

epibathra: *see* **machines and mechanical systems (III)**

epistylium: *see* **architrave**

eurythmia: *see* **architecture**; **harmony**

eustyle (*eustylos*): of a colonnade with an intercolumniation of two and a quarter diameters or four and a half modules (3.3.1, 6, 10; 3.4.3; Fig. 6.5)

exedra (*exhedra*): hall or room in a Greek or Roman house or gym not four walls (5.11.2; 6.3.6; 6.7.3; 7.3.4; 7.5.2; 7.9.2; Fig. 23.9); V. gives practically no information about its structure but Renaissance illustrators enjoyed giving it various configurations (Pl. 30): *see* **side-room**

expolitio: the process of finishing buildings by applying various surfaces such as pavements and the revetments of walls and vaults (6.8.10; 7.Introd.18; 7.1.1); use of lime in revetments (7.2): also a revetment or a painted surface (7.5.1; 7.9.2–3): *politio* is used for a revetment or a surface of a revetment (7.7.1; painting 7.9.4; also *corium*, 7.3.6,8): *see* ***albarium opus***; **concrete**; **revetment**; ***sectile***; ***spica***; ***testaceum***; **walls**

eye (*oculus*): *see* **volute**

façade (*frons*): of temples; of stage-set (*scaenae frons*, Pl. 26; Figs. 14–16). V. uses *posticum* for the back façade or colonnade of Roman temples (3.2.4–8; 3.3.6–7; 3.5.4; 4.3.4), twice with the specific meaning of *opisthodomus* (back porch; a term not used by V.), in apposition to *pronaus* (3.2.8)

facing: of wall (*frons*, 2.8.4, 7; 6.8.6; 7.Introd.11, 13); at 2.8.7, V. seems to use the noun *frontatus* to mean 'wall surface', then, additionally ('praeterea'), a long bond-stone with two outer faces, a ***diatonos*** (Fig. 4.6); facing-stone of wall (*orthostata*, 2.8.4). He uses *truncus* for the face of the pedestal of a **podium** (3.4.5, *see* **shaft**), *corium* for a wall surface, evidently, and a layer of bricks (2.3.4; 2.8.5) and *antepagmentum* for the protective facing fixed to the end of a joist of a wooden Tuscan temple (4.7.5); *see* **walls**

farmhouse (*villa*): 6.6; bakery (*pistrina*, 6.6.5); bathroom (*balnearius*, *lavatio rustica*, 6.6.2); courtyard (*chortis*, 6.6.1); kitchen (*culina*, 6.6.1, 2, 5). Barns, cellars, storerooms: grain (*granarium*, 1.4.2;

6.6.4; *horreum* 6.5.2, 5); hay (*fenile*, 6.6.5); oil-cellar (*olearia cella*, 6.6.3); oil-press, oil-press room (*torcular*, 6.6.2–3); for spelt (*farrarium*, 6.6.5); wine-cellar (*vinaria cella*, 1.4.2; 6.6.2). Stables, sheds: for goats (*caprile*, 6.6.4); horses (*equile*, 6.6.4; 6.7.1; *stabulum*, 6.5.2); oxen (*bubile*, 6.6.1–2); sheep (*ovile*, 6.6.4); stall (*presepium*, 6.6.1)

fascia (*fascia*): of Ionic architrave (3.5.10–11; Pl. 10.1, 2, 3). V. also uses *corsa* for a fascia of the architrave or frame of Ionic doors (4.6.3, 6; Pl. 16.I, K, L)

fauces: *see* **corridor**

fern (*filex*): 7.1.2

figlinum, opus: pottery or **tile** surface applied to the soffit of the **vault** of a hot **bathroom** (5.10.3)

fillet, *listello*: horizontal moulding of vertical profile, for which V. uses various words: *anulus* for a fillet below the echinus of a **Doric** capital (4.3.4; Pl. 13.K; for the **Tuscan** capital, Pl. 17.G); *quadra* for the fillet of the ***scotia*** of the Attic base and of the **abacus** of the Ionic capital (3.5.2, 5, 7; Pl.13); *supercilium* for the upper fillet of the upper *scotia* of the **Ionic** base (3.5.3; Pl. 9)

fir (*abies*): 1.2.8; 2.9.5–6. V. contrasts the highland fir of the north-east and Adriatic zone (*supernas abies*) with the lowland fir of the south-east, Tyrrhenian sea zone (*infernas abies*, 2.10.2); *see* **deal**; **knotwood**

fired brick, tile, terracotta (*testa*, *testaceus*): powdered terracotta or fired brick (*testa tunsa*, 2.5.1): *see* ***testaceum***

float (*liaculum*): trowel, float used in plasterwork to smooth off surfaces (2.4.3; 7.3.7)

floor, wooden floor (*coaxatio*, *contabulatio*, *contignatio*): in Hegetor's tortoise, the *contabulatio* is also called *planities transtrorum* and *tabulatum* (10.15.3–4; Fig. 39A.5); *cenaculum* is an upper floor (2.8.17); *suspensurae* are suspended floors, in this case made of tiles supported by piers in hot baths (5.10.2)

floors and walls: *see* **concrete**; ***expolitio***; **pavement**; **revetment**; **walls**

flute, fluting (*stria*, *striatura*, *strigilis*): flat or slightly concave segments or vertical channels of semicircular section on column-shafts; an arris or vertical fillet separating flutes is an *ancon* (3.5.14; disputed) or *angulus* (4.3.9)

forum (*forum*): Pl. 19

foundations (*fundamentum*, *fundatio*, *substructio*): Fig. 24; *see* **stereobate; stylobate**

frame, housing: *see* **housing**

fricatura: *see* **smoothing**

frieze: horizontal band, variously decorated according to the order, between the **architrave** and **cornice** in an **entablature** (*zophorus*, 3.5.10, 11, 13; 4.1.2; 4.8.1; 5.1.5); of Doric and Ionic doorways (*hyperthyrum*, 4.6.2, 4; Pll. 10.C; 14.N; 15.G; 16.C); see **ornament**

fulcrum (ὑπομόχλιον, *hypomochlion*): also *pressio* (10.3.2, 3): *see* Pl. 40

funnel (*infundibulum*): grain-hopper in watermill (10.5.2); referring to the hood (*paenula*) of the tank in the water-pump (10.7.2; Fig. 31.5)

furnace: under-floor furnace in baths (*hypocausis*, 5.10.1–2; Pl. 29); also *fornacula* and *fornax* (7.10.2)

gable, pediment (*fastigium*): of a temple (Pll. 10.L; 14.N); of primitive hut (2.1.3; 5.1.10); of siege engine (10.13.6; Fig. 36B.2); of tortoise-ram (10.13.6); half-gable, presumably meaning a split tympanum of some kind (*semifastigium*, 7.5.5); *see* **tympanum**

gallery, picture (*pinacotheca*): 1.2.7; 6.3.8; 6.4.2; 6.5.2; 6.7.3

gangway: *see* **stair**

ganosis (γάνωσις): process of applying a coat of Phoenician wax and oil, which is then melted by the heat from a brazier, smoothed off and polished, to protect marble or painted revetments, particularly those in cinnabar (7.9.3–4): *see* **candle; linen**

gnomon: metal pointer of a sundial, also called 'shadow-hunter' (σκιαθήρης, *skiatheres*, 1.6.6; Fig. 28)

granary: *see* **farmhouse**

gravel (*glarea*): 2.4.2; 2.6.5; 8.1.2; *see* **sand; stone**

green chalk (θεοδότιον, *theodotion*): green earth (*viridis*), so-called because a certain Theodotus owned the land in which it was discovered (7.7.4)

green malachite: *see* **malachite**

ground-plan (*ichnographia*): *see* **architecture; projection (II)**

guests' quarters (*hospitalia*): reached through the doors to left and right of the centre of the *scaenae frons* in the theatre (5.6.3, 8; Pll. 23; 26; Fig. 14.11–12)

gutta (drop): small truncated conical element (always in groups of six) attached below a small rectangular block (*regula*) under the triglyphs of the Doric entablature; also attached to the rectangular panels of the mutules in the soffit of the cornice in groups of 6 x 3 (4.1.2; 4.3.4, 6; Pl. 14.B)

gymnasium: possibly all the buildings and facilities for exercise and sport, of which the gymnastic complex (*palaestra*) was one (1.7.1; 6.Introd.1; 7.5.6)

gymnastic complex (*palaestra*): 5.11.1–2; Pl. 30; hall with punchbags (*coryceum*, 5.11.2; Pl. 30.B); dusting-room or room laid with sand for wrestling, etc. (*conisterium*, 5.11.2; Pl. 30.C); oiling-room (*elaeothesium*, 5.11.2; Pl. 30.E); *ephebeum* (5.11.2; Pl. 30.A) was a young men's hall of uncertain function, probably including seats and provision for eating; one source says that it was a 'locus constuprationis puerorum imberbium' ('place for debauching beardless boys'); *see* **bath**; ***diaulos***; **exedra**; ***paradromis***; ***signinum***; **Spartan hot room**; ***xystos***

gynaeconitis: *see* **women's quarters**

gypsum (*gypsum*): 7.3.3

hall (*oecus*): refers to the most imposing room in the house (6.3.8, 10; 6.7.2–5). The Corinthian hall had one storey of columns placed directly on the pavement or pedestals surmounted by an architrave with a wooden or plaster cornice supporting a segmental or semicircular coffered vault (*oecus corinthius*, 6.3.8–9); Cyzicene halls faced north and opened onto gardens (*cyzicenus oecus*, 6.3.10); Egyptian halls had two storeys of columns, with windows between the upper columns and an exterior gallery on the first floor (*aegyptius oecus*, 6.3.8–9); the **tetrastyle** hall was similar to Corinthian halls but had four columns (6.3.8); *see* ***andron*** **(II)**; **andronitis**; **courtyards**

handle (*manubrium*): of **stopcock** of water-organ (10.8.3; Fig. 32.18)

harbour, port (*portus*): construction of ports in naturally sheltered anchorages (5.12.1); construction of promontories (5.12.2); cofferdams (5.12.3); method of advancing promontory into the sea (5.12.3–4); double cofferdams (5.12.5–6); docks (15.12.7)

hard stone (*silex*): meaning uncertain; translated here as hard stone or hard limestone (1.5.8; 2.5.1; 2.8.4, 5; 8.1.2; 8.6.14)

harmonics: 5.4.1–9; and the **theatre** (5.5.1–8; 5.8.1–2; Figs. 12–13)

harmony (*eurythmia*): one of the six requisites of **architecture** (1.2.1; 3.4.5)

head-piece: *see* **main frame**

helepolis ('city-destroyer'): siege engine of Demetrius Poliorcetes (10.16.3–4, 7–8): *see* **machines and mechanical systems (III)**

hexastyle (*hexastylos*): of a temple with façades of six columns (3.3.7–8; 4.3.3, 7; Fig. 5.4)

hinge, joint (*verticula*): of piston-rod of water-organ (Fig. 32.8)

hinge-stile (*scapus cardinalis*): vertical side-post of a door with hinges (4.6.4–5; Doric, Pl. 15.QR); *see* **door**; **shaft**

hodometer: used for carts and carriages on land (10.9.1–4) and at sea (10.9.5–7; Pl. 44; Fig. 33; *see* Lewis, 2001, pp. 134–42)
hole, opening (*foramen*): in upper disc of hodometer (10.9.1; Fig. 33.7); referring to the diameter of the cylinder of the *ballista* and *scorpio* used as a module (1.2.4; 10.10.1–5; 10.11.2–9; 10.12.2; Figs. 34.20; 35.8); referring to the aperture for the arrow in the head-piece of the *scorpio* (Fig. 34.17)
homotonum: *see* **tension**
hood (*paenula*): of tank of water-pump (10.7.2–3; Fig. 31.5)
hook (*coracium*): metal spring or hook of key of water-organ (10.8.4; Fig. 32.21)
hot bathroom: *see* **bath; Spartan hot room**
houses: and climate (6.1); modularity and optics (6.2); house plans (6.3); orientation of rooms (6.4); social status and houses (6.5); Greek house (6.7); foundations and substructures (6.8): *see the Vitruvian house*, Fig. 23; *cf. the fantastic Renaissance reconstructions*, Pll. 31–3. Primitive houses, huts (*casa*, 2.1.2, 5, 7; and *tugurium*, 2.1.5); farmhouse (*aedes*, 6.5.2; also *villa*, 6.6.1, 4–6); residence, home (*domus*, *habitatio*); a house with some characteristics like those of urban houses (*pseudourbanus domus*, 6.5.3); palace (*regia domus*, 2.8.13; 8.3.4): *see* ***amphithalamos***; ***andron***; ***andronitis***; ***anta***; ***atrium***; **bath**; **bedroom**; ***compluvium***; ***conclave***; **corridor**; **courtyards**; ***diathyron***; **dining room**; ***displuviate***; **door**; **exedra**; **farmhouse**; **gallery**; **hall**; **library**; ***pastas***; **portico**; ***prothyron***; **Rhodian courtyard**; **side-room**; ***tab(u)linum***; ***thyroron***; ***vestibule***; **women's quarters**
housing, frame: V. uses *loculamentum* for the frame enclosing the vertical disc of the hodometer (10.9.2–6), also *theca*, implying a smaller frame or box, for the disc at the top (10.9.3; Fig. 33.8, 10); for the housing of the windlass at the end of the shaft of the *scorpio*, with the synonyms *scamillum* and *buccula* (10.10.3; Fig. 34.5); *operimentum* for the housing or cover of a socket-piece (*chelonium*) of the **ballista**, with the synonym *replum* at 10.11.8; Fig. 35.17
hub (*modiolus*): of four-wheel carriage or cart (10.9.2; Fig. 33.3)
hypaethral (*hypaethros*): of a temple with an unroofed *cella* (1.2.5; 3.2.1, 8; Pl. 8; Fig. 5.7); of an uncovered walk or promenade, also called a ***xystum*** in Latin (*ambulatio*, 5.9.5, 9; 6.7.5)
hypaethrum: *see* **aperture**
hypomochlion: *see* **fulcrum**

ichnographia: *see* **diagram; projection (II)**
image, representation (*imago*): *see* **diagram; projection (II); scale**
incertum*, *opus: of wall-surface with irregular stones (*incerta caementa*, 2.8.1; Fig. 4.2) which interlock (*imbricata caementa*, 2.8.1): *see* **walls**
indigo (*indicus*): 7.9.6; 7.10.4; 7.14.2
instrument, device (*organum*, *ratio*): instrumentally, by the use of an instrument (ὀργανικῶς, *organikos*, 10.1.2); of musical instruments (1.1.15; 5.3.8; 5.4.4; 6.1.5–6); of devices for raising water (10.4.1); of waterwheels and watermills (10.5); of the water-screw (10.6): *see* **machines and mechanical systems**
intercolumniation (*intercolumnium*, *intervallum*): V. does not mention interaxial distances, only intercolumniations; *see* **araeostyle; diastyle; eustyle; pycnostyle; systyle**; *see also* p. 370 n. 15; Fig. 6
intestinum*, *opus: *see* **carpentry**
invention (*inventio*): part of the intellectual process involved in **projection**, along with **analysis** (1.2.2)
Ionic order (Pll. 9–10; Fig. 7): **base** (3.5.13); **column** placement (3.5.4); **capital** (3.5.5–8); proportions of **entablature** and **tympanum** (3.5.8–13); **flutes** of columns (3.8.14); lions' heads in gutters (3.5.15)
isodomum, opus: wall-surface with courses of stones of uniform heights and, evidently, of variable lengths (2.8.5–6; Fig. 4.3); *see* **stone; walls**
ivy (*hedera*): 8.1.3

joint: in wood, pipes (*commissura*, 2.9.11; 8.6.6); *cubile* is a bed-joint or horizontal joint in brickwork and masonry (2.8.1, 4, 6; 4.4.4; 5.5.1), *coagmentum* a vertical joint (2.3.4; 2.8.1, 3–5, 7; 4.4.4; 5.10.3; 5.12.5; 6.8.4; 7.1.4, 6–7)
joist: *see* **beam; facing; woodwork in roofs**
juniper (*iuniperus*): 2.9.13; 7.3.1: *see* **timber**

kanon mousikos (κανὼν μουσικός): the principal element of the water-organ, the head-piece (*caput machinae*, 10.8.2–3; Fig. 32.15)
key (*pinna*): of water-organ (10.8.4, 6; Fig. 32.22)
kitchen: *see* **farmhouse**
knotwood (*fusterna*): the hard upper wood of fir trees, as opposed to the lower section which provides **deal** (2.9.7); *see* **timber**
koilia: *see* **'stomach'**

ladder, drawbridge (ἐπιβάθρα, *epibathra*): *see* **machines and mechanical systems (II)**

larch (*larigna, larix*): 2.9.14–16
lead (*plumbum*): 2.8.4; 7.12.1; 8.1.4; 8.3.5; 8.6.10–11; 9.8.3
levels, device for taking: *see* ***chorobates***; ***dioptra***; **water**
lever, handspike (*vectis*): of water-pump (10.7.3; Fig. 31.11); in water-organ (Fig. 32.7); of windlass of crane (10.2.2; Fig. 29.9); a *scutula* is a hand-spike of the windlass of a ***scorpio*** (10.10.5; Fig. 34.21)
library (*bibliotheca*): 1.2.7; 6.4.1; 6.5.2; 6.7.3; *see* **houses**
lime (*calx*): 2.5 in particular; *see* **concrete**; ***expolitio***; **pavement**; **revetment**; **screed**; ***signinum***; **stone**
lime, tree (*tilia*): 2.9.9; *see* **timber**
line (*cathetus*): perpendicular line used in the projection of the volutes of Ionic capitals (3.5.5–6; Fig. 7)
linen (*linteum*): used in the process of ***ganosis*** (7.9.3; 7.14.1)
lingula: *see* **ring**; **tongue**
lintel (*supercilium*): of door, of entrance to corridor of theatre (4.6.2, 4; 5.6.5; Pl. 15.I)
logeion (λογεῖον): *see* **theatre**
logotomus [reading disputed]: the line (GH) in the ***analemma*** parallel with the *axis* (9.7.6; Fig. 28, PQ)
lozenge (*scutula*): referring to the rhomboidal configuration of the holes in the upper and lower boards of the head-pieces of the *ballista* (also περίτρητος, *peritretos*, 10.10.2; 10.11.4; Fig. 35.8); *see* **perforated**
Lydian brick: *see* **brick**

machines (*machinae, rationes*) **and mechanical systems** (*machinationes*). (I) *Principles*: V. divides machines into three types (10.1.13): (i) for ascending or climbing (ἀκροβατικόν, *akrobatikon* = *scansorium*), essentially, **ladders**; (ii) for hauling (βαρουλκόν, *baroulkon*, *tractorium*) and raising loads, i.e. **cranes**; (iii) those operated by air pressure (πνευματικόν, *pneumatikon, spirabile*); the adverb for devices operated by machine is μηχανικῶς (*mechanikos*) and for those operated by an instrument, ὀργανίκῶς (*organikos*). (II) *Machines and instruments*: ladders (10.1.1); pneumatic devices (10.1.1–4); hauling, hoisting machines (10.1.2). *Cranes*: the *trispastos*, *pentaspastos* and *polyspastos* (10.2.1–9; Figs. 29–30); Chersiphron's device (10.2.11); Metagenes' device (10.2.12); Paconius' device (10.2.13–14); levers and **fulcrums** including balances, porters' poles, tillers, yokes of oxen (10.3.2–9). (III) *Military machines and devices*: the **scorpio** (10.10.1–6; Fig. 34); the **ballista** (10.11.1–9; Fig. 35); putting the *ballista* and *scorpio* under tension (10.12.1–2); siege engines (10.13); **battering-ram** and Pephrasmenos (10.13.1–5); **'raven'**, for demolishing walls (10.13.3); the tortoise-ram (10.13.3–6; Fig. 38); ladder, draw-bridge (ἐπιβάθρα, *epibathra*, 10.13.8); Diades' **drill** (10.13.7–8; Fig. 37); tortoise for filling ditches (10.14.1–3); **digging machine** (10.15.1); Hegetor's tortoise (10.15.2–7; Fig. 39). (IV) *Machinery for defence*: general (10.16.1–2); Diognetus, Callias and Demetrius' siege engine (10.16.3–4); scaling up models of machines (10.16.5); siege of Rhodes (10.16.6–8); sieges of Chios and Apollonia, and the engineer Trypho (10.16.9–10); siege of Marseilles (10.16.11–12): *see* **movement; water**

madder (*rubia*): 7.14.1: *see* **pigments and colours**

main frame, main component, head-piece (*caput*): of water-organ (*caput machinae*, κανὼν μουσικός, *kanon mousikos*; Fig. 32.15); of the ***scorpio*** and ***ballista*** (also called *capitulum*, 1.1.8; 10.10.1, 2, 6; 10.11.3; 10.12.12; Figs. 34.7; 35.7)

malachite green (*chrysocolla*): 7.5.8; 7.9.6; 7.14.2; *see* **pigments and colours**

marble (*marmor*): *see* **stone**

masonry (*structura*): there are occasions (e.g. *latericia structura*, 2.8.16: *structurae testaceae*, 2.8.17, disputed passage) when 'structure' or 'wall', not masonry, is the meaning: *see* **walls**

meeting place (*gerusia*): probably a place for political assemblies of the important elders (2.8.10)

megalographia: (7.4.4; 7.5.2); an emendation by Fra Giocondo (1511) for the *melographia* of the MSS: given the etymology implied, it could refer to (i) the dignity with which a subject is represented, or (ii) the moral altitude of the subject matter or (iii) the large scale of the representation within the picture frame: but in the context it is difficult to see how a particular treatment of the subject matter in a fresco should make it more or less susceptible than any other to blackening by smoke, like the fine stuccowork; the meaning of *melographia* (much disputed) open to the same objection, could be 'paintings of fruit' or 'festoons, garlands of fruit'

melinum: *see* **white**

menaeus: circle marking the months on the **analemma** (10.7.6; Fig. 28)

men's quarters: *see* ***andron*** **(II)**; ***andronitis***

mercury (*argentum vivum*): 7.8; 7.9.1

mesauloe: *see* ***andron*** **(I)**; **corridor**

metope (μετόπη): the square, recessed panel between triglyphs in Doric friezes (4.2.2, 4; 4.3.2, 4–6, 8; Pl. 14.G); the space between dentils (4.2.4). V. reports that in Greek the cavities housing joists are ὀπαί (*opai*; holes), called 'pigeon-holes' (*columbaria*) in Latin; hence the

Greeks call the spaces between joists μετόπαι (*metopai*, 4.2.4); half-metope (*semimetopium*, 4.3.5; Pl. 14.H); *see* **dentil**

mint (*aerarium*): 5.2.1; Pl. 19.D

modularity, commensurability (*symmetria*): 1.1.4 and *passim*; one of the six requisites of **architecture**, translated as *commensus* (1.3.2; 1.7.2; 3.1.2–4, 9; 5.Introd.2; 6.Introd.6; 6.1.12; 6.2.1); *commoditas* (1.2.2) and *commodulatio* (3.1.1) seem to be synonyms for *symmetria*; modularity originates in proportion (3.1.1); *symmetria* is often used by V. in the phrase *proportiones et symmetriae*: p. 363 n. 15: *see* **addition**; ***diapegma***; **entasis**; **hole**; ***scamillus***; **subtraction**

module (*modulus*): also *embater* (?) (1.2.4) or ἐμβάτης (*embates*, 4.3.3). The text at 1.2.4 is very problematic. It mentions using column diameters, or a triglyph or *embater* as modules. Apart from the problem of the word *embater* (p. 364 n. 16), the shift from the plural diameters to the singular triglyph is curious: and V. nowhere says that when planning a building the architect should start by selecting the breadth of the triglyph as the module, even though at 4.3.4 the triglyph is indeed one module wide; he always starts with subdivisions of the stylobate (3.3.7; 3.3.10; 4.3.3, 7) which generate the modules which govern column diameters, or in fact thinks of the column diameter first then divides up the stylobate accordingly. One wonders whether in fact both the words triglyph and *embater* should be excised as maladroit and inconsistent explanatory interpolations: 'Et primum in aedibus sacris [aut] e columnarum crassitudinibus [aut triglypho aut etiam embate]'; *see* **addition**; ***diapegma***; **entasis**; **hole**; **subtraction**

monopteral (*monopteros*): of a circular temple with a single row of columns and no *cella* (4.8.1); V. uses the term as a synynom for 'peripteral' in reference to Hermogenes' Temple of Father Liber at Teos (7.Introd.12); *see* **amphiprostyle**; **dipteral**; **façade**; **peripteral**; **prostyle**; **pseudodipteral**; **pseudoperipteral**

monotriglyph (*monotriglyphus*): refers to a **systyle** Doric temple with one **triglyph** over the centre of the intercolumniation (4.3.7): *see* p. 372 nn. 10, 12, 13

mortar (*materia*, *materies*): *see* **revetment**; **walls**

mortise (*pterygoma*): two wooden strips of triangular cross-section fixed to the sides of the channel of the *ballista* making a dovetailed mortise in which the trigger slides (10.11.7; Fig. 35.16)

mosaic: *see* ***sectile***

moulding (*cymatium*): refers to mouldings with more complex curvatures than those with vertical profiles (*anulus*, *quadra*, *supercilium*, ***taenia***) or convex and semicircular profiles (***astragalus***, ***torus***). V.

uses *cymatium* for the crowning moulding of architraves, door surrounds, capitals, cornices, dentilation, door-frames, friezes and triglyphs, but almost always without further definition: at 3.5.7 the *cymatium* or echinus of the Ionic capital is presumably decorated with egg-and-dart; at 4.6.2 he mentions a Lesbian moulding (*cymatium lesbium*), perhaps meaning a layer of heart-shaped leaves, and a Doric moulding (*cymatium doricum*), perhaps meaning a moulding with crows' beaks. V. also uses *unda* (wave) for a base moulding at 5.6.6; *lysis* for the final moulding of a podium of a temple and of the *scaenae frons* (3.4.5; 5.6.6); *sima* (3.5.11–12, 15) for the moulding in or above the cornice of Ionic temples (called ἐπαιετίς, *epaietis*, in Greek 3.5.12; Pl. 10.H). In English architectural vocabulary the words sima, syma, cima, cyma, cymatium and gola are apparently interchangeable, and usually refer to either a cyma recta, an S-curve with the concave curve above and convex curve below, or a cyma reversa, which is the opposite. Renaissance illustrators frequently insert cymas even when the type of profile V. has in mind is uncertain; Palladio often draws cymas (labelled *cimasia* or *gola*), and sometimes a *cavetto*, a quarter-circle moulding (Pll. 10.B, D, F; 13.G; 14.K, M; 15.O). I have made the nomenclature in the text and in the captions to Barbaro's plates uniform by writing 'moulding' for V.'s *cymatium* and Palladio's *cimasia* or *gola* since usually we do not know exactly what form it took; for the larger cornice mouldings, which V. calls *simae*, I have retained V.'s term: *see* **astragal; fillet; sima**

movement, motion: circular motion (κυκλικὴ κίνησις, ***kuklike kinesis***, 10.3.1; also called *rotundatio*, *versatio*) contrasted with rectilinear motion (*porrectum*, *porrectio*, ἐυθεῖα, *eutheia* = ἡ ἐυθεῖα γραμμή, 10.1.1; 10.3.1)

mud: *see* **clay**

music: *see* **harmonics**

mutule (*mutulus*): horizontal or slightly inclined rectangular panel with rows of ***guttae***, 6 x 3, in the soffits of Doric **cornices**; Pl. 14 (left)

nail (*clavus*): flat- or splay-headed nail (*muscarius clavus*, 7.3.11)

naos (ναός): in other texts refers to the ***cella*** of a temple but V. uses it once (3.2.1) to mean temple (*aedes sacra*) when he talks of the ναὸς ἐν παραστάσιν (*naos en parastasin*)

natron, flower of (*nitrum*): 7.11.1; *see* **pigments and colours**

nature, study of (φυσιολογία, *physiologia*): 1.1.7

neck, collar: at the top of a column (*hypotrachelium*, 3.3.12; 3.5.12; 4.3.4; Pll. 13.L, M; 17.H); perhaps includes the lower part of the

Tuscan capital (4.7.3–4); the neck of the regulator of the water-organ is called a *cervicula* (Fig. 32.13)

oak: common oak (*quercus*, 2.9.8; 7.1.2; 7.3.1); winter or Italian oak (*aesculinus quercus*, 2.9.9; 7.1.2); oak, perhaps Austrian or Russian (*robur*, 2.9.5; 7.3.1): *see* **timber**
ochre (*sil*, ὤχρα, *ochre*): yellow Attic ochre (7.7.1; 7.11.2; 7.14.1); burnt ochre (*usta*, 7.11.2): *see* **pigments and colours**
octastyle (*octastylos*): refers to temples with eight columns on the façades (3.2.7–8; 3.3.7–8; Fig. 5.5–6); *see* **pseudodipteral**
odeum: 5.9.1
oecus: *see* **hall**
oil-room: *see* **farmhouse**
oiling-room: *see* **gymnastic complex**
olive wood (*olea*): 7.3.1; *see* **timber**
opai (ὀπαί): *see* **metope**
opus: *see* ***albarium***; ***caementicium***; **carpentry**; ***craticium***; **earthwork**; ***figlinum***; ***incertum***; ***isodomum***; **pavement**; ***pseudisodomum***; ***reticulatum***; **revetment**; ***sectile***; ***signinum***; ***spicatum***; **stone**; ***testaceum***; **walls**
orchestra: in the Greek theatre the area in front of the stage reserved for the chorus; in the Roman theatre the semicircular area in front of the stage reserved for the seating of VIPs (5.6.1–4; 5.7; Pll. 23; 27; Figs. 14–16); *see* **theatre**
order, type, style of architecture (*genus*): p. 364 n. 17; *see* **Corinthian**; **Doric order**; **Ionic order**; **Tuscan**
ordinatio: *see* **architecture**; **planning**
organ: *see* **water**
orientation, of temples: 4.5
ornament (*ornamentum*): ornament in general; also refers frequently to the two elements in an entablature above the architrave, the **frieze** and **cornice**: p. 371 n. 4 for 4.2.1; *see* **architrave**; **entablature**
orpiment (*auripigmentum*, ἀρσενικόν, *arsenikon*): 7.7.5; *see* **pigments and colours**
orthographia: *see* **projection (II)**
oryx: *see* **digging machine or tortoise**; **machines and mechanical systems (III)**

paddle (*pinna*): of waterwheel, of naval hodometer (10.5.1; 10.9.5, 7)
painting (*pictura*, *pingere*): painted decoration, *trompe-l'œil* (6.2.2); painting on walls of a private house; Apaturius' stage decorations (7.5.2, 5–7); decadent painting (7.5)

palaestra: *see* **gymnastic complex**
palm (*palmeus*): of boards of palm-wood (10.14.3); *see* **timber**
palm of hand (δῶρον, *doron*): *see* **brick**
panel: in wooden door (*tympanum*, 4.6.4–5; Pll. 15.S; 16.M); framed panel-painting (*abacus*) on a **plinth** (7.3.10; 7.4.4); *see* **border**
paradromis (παραδρομίς): open-air track next to the Greek gymnastic complex; ***xystum*** in Latin (5.11.4; 6.7.5)
parastas (παραστάς): *see* **anta**
pastas: a vestibule in the Greek house at the south side of the peristyle of the *gynaeconitis* connected to the women's quarters; synonymous with *prostas*, according to V. (6.7.1); but in Greek *pastas* evidently means a portico connecting two separate parts of a house, and the *prostas* was the vestibule of the *gynaeconitis*
patron, architectural (*aedificans*, *aedificator*): 1.1.18; 2.10.3; 4.2.2; 6.Introd.6; 6.6.7
pavement: 7.1.3–7. (I) *Pavements under cover*: V.'s account may be condensed thus: once the site has been levelled (i) a bed of rubble comprising fist-sized stones, evidently without mortar, is spread on it (*statumen*, *statuminatio*); (ii) then a bed of **concrete** (*rudus*, *ruderatio*) comprising rubble, lime and sand mortar not less than nine inches deep is laid on top; if the rubble is new, then the proportions are 3 to 1 of lime, if reused, 5 to 2; (iii) then comes the **screed** (*nucleus*) comprising 1 part lime to 3 of crushed terracotta (**fired brick**) to a depth of at least six inches; (iv) the pavement of **brick** or **stone** in various configurations with its inclines is laid then smoothed off with pumice or **sand** to remove bumps and irregularities (*fricatura*), then rough polished (*levigatio*) and buffed or fine polished (*politura*); (v) finally, marble powder is sprinkled on, then a coat of lime and sand. (II) *Pavements in the open air*: once the rubble bed has been spread, a layer of concrete not less than a foot thick comprising the aggregate mixed 3 to 1 with powdered terracotta, with 2 to 5 parts of lime added to the result; the screed is laid on, then the pavement of large **tiles**. For greater impermeability, two-foot tiles should be laid on a bed of mortar on top of the concrete, then the screed and final surface provided by large tiles or bricks; *see* ***sectile***; **smoothing**; ***spicatum***
pedestal: *see* **podium**
pediment: *see* **gable**
pentadoron: *see* **brick**
pentaspastos: *see* **crane**
pepper (*piper*): 8.3.13

perfect number (τέλεον, *teleon*): 3.1.5
perforated (περίτρητος, *peritretos*): refers to the upper and lower boards of the head-pieces of the *scorpio* and *ballista* (1.2.4; 10.10.2; 10.11.4; Figs. 34.20; 35.8); at 1.2.4 V. says, apparently wrongly, that the Greeks call the hole in the board of the head-piece περίτρημα
periaktos: *see* **stage**
peripteral (*peripteros*): of a temple with one row of **columns** on all sides (Pll. 6–7; Fig. 5.4); of a circular columniated temple with a ***cella*** (4.8.1–2: Pl. 18): *see* **amphiprostyle; dipteral; façade; monopteral; prostyle; pseudodipteral; pseudoperipteral**
peristylium: *see* **courtyards**
peritrema (περίτρημα): *see* **perforated**
perspective drawing: *see* **projection (II)**
Phoenician wax (*cera punica*): *see* ***ganosis***
pier, pillar (*pila*): of stone, brick, tile; pillar placed above the beams over the capitals in the **Basilica** at Fano (5.1.9: Pl. 21.7; Fig. 11); *see* ***anta***; **strut**
pigments and colours: natural pigments (7.7); **cinnabar** and **mercury** (7.8–9); **black** (7.10); **blue** and burnt **ochre** (7.11); **white lead, verdigris** and **sandarach** (7.12); **purple** (7.13); artificial purple, yellow ochre, **malachite green** and **indigo** (7.14); *see also* **dyer's weed; green chalk; madder; natron; red clay; woad**
pilaster: *see* ***anta***; **pier; strut**
pillow-shaped: *see* **pulvinate**
pinax: *see* **board**
pincers: iron pincers (*ferrei forfices*, 10.2.2; Fig. 29.10)
pine (*pinus*): *see* **timber**
pipe (*fistula*): pipe called a 'trumpet' (*tuba*) leading out of the tank of the water-pump (Fig. 31.2, 7); in the water-organ (Fig. 32.25); V. calls an organ pipe an *organum* at 10.8.4 (Fig. 32.24)
piston: in water-pump (*embolus masculus*, 10.7.3; Fig. 31.9); in water-organ (*ambulatilis fundulus, fundus*, 10.8.1, 5; Fig. 32.5)
piston-rod: of water-pump (*regula*, 10.7.3; Fig. 31.10); of water-organ (*ancon*, 10.8.1,5; Fig. 32.6)
pivot: *see* **axle**
plank in wooden floor (*axis*): 4.2.1–2; 7.1.2: *see* **floor**
planning (*ordinatio*, τάξις, *taxis*): one of the six requisites of **architecture**
plaster: *see* **revetment; walls**
platform: *see* **base; theatre; tribunal**
plinth: the square block below Attic and Ionic bases, the cylindrical element below Tuscan bases (*plinthus*, 3.5.1–3; 4.7.3; 5.9.4; also

plinthis, 3.3.2); Doric, Pl. 13.A; Ionic, Pl. 9.A; Tuscan, Pl. 17C; plinth of the podium of a temple (*quadra*, 3.4.5; also *crepido*, 3.3.7); of *scorpio* (*plinthis*, 10.10.4); 'little plinths', small blocks which regulate the passage of air through to the pipes of the water-organ (*plinthides*, 10.8.3, also called *regulae*, Fig. 32.20); *see* **abacus**

plumb-bob (*perpendiculum*): Fig. 27.2

pneumatic devices: *see* **machines and mechanical systems (I)**

podium (*podium*): of Roman temples, of the Corinthian *oecus* (3.4.5; 6.3.9); pedestals of the columns of the *scaenae frons* (5.6.6; cf. Pl. 26; Fig. 15A); plinth below panel-paintings (7.4.4)

polishing (*politura*): the final operation after sanding and smoothing a floor (7.1.4); the final polishing of a revetment is *politio* (7.3.8–9): *see* **pavement; revetment**

polyspastos: *see* **crane.**

poplar (*populus*): white and black (1.2.8; 2.Introd.1; 2.9.5; 2.9.9): *see* **timber**

port: *see* **harbour**

portico (*porticus*): colonnade, portico of basilicas (Pl. 25), gymnastic complexes (Pl. 30), temples (Pll. 7–8, etc.), theatres (Pl. 25), urban and rural houses; and used of independent colonnaded structures built outside theatres, gymnastic complexes, squares, etc. (5.9); Caryatid portico (*opus*, 1.1.5; Pl. 1); Persian portico with statues of prisoners supporting the roof (*porticus* 1.1.6; Pl. 2); V. also uses *pteroma* for the colonnade or portico surrounding a temple (3.3.9; 4.4.1; 4.8.6); *columnatus* for that of a **monopteral** temple (4.8.1); *see also* **amphiprostyle; dipteral; façade; peripteral; prostyle; pseudodipteral; pseudoperipteral**

posticum: *see* **façade**

pozzolana (*puteolanus pulvis*): powder from Pozzuoli (5.12.2)

prison (*carcer*): 5.2.1

projection: (I) *proiectura*: of a base (= ἐκφορά, *ekphora*, 3.5.1); of a cornice moulding in an Ionic temple (3.5.11); illusion of projection given by mutules painted in *trompe-l'œil* (6.2.2). (II) *dispositio*, διάθεσις, *diathesis*: 1.2.2 in particular; projection is one of the six requisites of **architecture,** manifested by three types or forms (ἰδέαι, *ideai*), which are: (i) the ground-plan drawn to scale (*ichnographia*; Pl. 3); (ii) the orthogonal elevation (*orthographia*; Pl. 4; *see* **scale**); (iii) the perspectival drawing (***scaenographia***), which are the results of **analysis** (*cogitatio*) and **invention** (*inventio*)

pronaus: vestibule between the door of the *cella* of a temple and the front columns

proportion (ἀναλογία, *analogia*): 3.1.1; *see* **modularity**

proscaenium: the platform of the stage and its wall facing the *orchestra* (5.6.1; 5.7.1; Figs. 14–16)
prostas: *see* ***pastas***
prostyle (*prostylos*): of a temple with columns only at the front (3.2.1, 3; 3.2.4; 7.Introd.12; Fig. 5.2); *see also* **amphiprostyle; dipteral; façade; monopteral; peripteral; pseudodipteral; pseudoperipteral**
prothyron (πρόθυρον): a vestibule in front of the door of the Greek house; but in Latin usage, according to V., *prothyrum* means the same as διάθυρον (*diathyron*), the space between the front and back doors of the entrance corridor (6.7.5)
pseudisodomum*, *opus: masonry or wall-surface with courses of stone of dissimilar heights and, evidently, of variable lengths (2.8.5–6; Fig. 4.4); see ***isodomum***; **stone**
pseudodipteral (*pseudodipteros*): of a temple with space for two colonnades all round the *cella*, but the inner colonnade is omitted (3.2.1, 6; 3.3.8–9; 7.Introd.12: Fig. 5.5): at 3.3.8 (readings much disputed), V. says that Hermogenes removed 34 columns from a dipteral octastyle plan to create the pseudodipteral octastyle type; if such a temple had 8 × 15 or 42 columns in the exterior colonnade, like the architect's own Temple of Artemis at Magnesia, then the interior colonnade, which he dispensed with, would have included 6 x 13 columns = 34; *see* **amphiprostyle; dipteral; façade; monopteral; peripteral; prostyle; pseudoperipteral**
pseudoperipteral (*pseudoperipteros*): of a temple, notionally peripteral, of which the colonnade is not free-standing but engaged to the *cella* walls in the form of half columns or pilasters (4.8.6); *see* **amphiprostyle; dipteral; monopteral; peripteral; *posticum*; prostyle; pseudodipteral**
pteroma: *see* **portico**
pulley (*orbiculus*): 10.2.1, 3–4; 10.6.8–10; 10.3.2; Figs. 29.4; 30.7
pulley-block: V. employs various terms: *troclea* for the pulley-block of a crane (also called *rechamus* 10.2.1, 2; Figs. 29.3; 30.4, 6); *artemon* for the third pulley-block at the foot of the crane, ἐπάγων (*epagon*) in Greek (10.2.9; Fig. 30.8); *see* **crane; machines and mechanical systems (I)**
pulpitum: *see* **theatre**
pulvinate, cushion-shaped (*pulvinatus*, *pulvinus*): of Ionic capitals, which are provided with cushions, bolsters or rolls at right angles to the volutes (1.2.6; 3.5.5, 7; 4.1.12); *see* ***balteus***
pumice (*pumex*): 2.3.4; 2.6.2
punchbags: *see* **gymnastic complex**

purlin: *see* **beam; woodwork in roofs**
purple (*ostrum*, *purpura*): 7.5.8; 7.13; 7.14.1
pycnostyle (*pycnostylos*): of a colonnade with an intercolumniation of one and a half diameters or three modules (3.3.1–3, 10–11; 3.4.3; Fig. 6.1)

quantity (ποσότης, *posotes*): an aspect of **planning**; *see* **architecture**
quarry, mine: mine (*metallum*, 7.7.3, 5; 7.9.4, 6; 7.12.2); sand quarry (*arenarium*, 2.4.2–3; 2.6.5); stone quarry (*lapidicina*, 2.7.2–5; 8.3.9; 10.2.11–13, 15); silver mine (*argentifodina*, 7.7.1); vein, layer, stratum of stone or metal (*vena*, 7.1.1; 7.8.1; 7.9.4): *see* **stone**; *see also Index of Names and Places*, s.v. quarries

rafter, common: *see* **beam; woodwork in roofs**
'raven' (*corvus demolitor*): a siege engine for demolishing walls, also called a crane (*grus*, 10.13.3); *see* **machines and mechanical systems (III)**
receptacle, bronze (*vas aeneum*): for collecting stones in hodometer (10.9.3, 6; Fig. 33.9)
red clay (*rubrica*): used as a pigment and to make bricks (2.3.1; 7.7.2)
reed, cane (*arundo*, *canna*): 2.1.3, 5; 7.3.2 (Greek reed), 11; 8.1.3–4; 8.3.9
regula: small rectangular block below the **taenia** to which the ***guttae*** are attached in the Doric entablature (4.3.4; Pl. 14.C)
regulator (*pnigeus*): a conical container controlling the flow of air in the water-organ (10.8.2, 4–5; Fig. 32.11)
retaining-hook, claw (*epitoxis*): the claw which holds back the cord of the *scorpio* (10.10.4; Fig. 34.15)
reticulatum*, *opus: wall composed of truncated conical or pyramidal stones with square ends or bases inserted horizontally leaving a network pattern on the outer surface (2.8.1; Fig. 4.1); *see* **walls**
revetment (*opus tectorium*, *tectorius*): *see especially* 7.3.5–11; 7.4.1; refers to all the layers of a revetment as well as the final coat with its painted decoration; a *tector* is a revetment specialist, so a plasterer or painter (7.3.10; 7.10.2, 3; 7.14.1). V. also uses *politio* for a revetment or a layer of a revetment (7.2.1; 7.3.7; 7.4.1, 4; 7.7.1), for a painting or fresco (7.9.4; also *opus*, 7.6.1; 7.7.1; 7.10.1, 4), and *corium* for a layer. The process of applying a revetment may be summarized thus: (i) the application of a layer of rough sand mortar or mortar of powdered terracotta to cover the joints in the **masonry** and create a smooth surface (*trullisatio*); (ii) three layers of sand mortar (*arenatio, arenatum opus*); (iii) three layers of marble mortar

(*marmor*). The various layers had to be aligned (*directio*) by checking the length with string, the height with a plumb-bob, the angles with a set-square, then smoothed off with a **float**; the result was *directura* (7.3.5), a perfectly aligned and level surface: *see* **albarium**; **walls**; **wax**

revolving platform or crane (*carchesium*): a device for lifting besiegers up to the walls (10.16.1). The *carchesium versatile* may have been a revolving platform rather than a revolving yardarm on a mast (10.2.10; 10.16.3); *see* **machines and mechanical systems (III)**

Rhodian courtyard (*rhodiacus porticus*): courtyard with four porticoes, with that on the south side taller than the others (6.7.3); *see* **courtyards**

ridge-piece: *see* **woodwork in roofs**

rim: *see* **volute**

ring (*anulus*): ring holding the 'little tongue' (*lingula*) of a pipe of the water-organ in position (10.8.4; Fig. 32.23)

roof (*tectum*, *testudo*): in V.'s Basilica at Fano, *testudo* refers to the roof, but also to the area covered by it, and so in effect means 'nave' (5.1.6, 7); of Hegetor's tortoise (*contabulatio*, Fig. 39A.11). V. uses *tholus* for the roof of a round temple at 4.8.3; *see* **dome**; **woodwork in roofs**

roofed (*testudinatum*): of a courtyard entirely roofed without a *compluvium* (6.3.1–2; Fig. 22); *see* **courtyards**

rope (*funis*, *rudens*): support-rope, guy (*antarius funis*, 10.2.3; Fig. 29.12); traction-rope (*ductarius funis*); control rope, retaining rope (*retinaculum*, Figs. 29.13; 30.2); ropes under tension in head-pieces of the ***scorpio*** (*nervi torti*, Fig. 34.23) and the ***ballista*** (*funes*, Fig. 35.9). A small rope or cord (*resticula*) kept the traction ropes of some cranes together where they left the pulley so that they did not get entangled with the others (10.2.6)

round temple (*tholus*): at Delphi (7.Introd.12); round temples represented in Apaturius' stage scenery (7.5.5), also *aedes rotunda* (4.8.1); *see* **monopteral**; **peripteral**

rubble (*caementum*, *rudus*, *ruderatio*, *statumen*, *statuminatio*): *see* **concrete**; ***expolitio***; **pavement**; **walls**

rule (θεματισμός, *thematismos*): obedience to a rule is an aspect of architectural **appropriateness** (1.2.5)

ruler, straight edge (*regula*): *see* **set-square**

sambuca: device for storming walls from ships (10.16.9); triangular configuration of the world (σαμβύκη, *sambuke*, 6.1.5; Fig. 17)

sanctuary (*fanum*, *templum*): *fanum* usually means a sanctuary as distinct from the temple (*aedes sacra*) or temples within it (1.3.1); but at 1.7.1, where V. talks first about the founding of the *aedes sacrae* of the chief protectors of the city, and then of the *fana* of Venus, Vulcan and Mars, *fanum* is apparently synonymous with temple; at 1.2.7 *templum* evidently means sanctuary or sacred enclosure and *fanum* means temple

sand (*arena*, *sabulo*): river sand, black, white, red, etc. (*fluviatica arena*); quarry-sand (*fossicia arena*); marine sand (*marina arena*). V. also uses the terms *masculus sabulo* for compact sand and *solutus sabulo* for loose sand (2.3.1; 8.1.2)

sandarach, red arsenic (*sandaraca*): 7.7.5; 7.12.2; 8.3.11; *see* **pigments and colours**

sanding: *see* **pavement**

sapwood (*torulus*): 2.9.3, 7: *see* **timber**

scaenographia: drawing or painting in perspective; *see* **architecture; projection (II)**

scaffolding (*machina*): 7.2.2

scale: of a drawing made to scale (*imago modice picta*, 1.2.2); for the problem of scaling up models of machines, 10.16.5; *see also* **diagram; projection (II)**

scamillus: small block of wood or a stone projection on a block of the stylobate used for projecting the curvature ? (3.4.5; 5.9.4; *see* p. 370 n. 16); *scamillum* is also used as a synonym for *buccula* and *loculamentum*, in reference to the block or box of wood housing the windlass at the end of the shaft of the *scorpio* (10.10.3; Fig. 34.5); *see* **housing**

scorpio: *see* **catapult**

scotia (τροχίλος, *trochilos*): 3.5.2–3; Pll. 9.B; 13.C; V. also uses *scotia* of a moulding or groove at the edge of the cornice of a Doric diastyle temple (4.3.6)

screed (*nucleus*): the layer of mortar made of lime and ground-up terracotta (**fired brick**) spread above the bed of **concrete** in a **pavement** (7.1.3, 6, 7)

seats, seating: a wedge or block of seating in the theatre is a *cuneus*, probably referring to its triangular vertical section rather than its truncated conical plan (5.6.2; 5.7.2; Figs. 14.16; 15; 16); tiers of seats (*gradationes*, 5.3.3, 8; 5.6.4); *subsellium* is a long bench-like seat in the theatre (5.6.3); *tribunal* is honorific seating in the theatre (5.6.7); *see* **stair**

sectile, opus: of pavements surfaced with cut stones of various configurations: the hexagon (*favus*), lozenge (*scutula*), triangle (*tri-*

gonum), cube (*tessera*) and square pieces of stone (*quadratus*, 7.1.3–4, 6–7)

senate house (*curia*): 5.2.1

set-square (*norma*): 1.1.4; 3.1.3; 3.5.13–14; 4.3.5; 7.3.5; 8.5.1; 9.Intro.6; 9.7.2

shadow-hunter (σκιαθήρης, *skiatheres*): Greek name for the pointer or **gnomon** of a **sundial** (1.6.6)

shaft (*scapus*): of column; V. uses *truncus* at 4.1.7 for the shaft of a fluted Ionic column presumably to indicate its origin in trees (also *corpus* at 5.1.9); vertical post or stile of door (4.6.4–5); main beam of *ballista* (*scapus climacidos*, Fig. 35.12); at 3.4.5, V. uses *truncus* to describe the shafts or tall vertical faces of pedestals

shop, workshop: money changer's, bankers (*argentaria taberna*, 5.1.2); workshop (*officina*, 7.8.2; 7.9.4); painter's studio (*officina pictorum*, 6.4.2); workshop for weaving (*textrinum*, 6.4.2).

side-room (*ala*): small rectangular room with three, not four walls, like an **exedra**, at left and right of the ***atrium*** and near the ***tablinum*** (6.3.4–6; Fig. 23.5); *see* **corridor**

siege engines: *see* **machines and mechanical systems (III)**

signinum*, *opus: waterproof revetment of Signian type made of rubble (*caementum de silice*) and a mortar of lime and coarse sand (5.11.4; 8.6.14); Signian sand (2.4.3)

silex: *see* **hard stone**

sima: where V. uses the word 'sima' for large curved mouldings with an S curve or part of one, the English 'sima' rather than simply 'moulding' is perhaps justified; *see* **moulding**

smoothing (*fricatura*): rubbing down to produce a smooth surface on tile or stone **pavements** (7.1.4), followed by *levigatio* (7.1.4; 7.3.6 for revetments) then **polishing** (*politura*, 7.1.4)

socket-piece (*chelonium*): socket-piece housing the axle of the windlass of a crane (10.2.2, 8; 10.3.2; Figs. 29.7; 30.3); part of the mechanism for tightening up the head-pieces of *scorpiones* and *ballistae* (10.12.1); component of Hegetor's tortoise (10.15.4); the socket-piece above the channel of the *scorpio* into which the retaining hook (*epitoxis*) and trigger (*chele*, *manucla*) were fixed (10.10.4); socket-piece, also called a 'cushion' (*pulvinus*, 10.10.5) for the small counter-strut of the *scorpio* (Figs. 34.1, 12; 34.12)

sounding vessel (ἠχεῖον, *echeion*): made of bronze and horn (1.1.9; 5.3.1, 8; 5.5.2, 7; Figs. 12; 13)

space, gap (*intervallum*): between the walls of a ***cella*** and **columns**, between columns, between dentils or triglyphs, between the uprights of the head-piece of *ballistae* and *scorpiones* through which the missile was discharged (10.10.3; 10.11.7; Fig. 34.17); the breadth of a tortoise-ram (10.13.6; Fig. 36B.1); the length of a lever (10.3.9); *see* **intercolumniation**

Spartan hot room (*laconicum*): a hot room like a sauna heated by a brazier in the middle and not supplied with hot water (5.10.5: 5.11.2; Pll. 28.I; 29.d–; *laconicum* refers to a furnace or stove at 7.10.2

spicatum, opus, spica: brick work in herringbone pattern recommended by V. for floors open to the elements (7.1.7); tiled pavement in the Tivolean manner (*testacea spicata tiburtina*, 7.1.4)

spleen, absence of (ἄσπληνον, *asplenon*, 1.4.10): the name of the herb used to cure the condition

stable: *see* **farmhouse**

stadium: at Tralles (5.9.1); covered stadium next to the **gymnastic complex** (*palaestra*) for exercise in the winter (5.11.4; Pl. 30.Q): *see* **gymnastic complex**; ***xystos***

stage: Roman, 5.6; Greek, 5.7 (Pll. 23–7; Figs. 14–16): (I) *Elements, components of the stage*: stage-set, used here to describe all the structures at the back of the stage (*scaena*); façade of stage-set (*scaenae frons*); a level or storey of the façade of the stage-set (*episcaenium*, 5.6.6; 7.5.5); guests' quarters, indicated by doors left and right of the centre of the façade of the stage-set (*hospitalia*, 5.6.3); platform of the stage (*pulpitum*, 5.6.1–2, 7; a translation of λογεῖον, *logeion*, at 5.7.2); *periaktos* (περίακτος), vertical, revolving triangular device mounted on a pole of which the three faces were decorated with different scenes (5.6.8); the area in front of the stage-set with its wall facing onto the orchestra (*proscaenium*, 5.6.1; 5.7.1); the central doorway in the façade of the stage-set indicating the entrance to royal quarters is called the *regiae valvae* (5.6.3); the wings of the stage to left and right (*versurae*). (II) *Types of scenery and stage-set*: comic stage-set with façades of private houses and balconies (*comica scaena*, 5.6.9; 7.5.2); satiric stage-set with representations of the countryside, hills, grottoes, herds of animals, shepherds (*satyrica scaena*, 5.6.9; 7.5.2); tragic stage-set with palaces and grand architecture, tympanums (*tragica scaena*, 5.6.9; 7.5.2); *see* **aisle**; **auditorium**; **corridor**; ***orchestra***; **sounding vessel**; **stair**; **theatre**; *see also Index of Names and Places*, s.v. Apaturius; Tralles

stair, staircase, steps: V. uses various terms: the generic term is *scala* in theatres, private houses (5.6.2; 5.7.2; 6.6.7; 9.Introd.7–8); *gradus* is a generic term for a step in a temple, theatre, house. For the flights of stairs in the theatre he also uses *ascensus* (5.6.2, 7; at 4.8.1 of the staircase leading up to the stylobate of a monopteral temple), also *gradatio* (5.7.2, which can also mean seats, 5.3.3, 8; 5.6.4) and *scalarium* (5.6.3); he uses *spectaculum* for a place of spectacle (5.1.2; 5.3.5), hence seats or steps in a theatre (*cunei spectaculorum*, 5.6.2–3)

stake (*palus*): used to secure the stay-ropes of cranes to the ground (10.2.3–4; Fig. 29.14)

stalk (*coliculus*): stalk springing from the second row of leaves of the Corinthian capital supporting the volutes (4.1.9, 12; 7.5.3–4; Pl. 11)

step (*gradus*): *see* **stair**

stereobate (*stereobates*): rectangular system of containing walls above foundations comprising vaulted and filled-in chambers surrounded by steps or a podium on which the stylobates of temples rest (3.4.1; cf. Pl. 4)

stile: *see* **cover-joint; hinge-stile**

'stomach' (*venter*, κοιλία, *koilia*): a giant inverted syphon or system of pipes conveying water across the bottoms of valleys between hills or mountains (8.6.5–8)

stone (*saxum*, *lapis*): 2.7–8; square or rectangular dressed stone, ashlar (*quadratum saxum*, *quadratus lapis*), contrasted with quarry-stone, rough stone (*caementa*); at 2.8.5, V. says that when the Greeks did not use ashlar, meaning rectangular or square blocks of dressed stone equal in height and length in a given wall surface, they used hard stone laid like bricks with alternating vertical joints instead. This resulted in two types of surface: *opus isodomum*, with courses of stones of equal height but not necessarily equal length, a necessary hypothesis, it seems, else *opus isodomum* would be identical in principle to walls built with *opus quadratum*, though constructed of stones which were less carefully dressed; and *pseudisodomum*, with courses of stones of varying length and varying heights (which is probably the meaning of 'inpares et inaequales ordines coriorum'): Fig. 4.3–4 shows the most prevalent interpretation. In the case of *pseudisodomum* V. may also have intended to imply that it need not have courses of just *two* different, alternating heights, as it is often reconstructed, but courses of several different heights in one wall; presumably both *isodomum* and *pseudisodomum* would have included stones of variable lengths (similarly Rowland, 1999, p. 183); *see* ***diatonos***; **facing; hard stone; marble; pavement; quarry;** ***sectile***; **walls**

stopcock (*epitonium*): in water clock (9.8.11); in water-organ (10.8.3, 5; Fig. 32.17)

storerooms: *see* **farmhouse**
straight line: 10.3.1; *see* **movement**
street, side street (*angiportum*): 1.6.1, 7–8, 13; 1.7.1; cf. **avenue**
strut, support: V. has various terms: struts or rafters (*capreolus*, 10.14.2; 10.15.1,3); struts and cross-beams (*capreoli et transtra*, 4.2.1; *transtra cum capreolis*, 5.1.9); compound strut or post (*trabs compactilis*, 4.7.4; Fig. 8.1; also *postis compactilis*, 10.14.2; Fig. 38.4); vertical wooden support (*postis*, 6.8.2; Fig. 39A.6); side support of water-organ (*regula*, 10.8.1; Fig. 32.3); central and side support of head-piece of *scorpio* (*parastata, parastatica*, 10.10.2; Figs. 34.18–19; 35.18); front strut of *scorpio* (*capreolus*, Fig. 34.8); back strut of *scorpio* (*subiectio*, 10.10.5; Fig. 34.22); small counter-strut of *scorpio* (*posterior minor columna*, called ἀντίβασις, *antibasis*, 10.10.5; Fig. 34.1); front strut of *ballista* (*anteris*, 10.11.9; Fig. 35.2); *columna* and *columella* of vertical support of *scorpio* and *ballista* (10.10.5; 10.11.9; Figs. 34.13; 35.21): *see* **woodwork in roofs**
stucco: *see* ***albarium***; **walls**
stylobate (*stylobata*): platform of stone blocks above the **stereobate** on which the columns of temples were placed (3.4.2–3, 5, 8; 4.8.1–2; 5.9.4)
subtraction (*detractio*): from the modular system (*detractio modulorum*, 6.2.1, 4; 6.3.11; 10.10.6); *see* **addition; entasis; hole; module**
sundial and clock (*horologium*): the sundial, the ***analemma*** and its construction (9.7); sundials and water clocks (9.8); in principle, sundials are summer clocks, water clocks are winter clocks: *see* **gnomon**
sweating-rooms: *see* **bath; Spartan hot room**
symmetria: *see* **architecture; modularity**
systyle (*systylos*): of a colonnade with an intercolumniation of two diameters or four modules (3.3.1–2, 10; 3.4.3; 4.3.7; Fig. 6.2)

tab(u)linum: large room of disputed function placed centrally between the *atrium* and peristyle. V.'s phraseology at 6.3.6 misled some interpreters to assume that the statues of illustrious ancestors were placed in the *tablinum* rather than in the *atrium* (6.3.5–6; 6.5.1; Fig. 23.6; cf. Pl. 31)
taenia: small continuous horizontal moulding of vertical profile dividing the top of the architrave from the bottom of the frieze in Doric entablatures (4.3.4; Pll. 13.q; 14.D)
tank: in water-pump (*catinus*, 10.7.1–3; Fig. 31.3); also *castellum* (8.6.1, 4, 7; 9.8.11; 10.4.3; 10.7.3; Fig. 31.12); water-tanks in baths

for cold, warm and hot water (therefore, boiler: *vasarium*, 5.10.1; Pl. 29.a, b, c)

telamon (ἄτλας, *atlas*): male statue with raised arms bent at the elbow supporting mutules (i.e. in a Doric context) or a cornice (6.7.5,6); *see* Pl. 37

teleon: *see* **perfect number**

temple (*aedes, templum*): the word *aedes* (building) is often qualified by V. with *sacra* and *deorum immortalium* to distinguish it from non-religious *aedes*; 3.2–4 for plans, column spacing and foundations: here, 'temples of the immortal Gods' or 'sacred temples' are retained, despite the tautology; *see* **amphiprostyle; *anta*; araeostyle; diastyle; dipteral; eustyle; façade; hypaethral; monopteral; order; peripteral; portico; *pronaus*; prostyle; pseudodipteral; pseudoperipteral; pycnostyle; round temple; sanctuary; stereobate; stylobate; systyle**: *see Index of Names and Places*, s.v. Alabanda; Argos; Athens; Claros; Delphi; Eleusis; Ephesus; Fano; Halicarnassus; Magnesia; Miletus; Patras; Priene; Rome; Samos; Sunium; Teos; Tralles

tension: high-tensioned, referring to the head-piece of a *scorpio* that is taller than it is wide (*anatonum capitulum*, 10.10.6); low-tensioned, referring to the head-piece of a *scorpio* that is wider than it is tall (*catatonum capitulum*, 10.10.6); when both sets of ropes in the cylinders of the *scorpio* and *ballista* are under the same tension the firing arms inserted in them are described as 'identically tuned' (*homotona brachia*, 1.1.8)

tension-bolt (*epizygis*): (or torsion-bolt) at the top of the cylinder of the *ballista* and *scorpio* maintaining the ropes under tension (10.11.4; also *cuneolus* at 10.12.1; Figs. 34.14; 35.13)

terrenum, opus: *see* **earthwork**

testaceum, opus: pavement or wall-surface with **tiles** or fired **brick** (2.8.17–18, 20; 7.1.4, 7; 7.4.3, 5; 8.3.8); the phrase *testaceum pavimentum* at 7.4.5 refers to the layer of tiles or bricks used in the Greek **pavement**; brick pavement laid in herringbone pattern in the Tivolian manner (*testacea spicata tiburtina*, 7.1.4): *see* ***spicatum***

testudinatum: *see* **courtyards; roofed**

tetradoron: *see* **brick**

tetrastyle (*tetrastylos*): of temples with four columns at the front (3.3.7; 4.3.3, 7; Fig. 5.2); *for tetrastyle courtyards, see* p. 376 n. 7; Fig. 19; *see* **courtyards; hall; temple**

theatre (*theatrum*): 5.3; **acoustics** (5.4); **sounding vessels** (5.5; Figs. 12–13); Roman theatre (5.6; Pl. 23; Figs. 14–15); Greek theatre (5.7; Pl. 27; Fig. 16); acoustics (5.8; Figs. 12–13); the Stone Theatre

(3.3.2): *see* aisle; auditorium; corridor; ***ekklesiasterion***; harmonics; ***orchestra***; stage; stair

thematismos (θεματισμός): *see* appropriateness; architecture

theodotion: *see* green chalk

'thigh' (*femur*, μηρός, *meros*): vertical surface between the flutes or channels of triglyphs (4.3.5,6; Pl. 14.E)

tholus: *see* dome; round temple

thyroron (θυρωρών): doormen's quarters between entrance door and interior door of the Greek house (6.7.1)

tile (*tegula*): two-foot tile (*bipedalis tegula*, 5.10.2); tile of a foot and a half (*sesquipedalis tegula*, 5.10.2); tile with corner hooks (*hamata tegula*, 7.4.2; Fig. 26): *see* brick; ***testaceum***

tilework: *see* ***figlinum***

tiller (*ansa*, οἴαξ, *oiax*): 10.3.5

timber, wood (*materia*, also *materies*): *see* alder; ash; beech; boxwood; carpentry; cedar; cypress; deal; elm; fir; juniper; knotwood; larch; lime; oak; olive wood; palm; poplar; sapwood; turkey oak; willow; woodwork in roofs; yoke-wood

Tivoli, Tivolian: travertine (*tiburtina saxa*, 2.7.2); of fired bricks, like those made in Tivoli (*testacea spicata tiburtina*, 7.1.4): *see* ***spicatum***

tongue (*lingula*): 'little tongue', evidently referring to the tapered ends of the pipes of the water-organ (10.8.4); the hand or pointer of a clock (9.8.12–13); front end or point of a lever or pole used with a fulcrum (10.3.3)

tooth (*denticulus*): small tooth on the disc of a hodometer (10.9.2–3, 5–7; Fig. 33.4–6, 11–12); on the disc of a water clock (9.8.5)

tortoise-ram: *see* battering-ram

torus (*torus*): large circular moulding of a base with convex profile (3.5.2–3; 4.7.3; Pll. 9.C; 13.B, D)

tower (*turricula*): small tower on the tortoise-ram and Hegetor's tortoise (10.13.6; 10.15.5; Fig. 36B.5)

treasury (*aerarium*): 5.2.1

tribunal: platform on which Alexander sat (2.Introd.1); platform of monopteral temple (4.8.1); in Temple of Augustus at Basilica at Fano (5.1.8); in theatre for seats of magistrates (5.6.8)

triclinium: *see* dining room

trigger, trigger handle (*chele*): of *scorpio* and *ballista* (10.10.4 = *manucla*; 10.11.7–8; Figs. 34.11; 35.10)

triglyph (*triglyphus*): vertical rectangular element in Doric frieze, representing the end of a joist in V.'s theory, supplied with two vertical grooves in the middle and two half-grooves at the corners (Pll. 13R; 14.E, F, I); half-triglyph (*hemitriglyphus*, 4.3.8; Pl. 14.H); at 4.2.2, a *tabella* is a wooden precursor of the triglyph, painted blue and nailed onto the ends of the joists in Doric temples to protect them (4.2.2): *see* **capital**; Doric order

trispastos: *see* **crane**

trowel (*liaculum*): *see* **float**

trumpet: *see* **pipe**

tube: *see* **pipe**

turkey oak (*cerrus*): 2.9.9; 7.1.2

Tuscan (*tuscanicus*): 3.3.5; 4.6.6; 4.8.5; Tuscan order (*tuscanicae dispositiones*, 4.6.6; 4.7.1–5; Fig. 8); Tuscan **courtyard** (p. 376 n. 6; 6.3.1: Fig. 18): hybrid Tuscan-Corinthian and Tuscan-Ionic temples (4.8.5)

tympanum: the triangular surface enclosed by a gable or pediment (*tympanum . . . in fastigio*, 3.5.12–13; 4.3.6; *tympanum fastigii*, 4.7.5; Pll. 10.K; 14.N)

underground structure (*hypogeum*): 6.8.1

valve: in tank of water-pump (*assis*, 10.7.1; Fig. 31.4, 8); cymbal-shaped valve of water-organ (*cymbalum*, 10.8.1,5; Fig. 32.10)

vault: V. uses *camera* for an arch and vault in a bath (Books V and VI), in houses (Book VII); *concameratio*, sometimes referring to a vault of masonry (5.10.3; of a *sudatio*, 5.11.2; in underground chambers, 6.8.1; also 2.4.2–3), sometimes of wood (5.10.3); *confornicatio* is a vault in the corridor of a theatre (5.6.5). V. uses *caelum* for the soffit of a vault (7.3.3) and perhaps at 8.2.4; *curvatura* for the vault of a Spartan bathroom and of a kiln (5.10.5; 7.10.2; 8.2.4); at 6.3.9, *lacunar* is used to refer to the curved, coffered vault in a Corinthian hall; *see also* **arch**

verdigris (*aerugo*): also called *aeruca* (7.12.1; 8.3.19): *see* **pigments and colours**

vermilion: *see* **cinnabar**; **pigments and colours**

vestibule (*vestibulum*): 1.2.6; 6.5.1–2; 6.7.3, 5; 7.Introd.17: Fig. 23.1; *see also* **houses**

via: perhaps referring to the spaces between **mutules** rather than the spaces between the ***guttae*** of the mutules (4.3.6)

vision (λόγος ὀπτικός, *logos optikos*): the subject of vision (1.1.16)

volute (*voluta*): of Ionic and Corinthian capitals; construction of Ionic volute (3.5.5–7; Fig. 7); eye of the volute (*oculus*, 3.5.6–7; Pl. 9.p); the outer rim of an Ionic volute is the *axis* (3.5.7; Pl. 9.n); *see also* **capital**

voussoir (*cuneus*): 6.8.3–4; *see* **arch**

Index of Names and Places

The Index includes names and places referred to in the Introduction to this volume as well in Vitruvius' text. Some of the Latin names for constellations (Aries, Capricorn, Gemini, Taurus) are so familiar in English that they have been left in their original form; the rest are translated.